Neutron Induced Reactions

SLOVAK ACADEMY OF SCIENCES

Institute of Physics, Electro-Physical Research Centre of the Slovak Academy of Sciences

Reviewer:

Ing. Š. Luby, DrSc.

NEUTRON INDUCED REACTIONS

Proceedings of the 4th International Symposium
Smolenice, Czechoslovakia, June 1985

Edited by

J. Krištiak and E. Běták

Institute of Physics, Electro-Physical Research Centre,
Slovak Academy of Sciences, Bratislava, Czechoslovakia

D. Reidel Publishing Company

A MEMBER OF THE KLUWER ACADEMIC PUBLISHERS GROUP

Dordrecht / Boston / Lancaster / Tokyo

Library of Congress Cataloging-in-Publication Data

Neutron Induced Reactions.

1. Nuclear reactions – Congresses. 2. Neutrons – Congresses. I. Krištiak, J. (Jozef), 1942 – . II. Běták, E. III. International Symposium on Neutron Induced Reactions (4th: 1985: Smolenice, Czechoslovakia). IV. Title.
QC793.9.N47   1986   539.7'5   86–3227
ISBN-13:978-94-010-8561-8   e-ISBN-13:978-94-009-4636-1
DOI: 10.1007/978-94-009-4636-1

Distributors for the U.S.A. and Canada
Kluwer Academic Publishers,
101 Philip Drive, Assinippi Park, Norwell, MA 02061, U.S.A.

Distributors for Albania, Bulgaria, People's Republic of China,
Cuba, Czechoslovakia, German Democratic Republic, Hungary,
Korean People's Democratic Republic, Mongolia, Poland, Rumania,
U.S.S.R., Vietnam, and Yugoslavia
VEDA, Publishing House of the Slovak Academy of Sciences,
Bratislava, Czechoslovakia.

Distributors of all remaining countries
Kluwer Academic Publishers Group,
P.O. Box 322, 3300 AH Dordrecht, Holland.

First edition published in 1986 by VEDA, Bratislava,
in co-edition with D. Reidel Publishing Company, Dordrecht, Holland.

CONTENTS

STRUCTURE AND INTERACTIONS

Invited talks

Contributions

RESONANCES AND GAMMA EMISSION

Invited talks

Contributions

FISSION

Invited talks

Contributions

INSTRUMENTATION

Invited talk

Contributions

PREFACE

The Fourth International Symposium on Neutron Induced Reactions
was held at the Smolenice Castle, Czechoslovakia, June 17–21,
1985.

It was sponsored by the European Physical Society and the
Union of Slovak Mathematicians and Physicists.

The primary aim of the Symposium was to provide space for
reports on developments in the field of nuclear reaction theory,
studies of decay of unbound nuclear states excited by simple
projectile as well as on methodical and instrumental progress
in the low energy nuclear physics.

Ample time was deliberately reserved for discussions and
the exchange of ideas among the participants.

Two committees have been involved in organizing the Symposium: the International Advisory Committee has been advising
on topics and selection of speakers and the Organizing Committee
has taken a burden of usual daily work.

The members of both committees are listed below.

The International Advisory Committee:

H. H. Barschall (USA), H. Feshbach (USA), J. Formánek (Czechoslovakia), I. M. Frank (USSR), E. Gadioli (Italy), M. K. Mehta
(India), S. Raman (USA), I. Ribanský (Czechoslovakia), D. Seeliger (GDR), C. Wagemans (Belgium), H. A. Weidemüller (FRG),
Z. Wilhelmi (Poland).

The Organizing Committee:

R. Antalík, E. Běták, Š. Gmuca, S. Hlaváč, J. Krištiak (Secre-
tary), Š. Luby, P. Obložinský, J. Pivarč, I. Ribanský, J. Šácha
(all from the Institute of Physics, Electro-Physical Research

Centre of the Slovak Academy of Sciences, Bratislava), J. Dobeš (Institute of Nuclear Physics, Řež near Prague), I. Wilhelm (Charles University, Prague).

The Symposium was organized in seven sessions with 13 invited talks and 11 orally presented short contributed papers. Three poster sessions have been arranged so that all other contributed papers could be presented as posters. The chairmen of the sessions, M. K. Mehta, R. E. Chrien, I. Ribanský, H. A. Weidenmüller, C. Wagemans, V. G. Soloviev, J. Rapaport, were excellent in their guidance of the discussions which were, in many cases, very lively.

These Proceedings contain texts of almost all papers presented at the Symposium. Unfortunately, it came that all manuscripts had to be edited and retyped. They were carefully compared with originals, however, such a procedure is not guaranteed to be an error-free one and we ask an excuse for occurred errors. We also take opportunity to redraw some of the figures.

In this book, the papers have been grouped by subject in somewhat different order from that in the conference programme.

The major credit for the success of the Symposium goes to all the speakers for their stimulating and excellent talks and to attendance for their active participations in the discussions and social events.

Anyone (including us) who thinks that planning and running a conference is entertainment is relieved of this fancy by actually running it. Therefore, we want to thank the members of the International Advisory Committee for their support and help. We should like to acknowledge the work of the members of the Organizing Committee, who cared for transportations, technical aids and poster sessions.

From early stages of the Symposium its secretaries Miss E. Kubincová and Mrs. D. Strapcová dedicated a sizeable fraction of their time to organization. We highly appreciate them for their competence and charm.

Once more, many thanks to all those who made everything run smoothly.

Bratislava, August 30, 1985   J. Krištiak, E. Běták

PRE-EQUILIBRIUM PROCESSES IN NUCLEAR REACTIONS

P. E. Hodgson

Nuclear Physics Laboratory, Oxford, U.K.

ABSTRACT. The quantum-mechanical theory of pre-equilibrium reactions due to Feshbach, Kerman and Koonin is reviewed and applied to the analysis of reactions of neutrons on ^{59}Co. The semi-classical and quantum-mechanical theories of pre-equilibrium reactions are compared, and their future development discussed.

1. INTRODUCTION

During the nineteen sixties, evidence accumulated indicating that it is possible for particles to be emitted after the first stage of nuclear interaction but long before the attainment of statistical equilibrium. These are the pre-equilibrium or pre-compound reactions.

Many attempts have been made to understand pre-equilibrium processes in terms of a series of nucleon-nucleon interactions within the target nucleus; the first interaction gives the direct process, and the following stages the pre-equilibrium processes. Starting with the pioneer work of Griffin /1/, a series of semi-classical or exciton models of varying complexity has been developed, and with appropriate choice of parameters these are often able to fit the observed energy and angular distributions of the emitted particles. More recently, several quantum-mechanical theories have been proposed, and these provide in principle a way of calculating the cross-sections of pre-equilibrium processes without the uncertainties of the semi-classical approximations.

Pre-equilibrium processes make substantial and in some cases dominant contributions to the cross-sections of reactions ini-

1

tiated by neutrons from 10 to 20 MeV. The multistep compound
(MSC) and multistep direct (MSD) theories of Feshbach, Kerman
and Koonin are described in section 2, and applied to some neutron reactions on ^{59}Co. Some conclusions are given in section 3.

2. THE QUANTUM-MECHANICAL THEORY OF FESHBACH, KERMAN AND KOONIN

The basic physical picture underlying the quantum-mechanical
multistep theory is the same as for the exciton model. It is
assumed that the interaction between the incident nucleon and
the target nucleus takes place in a number of stages of increasing complexity. The projectile enters the nucleus and collides
with a nucleon, producing a two-particle one-hole excitation.
The secondary particles can themselves interact, producing three-
particle two-hole excitations and so on. At each stage there is
a finite probability that the reaction proceeds to the next stage,
returns to a previous stage or goes directly to the continuum;
this latter possibility corresponds to pre-equilibrium emission
/2/.

2.1. The Multistep Compound Theory

To make the computation practicable, it is assumed as in EM that
the transitions within a chain of excitations can only be made
one step at a time; this is called the chaining hypothesis and
is equivalent to the assumption of binary collisions only. The
stage number N is related to the exciton number n by n = 2N+1.
Since the probability of any transition is proportional to the
level density in the final stage, and since the level density
increases very rapidly with the number of particle-hole pairs,
transitions back to a preceding stage are very unlikely. For
simplicity it is assumed that the probability of such transitions is negligible, this is the "never come back" hypothesis.
    In the context of this physical picture of the reaction, the
cross-section for pre-equilibrium emission as in EM is the product of three factors, namely the cross-section for the formation of the composite system multiplied by the sum over all

2

stages of the probability of precompound emission from the N-th
stage times the probability of reaching that stage.

The probability of emission of a particle into the continuum
from the N-th stage is given by the sum of the probabilities of
all possible emission processes divided by that of the sum of
all processes. The former are proportional to the sum of all
products of emission widths $\Gamma_{NJ}^{\uparrow lsv}(U)$ and the level densities
$9_J^v(U)$ of the final states in stage v at excitation energy U,
and the latter are the total widths Γ_{NJ}.

The probability of reaching the N-th stage without pre-equilibrium emission is given by the product of the probabilities
of surviving the m-th stage; this may be compared with the de-
pletion factor used in EM following the "never-come back" assumption. Collecting these factors together gives for the double
differential cross-section for pre-equilibrium emission

$$\frac{d^2 G}{d\Omega\, d\varepsilon} = \pi\lambda^2 \sum_J (2J+1)\left[\sum_{N=1}^{r} \sum_{ls\lambda v} C_{lsJ}^\lambda P_\lambda(\cos\theta)\,\right].$$

$$\cdot \sum_v \frac{<\Gamma_{NJ}^{\uparrow lsv}(U)9_J^N(U)>}{<\Gamma_{NJ}>}\right]\left(\prod_{m=1}^{N-1} \frac{<\Gamma_{mJ}^{\downarrow}>}{<\Gamma_{mJ}>}\right)T_J.$$

All the factors in the above expressions are calculated quan-
tum-mechanically or, as in the case of the level density function, obtained from the known systematics of nuclear properties.
The cross-section for the formation of the compound nucleus is
obtained from the optical model.

2.2. The Multistep Direct Theory

At higher energies it is increasingly likely that throughout the
reaction there is always at least one particle in the continuum.
As before it is convenient to consider the reaction as taking
place in a number of stages. The total emission cross-section
is the sum of emissions in all stages, and may be written as the
sum of one-step and multistep processes. At each stage we must
integrate over all angles and momenta so the full expression for
the multistep cross-section is

$$\frac{d^2\sigma_{if}}{dU\,d\Omega}\bigg|_{multistep} = \sum_n \sum_{m=n-1}^{n+1} \int \frac{dk_1}{(2\pi)^3} \int \frac{dk_2}{(2\pi)^3} \cdots \int \frac{dk_N}{(2\pi)^3} \cdot$$

$$\cdot \frac{d^2W_{mN}(k_f,k_N)}{dU_f\,d\Omega_f} \frac{d^2W_{N,N-1}(k_N,k_{N-1})}{dU_N\,d\Omega_N} \cdots$$

$$\cdots \frac{d^2W_{2,1}(k_2,k_1)}{dU_2\,d\Omega_2} \frac{d^2\sigma_{if}}{dU_1\,d\Omega_1}\bigg|_{singlestep} ,$$

where transition matrix element

$$\frac{d^2W_{N,N-1}(k_N,k_{N-1})}{dU_N\,d\Omega_N} = 2\pi^2 \varrho(k_N)\varrho_N(U) <|v_{N,N-1}(k_N,k_{N-1})|^2> ,$$

where $\varrho(k_N) = mk/(2\pi)^3\hbar^3$ is the density of particle states in
the continuum, $\varrho(N,U)$ the level density of the residual nucleus
at excitation energy U, and $v_{N,N-1}(k_N,k_{N-1})$ is the matrix element describing the transition from a state (N-1) to a state N
when the particle in the continuum changes its momentum from
k_{N-1} to k_N. This matrix element is given by the distorted wave
Born approximation expression

$$v_{a,b}(k_i,k_f) = \int \chi_a^{(-)*} <\Psi_f|V(r)|\Psi_i> \chi_b^{(+)}\,dr,$$

where $V(r)$ is the effective interaction for the transition $\chi_a^{(-)}$
and $\chi_b^{(+)}$ the incoming and outgoing distorted waves and Ψ_i and
Ψ_f refer to the initial and final nuclear states. To include
all the transition strength, the spectroscopic factors are always taken to be unity.

3.3. Multistep Analyses of Neutron-induced Reactions

The Milan group has already made many calculations using the
multistep compound and multistep direct theories /3/. More recently, the multistep compound theory has been used to calculate the inelastic spectrum of neutrons emitted when 14 MeV
neutrons interact with ^{59}Co and ^{93}Nb /4, 12/. These reactions

are particularly suitable to test the theory as much data are available, and they are also of practical importance.

At these energies, there are rather few open reaction channels and so it is possible to make a relatively complete analysis. In particular, the effective interaction strength can be fixed so that the sum of all the non-elastic cross-sections is the reaction cross-section obtained from the optical model calculation.

Preliminary calculations using the statistical theory of Weisskopf and Ewing /5/ showed that the (n,2n) cross-section dominates, (n,n´) makes a substantial contribution, and (n,np) and (n,pn) also contribute significantly. The calculation of V_o was

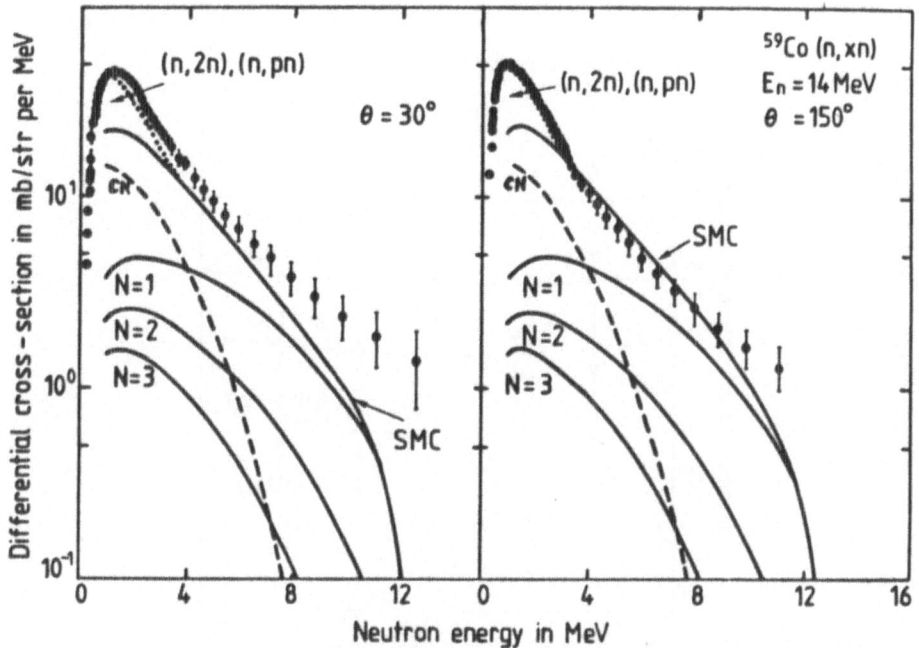

FIG. 1. Energy spectra of neutrons emitted at 30° and 150° from ^{59}Co at an incident neutron energy of 14 MeV /13/ compared with statistical multistep compound (SMC) calculations. The curves labelled with the value of N show the contributions of N-step processes and that labelled CN the final statistical compound nucleus cross-section. The SMC curve gives the sum of all these processes, and the dotted curve shows the effect of adding the neutron cross-section attributable to (n,2n) and (n,pn) reactions /4/.

made accordingly, using the experimental value of the (n,2n) cross-section and the Weisskopf—Ewing values of the smaller cross-sections. The resulting energy distributions of the neutrons at 30° and 150° are compared with the experimental data /13/ in Figs. 1 and 2. It is notable that the angular distribution is nearly isotropic up to about 6 MeV, confirming that most of the cross-section is attributable to multistep compound processes. There is evidence for forward peaking at the higher neutron energies in the 30° spectrum which is probably attributable to multistep direct reactions. At all angles there are substantially more higher energy neutrons than would be expected from the Hauser—Feshbach theory and this excess is accounted for by the FKK theory. Al lower neutron energies there is an excess of measured cross-section over that given by the FKK theory, and

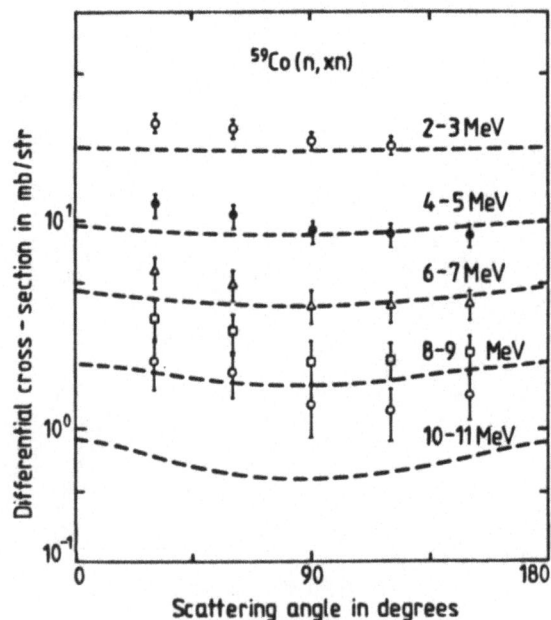

FIG. 2. Angular distributions of neutrons in selected energy ranges emitted from ^{59}Co at,an incident energy of 14 MeV /13/ compared with the sum of the multistep compound (n,n) cross-sections and the contributions of the (n,2n) and (n,np) cross-sections evaluated by the Weisskopf—Ewing theory. The excess cross-section at forward angles and at higher outgoing energies is attributable to multistep direct processes /4/.

this is attributable to the additional second-stage emission of neutrons from the (n,2n) and (n,pn) reactions. The MSC theory is very suitable for calculations of these processes, because direct terms are only important in first-particle emission at high outgoing energies /12/. Calculations of the cross-sections of these processes show that they account for the observed excess. Another check on the correctness of the calculation is provided by the (n,p) cross-section at backward angles, and this is compared with the experimental data in Fig. 3.

The Validity of the FKK Theory. It was pointed out by Kawai /6/ (1980) that the FKK multistep direct theory expresses the cross-section in terms of non-DWBA matrix elements, whereas the Milan group used DWBA matrix elements in their calculations. Bonetti therefore re-programmed the calculations using the non-DWBA matrix elements required by FKK, but found that it was no longer possible to obtain acceptable fits to the data.

The problem was finally solved by Feshbach /8/ who realized that the energy-averaging had been incorrectly carried out in

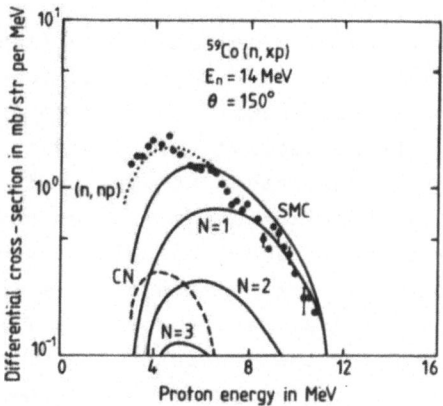

FIG. 3. Energy spectrum of protons emitted at 150° from ^{59}Co at an incident energy of 14 MeV. The experimental data /7/ are compared with the statistical multistep compound (SMC) calculations. The curves labelled with the values of N show the contributions of N-step processes and that labelled CN the final statistical compound nucleus cross-section. The full curve gives the sum of these processes and the dotted curve shows the effect of adding the cross-section attributable to the (n,np) reactions /4/.

the FKK paper, and that when this is done correctly it has the effect of transforming the non-DWBA matrix elements into DWBA matrix elements. Thus the original calculations of the Milan group are correct after all, and some critical comments /9, 10/ are no longer relevant.

3. CONCLUSIONS

The various pre-equilibrium theories differ appreciably in their flexibility, in several different respects. The semi-classical theories have been applied to a much wider range of reactions than the quantum-mechanical theories, in particular to those initiated by complex particles and those leading to the emission of many particles. By contrast, the quantum-mechanical theories have so far been confined to nucleon interactions with not more than two emergent particles. In the next few years the quantum-mechanical theories will certainly be applied to a wider range of reactions, but at present the semi-classical theories are the only ones that can be used for many of the more complicated reactions.

The semi-classical theories are also more flexible in that they have more model parameters than the quantum-mechanical theories. Here one must distinguish between internal parameters that are special to the particular theory and those that are fixed by some external constraint. As an example of the former we may mention the parametrization of the residual matrix element in EM which at present cannot be calculated from a more fundamental theory. In the latter category are the optical model and level density parameters. Here there is a further distinction that is important for the predictive power of these theories: reliable global optical potentials are now available for nucleons so that the cross-sections can be calculated from them for any nucleons with good accuracy. The particle-hole level density parameters, on the other hand, cannot be represented with sufficient accuracy by global formulae and only the total level density could be fitted to the experimental data for each nucleus. Therefore there is a strong need to obtain reliable expressions for these particle-hole level densities, in parti-

cular for the lowest stages. These considerations apply both to
the semi-classical and to the quantum-mechanical theories.

If all that is required is a fit to a particular data set,
this can be achieved by both the semi-classical and the quantum-
mechanical theories for the energy distributions of the emerging
particles. The semi-classical calculation may require some adjustment of an "internal" parameter, and both types are subject
to the above remarks about "external" parameters, particularly
those relating to particle-hole level densities. In practice
the range of applicability of the semi-classical models is often
well known and it is not necessary to adjust "internal" model
parameters in each computation. The differences between the results are therefore quite small and further calculations are
needed to detect deviations in other energy and mass ranges.

With respect to <u>angular distributions</u> the quantum-mechanical
effects may be more important. The most recent descriptions in
EM and GDH are based upon the scattering in infinite nuclear
matter with quasi-classical descriptions of refraction and/or
finite-size effects. Conceptually the QM theories are of course
superior. In particular, there are difficulties in the description of back-angle cross-sections with the semi-classical theories that do not exist in QM theories. This was illustrated
recently by Holler et al. /11/ in a comparison of semi-classical
and quantum-mechanical pre-equilibrium calculations for the
$^{65}Cu(p,\alpha n)$ reaction at 26.7 MeV. The GDH model of Blann and
Vonach /14/ gives a good overall fit to the energy distribution
of the emitted neutrons but is unable to fit the angular distribution in the very forward and backward directions. As shown in
Fig. 4a the back-angle discrepancy persists when refraction and
finite-size effects are included in the calculations. Similar
calculations have been made by the present authors using the
PRANG code. With the same calculational method as reported in
section 2 we obtained the dotted curve in Fig. 4a. The differences are due to a truncation of the Legendre polynomial expansion to l_{max} = 6 and to an adjustment of f_2 by 20% just to avoid
negative values of the scattering kernel. The resulting curve
is still below the data at backward angles. Quantum-mechanical
calculations with the FKK theory are, however, able to fit the

data over the whole angular range, as shown in Fig. 4b. This
shows that the quantum-mechanical theories are able to evaluate
interference effects that are beyond the scope of the semi-classical theories. In addition, the description of finite-size
effects, leads to serious difficulties in the semi-classical
models, which perhaps could be solved in a pragmatic way by

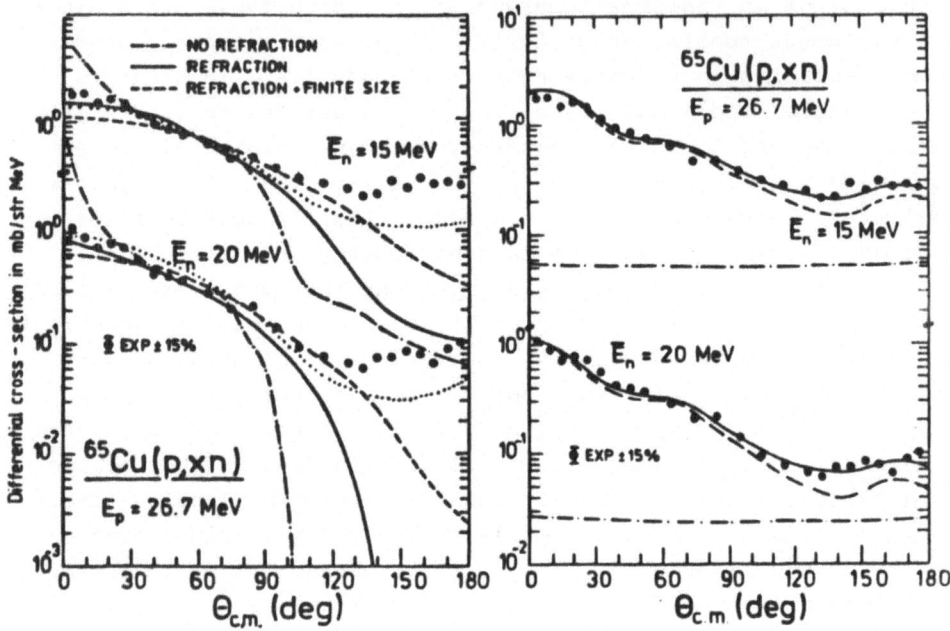

FIG. 4. a - Angular distributions for the $^{65}Cu(p,xn)$ reaction
at 26.7 MeV compared with geometry-dependent hybrid model calculations. The data are plotted in ΔE_n = 1 MeV bins centred at
energies E_n. The calculations are shown for pure NN-scattering
(dot-dash), NN-scattering with entrance channel refraction
(solid) and with refraction plus finite size correction (dashed)
/11/. The dotted curve has been added and shows the results of
using the same model (PRANG). If refraction effects are not
included and a very large number of Legendre coefficients are
used the dot-dash curve is reproduced. Introduction of entrance
and exit refraction and truncation of l_{max} = 6 gives slightly
better results as the dashed curve. Correcting for the negative
values of the scattering kernel by increasing f_2 by 20% gives
the dotted curve. A further increase of f_2 would fit the data.
b - Angular distributions for the $^{65}Cu(p,xn)$ reaction at 26.7
MeV compared with FKK quantum-mechanical calculations. The statistical multistep direct contributions are shown by the dashed
line, the statistical multistep compound by the dash-dot line
and their sum by the solid line /11/.

utilizing the results for a systematic study of precompound angular distribution using QM theories.

This work is part of a programme of quantum-mechanical studies of pre-equilibrium reactions that is being carried out in collaboration with Dr R. Bonetti (Milan), and Miss G. M. Field (Oxford) and I am grateful to them for permitting me to present their results. These results have also been presented, together with a survey of semi-classical theories by Dr H. Gruppelaar (Petten) and Dr P. Nagel (NEA Data Bank), to the International Conference on Nuclear Data for Basic and Applied Science held in Santa Fé, New Mexico in May 1985. A fuller account of all this work will be published in Revista del Nuovo Cimento.

REFERENCES

1. J. J. Griffin, Phys. Rev. Lett. $\underline{17}$ (1966) 478.
2. H. Feshbach, Proceedings of International Conference on Nuclear Reaction Mechanisms, Varenna, 1977.
   H. Feshbach, A. Kerman and S. Koonin, Ann. Phys. (N.Y.) $\underline{125}$ (1980) 429.
3. L. Avaldi, R. Bonetti and L. Colli-Milazzo, Phys. Lett. $\underline{94B}$ (1980) 463.
   R. Bonetti, M. Camnasio, L. Colli-Milazzo and P. E. Hodgson, Phys. Rev. $\underline{C24}$ (1981) 71.
   R. Bonetti, L. Colli-Milazzo, I. Doda and P. E. Hodgson, Phys. Rev. $\underline{C26}$ (1982) 2417.
   R. Bonetti and L. Colombo, Phys. Rev. $\underline{C28}$ (1983) 980.
   R. Bonetti, L. Colli-Milazzo and M. Melanotte, Phys. Rev. $\underline{C27}$ (1983) 1003.
   R. Bonetti, L. Colli-Milazzo, A. De Rosa, G. Inglima, E. Perillo, M. Sandoli and F. Shahin, Phys. Rev. $\underline{C21}$ (1980) 816.
   R. Bonetti, L. Colli-Milazzo, M. Melanotte, A. De Rosa, G. Inglima, E. Perillo, M. Sandoli, V. Russo, N. Saurier and F. Shahin, Phys. Rev. $\underline{C25}$ (1982) 717.
4. G. M. Field, R. Bonetti and P. E. Hodgson (1985), in preparation.
5. V. F. Weisskopf and D. H. Ewing, Phys. Rev. $\underline{57}$ (1940) 472.
6. M. Kawai (1980), private communication to H. Feshbach and A. Kerman.
7. L. Colli, S. Micheletti and M. Pignanelli, Nuovo Cim. $\underline{21}$ (1961) 966.
8. H. Feshbach, Ann. Phys. (N.Y.) $\underline{159}$ (1985) 150.
9. T. Udagawa, K. S. Low and T. Tamura, Phys. Rev. $\underline{C28}$ (1983) 1033.
10. I. Kumak, H. Haruta, M. Hyakutake and M. Matoba, Phys. Lett. $\underline{140B}$ (1984) 272.

11. Y. Holler, A. Kaminsky, R. Langkau, W. Scobel, M. Trabandt, and R. Bonetti, Preprint, 1985.
12. M. Herman, A. Marcinkowski and K. Stanklewicz, Nucl. Phys. A430 (1984) 69.
13. O. A. Salnikov, G. N. Lovchikova, G. V. Kotelnikova, A. M. Trufanov and N. I. Fetisov, Yad. Fiz. 12 (1970) 1132; Sov. J. Nucl. Phys. 122 (1971) 620.
14. M. Blann and H. K. Vonach, Phys. Rev. C28 (1983) 1475.

NUCLEONS CLUSTERIZATION

A. Iwamoto, *K. Harada and *K. Sato

Department of Physics, Japan Atomic Energy Research
Institute, Tokai, Naka-Gun, Ibaraki, Japan
*Division of Physics, Tohoku College of Pharmacy,
Komatsushima, Seendai, Japan

ABSTRACT. Models for the pre-equilibrium emission of fast light
particles are given based on exciton model. For the light ion
reaction, energy spectra of p, d, ^{3}He and alpha emission are
given together with the calculation of the angular distribution
of nucleon and alpha. A new model for the fast particle emission
in heavy-ion reaction is introduced and applied to data analysis.

1. INTRODUCTION

For the calculation of the high-energy tail of particle energy
spectra, exciton model proposed by Griffin /1/ has been used
extensively and reasonable fits have been obtained for the low-
energy nuclear reaction. This model proved to be very effective
to describe in a simple manner the cross-section for nucleon
and other light ion induced reaction. For the emission of light
composite particle like d, t, ^{3}He and alpha, however, the applicability of it is not so clear. Direct application of this model
to composite particle emission leads to the cross-section much
smaller than that of nucleon emission, because the removal of
more than one particle from particle state reduces the level
density much. It is much smaller than the experimental data. To
solve this problem in the framework of exciton model, several
models have been proposed among which the model proposed by
Ribanský and Obložinský /2/ and that of Milan group /3/ have
been applied to many data and both have proved to get reasonable good fitting to data. We found some problems in these
treatments and in these few years developed a new model of composite particle emission on the basis of exciton model. It takes
into account the intrinsic structure of emitted particle in its

13

formulation /4/. Application of it to light ion induced reaction showed a good result for d, t, ^{3}He and alpha-energy spectra. We developed afterwards the model /5/ which can calculate the angular distribution based on the idea of generalized exciton model /6--8/ where the two-body scattering kernels are treated more correctly taking into account the Fermi motion of target nucleons. Since this extension of generalized exciton model were found to be very effective to calculate the angular distribution, we applied the idea also to alpha-particle emission. The resulting angular distribution fitted the experimental data very well.

Our model described so far seems to be a good tool for the description of composite particle emission with incident energy of several tens of MeV. Especially the ratio of cross-section between nucleon emission and composite particle emission is described reasonably well. In the calculation of angular distribution in our model, the cross-section at very backward angle is too small compared to the experimental data. For the solution of this problem, we have recently developed a model /9/ in which the Fermi gas was replaced by the gas in the harmonic oscillator potential. It has really brought an effect of enhancing the backward cross-section for fast particle emission.

Another challenging field is the applicability of our model of composite particle emission in the case of heavy ion reactions. In this case, the ratio of fast alpha particle relative to the nucleon is much greater than in the case of light ion reaction. Thus we expect it very difficult to get this ratio in the exciton model. Of course the nucleon emission itself is not established for heavy ion reaction /10/ especially the set up of an initial state of the reaction is ambiguous. Very recently we developed a model to handle it and have made the preliminary calculations. The result was very promising with the idea of pick-up type component as was used in light ion reaction. Also the new factor coming from the linear momentum constraint plays the crucial role to enhance the cross-section of alpha emission.

In the next section, we briefly review our model of light ion reaction and describe also the recent work of modification of the angular distribution based on the harmonic oscillator model. In section 3, we give our model of heavy ion reaction.

2. LIGHT ION REACTION

We concentrate on the high-energy component of the energy spectra and thus use the most simple version of the exciton model, that is, we use the never-comeback approximation. The nucleon emission part is taken as the standard treatment and we describe only the change in describing the composite particles. Emission rate of the particle of energy ε is given for the nucleon emission assuming the detailed balance argument as

$$W_n(\varepsilon) = \frac{2s+1}{\pi^2 \hbar^3} \mu\varepsilon\sigma_{inv} \frac{\omega(p-1,h,U)}{\omega(p,h,E)} \qquad (1)$$

where s is the spin, μ the reduced mass and E and U, the excitation energy of composite system and residual nucleus, respectively. Particle hole level density ω is given by the Ericson's formula. In the case of composite particle emission, we assumed instead of eq. (1) the form given by

$$W_{n(1,m)}(\varepsilon) = \frac{2s+1}{\pi^2 \hbar^3} \mu\varepsilon\sigma_{inv} F_{1,m}(\varepsilon) \frac{\omega^*_{n(1,m)}(U)}{\omega(p,h,E)} \qquad (2)$$

where $F_{1,m}$ is the formation factor given below. The quantity ω^* is the level density of the residual nucleus of excitation energy U. Introduction of $F_{1,m}$ is related to the inclusion of the pick-up type reaction for the case of composite particle emission. In Fig. 1, we show for the case of alpha emission the momentum space picture. Here the large sphere corresponds to the Fermi sphere of the parent nucleus and the smaller sphere is that of the alpha's. As is seen in this figure, the spread of the alpha momentum space is rather large and as a result, even for the relatively high energy, there is an overlap of two spheres. This overlap means that four nucleons which constitute alpha particle are not always the particle state but it contains contribution of particle inside the Fermi sphere. This figure leads us to the idea that for the case of composite particle emission, the constituent particles need not always be the one coming from the particle state but also come from the nucleon

15

inside the Fermi sphere. Inclusion of this kind of the process corresponds to the pick-up type reaction.

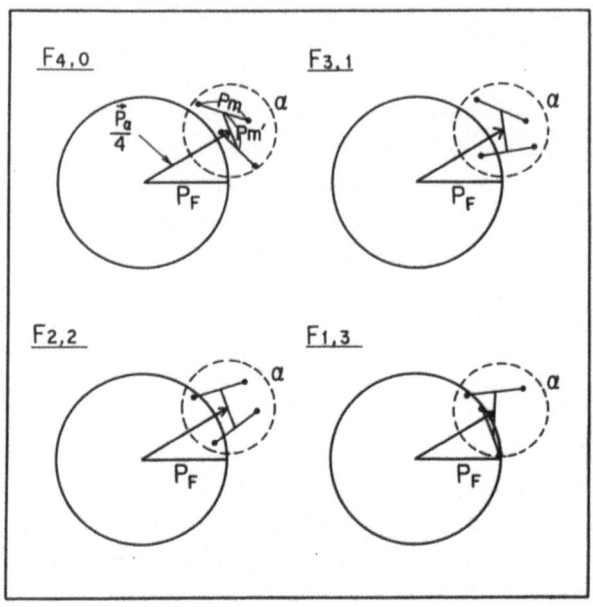

FIG. 1. Momentum space relation of the parent nucleus and alpha particle. P_α is the total momentum of alpha.

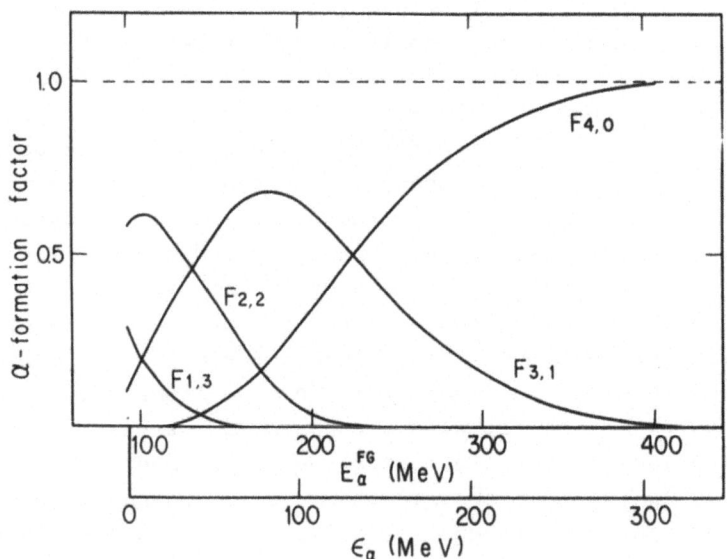

FIG. 2. Alpha particle formation factor $F_{1,m}$ as a function of alpha energy ε_α.

We calculated the portion of the contribution corresponding to four figures in Fig. 1 by using the simple Fermi gas model. The result is shown in Fig. 2. As is seen clearly in this figure, $F_{4,0}$ component which is usually used for the alpha emission in the exciton model occupies only a small part. Remaining part is occupied by various pick-up components, one-particle pick-up type $F_{3,1}$, two-particle pick-up type $F_{2,2}$ and three-particle pick-up type $F_{1,3}$. Since the intrinsic state of alpha is a pure state, the sum of all four contributions is equal to unity. Four-particle pick-up does not contribute to the positive alpha energy. At very high energy emission, as is expected, the pick-up components diminish and $F_{4,0}$ tends to one. But the diminishing of pick-up component occurs at an energy of more than 300 MeV and for alpha particle, which is due to the fact that the alpha particle is a tightly bound system and the spread of the momentum space for the intrinsic motion is large. As a result, the pick-up contribution remains finite even at very high energy. On the contrary, a weekly bound system like deuteron, pick-up component diminishes at rather low energy. Use of the formula given by eq. (2) brings about the enhancement of the emission rate. It is because for the pick-up type reaction, the particle number of the residual nucleus is larger than p-4 for alpha particle and reduction of the level density compared to that of the parent nucleus is smaller. The newly created holes from pick-up type reaction, are not to be assumed effective for the calculation of the level density of the residual nucleus. It comes from the fact that the intrinsic state of the composite particle is a pure state and though the strength splits up into many single particle states, the sum of these probability is one. We calculate the formation factor F for composite particles and then using the emission rate given by eq. (2), calculate the energy spectra. An example of the result is given in Fig. 3 for the ^{197}Au+p reaction at incident energy of 62 MeV. As is seen from this figure, the spectra of light composite particle are reproduced rather well except at very high-energy part. The fitting is best for the alpha particle and as we go to ^{3}He and d, the discrepancy at high energy becomes large. The high-energy tail of deuteron will arise from the pure direct reaction. For

the ^{3}He, the cross-section is about one order smaller than t
and alpha and spectra shape is not of the standard form, which
causes the fitting difficult. We calculated several other systems for proton to alpha emission at incident energy of several
tens of MeV and got the similar quality of fitting for all the
reactions, without changing any parameters. From this experience,
we conclude that our model of composite particle emission is
one promising way of describing such spectra.

Common feature of the composite particle emission is that
the pick-up component play the important role. Dominating term
comes from, in the case of alpha emission, three-particle pick-
up. The original component of $F_{4,0}$ contributes negligibly small.
Thus the inclusion of the pick-up type component is essential
in our model to reproduce the magnitude of the cross-section of
the composite particle emission.

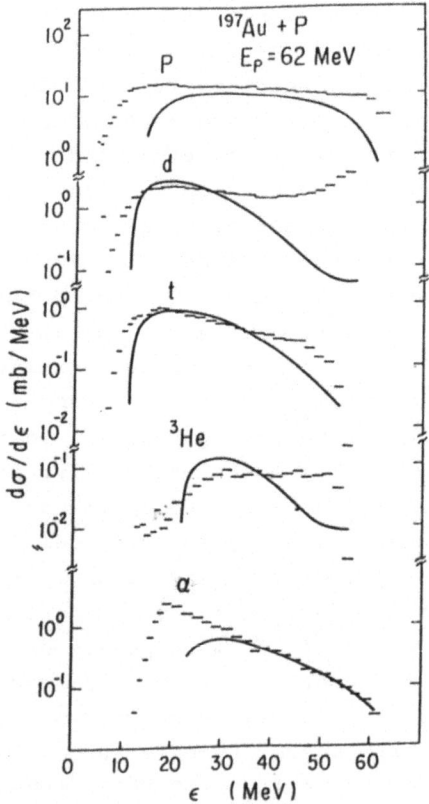

FIG. 3. Angle integrated energy spectra of p, d, t, ^{3}He and
alpha from 62 MeV p on ^{197}Au.

Next topic is the calculation of the angular distribution in exciton model. The model which is used often is the so-called generalized exciton model /6/, which is rather easy to calculate by use of the Legendre polynomial expansion method /7/. Introducing the effect of the Fermi motion, Ziyang et al. /8/ got an improvement in the angular distribution of backward angle. All such treatments use the assumption of "fast particle" and the direction of it is included as a new variable in the master equation. In order to further improve the assumption of "fast particle" and to include the Fermi motion effect correctly, we are lead to include the energy of "fast particle" also as a new variable. It is because the angular distribution in the nucleon-nucleon collision is very sensitive to the final energy of nucleon, that is, the high-energy nucleon is forward peaked, and the low-energy nucleon becomes more backward peaked. Previous treatments neglect this energy-angle correlation and use the same form of angle-dependent transition probability independent of the energy of the emitted particle.

We start from the master equation for the probability $q(n,\Omega,\varepsilon,t)$, where in addition to the n, Ω, and t of the generalized exciton model, the energy ε of "leading particle" is specified. It is given as

$$\frac{d}{dt} q(n,\Omega,\varepsilon,t) = \sum_m \int d\Omega' \int d\varepsilon' \, q(n,\Omega',\varepsilon',t) w_{m\to n}(\Omega',\varepsilon'\to\Omega,\varepsilon) -$$

$$- q(n,\Omega,\varepsilon,t)\left\{\sum_m \int d\Omega' \int d\varepsilon' w_{n\to m}(\Omega,\varepsilon\to\Omega',\varepsilon') + W_n\right\},$$

$$(3)$$

where we neglect the emission of secondary particle. Here W_n is the total emission rate of the particle and $w_{m\to n}(\Omega',\varepsilon\to\Omega,\varepsilon)$ is the transition rate from the m-exciton state to n-exciton state in association with the change of the particle variable from (Ω',ε') to (Ω,ε). The explicit form of $w_{m\to n}$ is written as

$$w_{n\to m}(\Omega',\varepsilon\to\Omega,\varepsilon\ ) = \lambda_{n\to m} G_m(\Omega',\varepsilon\to\Omega,\varepsilon),\qquad(4)$$

where $G_m(\Omega',\varepsilon\to\Omega,\varepsilon)$ is the two-body scattering kernel normalized with respect to integration of Ω and ε. $\lambda_{n\to m}$ denotes the total transition rate from n-exciton state to m-exciton state. Now we relate the quantity $q(n,\Omega,\varepsilon,t)$ of eq. (3) to double differential cross-section for the energetic nucleon emission. To this end, we postulate that it is written as

$$\frac{d^2\sigma}{d\Omega\,d\varepsilon} = \sigma_{abs} \sum_m \tau(n)W_n(\varepsilon) \frac{\tau(n,\Omega,\varepsilon)}{\int d\Omega \tau(n,\Omega,\varepsilon)} , \qquad (5)$$

where $\tau(n,\Omega,\varepsilon)$ is the time integration of $q(n,\Omega,\varepsilon,t)$ of eq. (3). In order to get this form, we assumed that the angle-integration of it coincides with the usual form of the exciton model. The structure of eq. (5) is very similar to that of the generalized exciton model. With our extension, however, the energy-angle correlation is taken into account.

Calculation of eq. (5) is rather easily done and for an example, we show in Fig. 4 the angular distribution of ^{197}Au+p

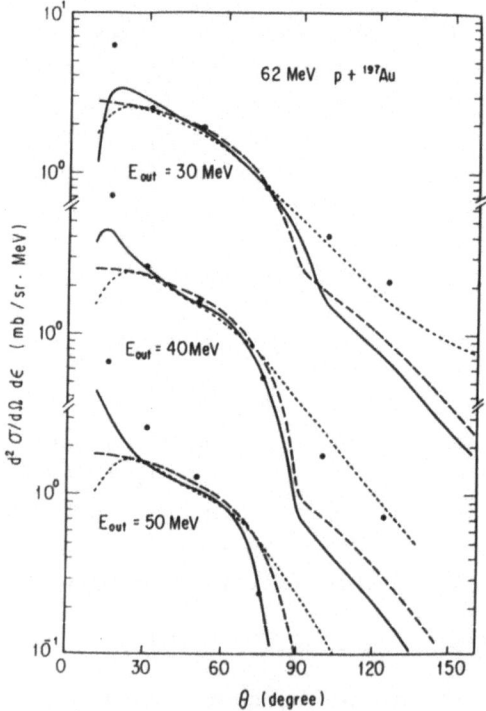

FIG. 4. ^{197}Au(p,p´) angular distribution for three outgoing energies.

reaction of incident energy 62 MeV for three different outgoing
energies. The solid line is our result and experimental data
are plotted together with the calculation under the method of
Mantzouranis et al. /6/ (dashed line) and of Ziyang et al. /8/
(dotted line). As is seen in this figure, our calculation reproduces nicely the angular distribution at forward angle. On
the other hand, at backward angles for 30 and 40 MeV, our results underestimate the data. In these outgoing energy region,
the method of Ziyang et al. gives good result, but it gives
too large backward cross-section at outgoing energy of 50 MeV.

The double differential cross-section of (p,α) reaction was
also calculated by using the above-mentioned method. The form
of the double differential cross-section is given parallel to
eq. (5), but because of the pick-up type component, it is a
little complicated and we will not give the detail of the formula /6/. Essential idea is that the angular distribution of
alpha particle is closely correlated to the "leading nucleon"
as calculated in eq. (5). It will be allowed because the dominating part of the alpha cross-section in our model comes from
the three-particle pick-up component and in such case, the scattering angle of one "leading particle" will play the crucial
role to the angle of alpha. In this way, the angular distribution of alpha is determined by the folding of the angular distribution of "leading particle" with respect to the intrinsic
momentum distribution of alpha particle. An example of the calculation is given in Fig. 5 for $^{209}Au(p,\alpha)$ reaction of incident
energy of 62 MeV for three different outgoing energies together
with the experimental data. Solid lines and dashed lines are
our calculation which differ with respect of the folding procedure. From this figure, we see a very good fit to data for both
forward and backward angles. We calculated 62 MeV $^{197}Au(p,\alpha)$,
39 and 62 MeV $^{120}Sn(p,\alpha)$ and 39 MeV $^{209}Bi(p,\alpha)$ reactions and
the same quality of fitting to data was obtained.

The last topic of the light ion reaction is the improvement
of the backward cross-section in (p,p'). Among the many possible
reasons for the lack of the backward cross-section, we think it
most important to include the effect of the finiteness of the
nucleus. For the Fermi gas, the nucleon-nucleon collision occurs

like a free collision apart from the limitation coming from the
Pauli principle. Collision of incident nucleon with the nucleon
in the finite well, on the other hand, is very different from
the free nucleon-nucleon scattering, because through the single
particle wave function, some many body effect is taken into
account automatically. For the backward cross-section, such
effect like the scattering with the target nucleus as a whole
is very important since it can cause a large momentum transfer.
Recently, Sato developed a model in which, roughly speaking,
the scattering kernel $w_{m \to n}(\Omega', \varepsilon' \to \Omega, \varepsilon)$ in eq. (4) is calculated
with the harmonic oscillator single particle wave function. For
the case of harmonic oscillator wave function interacting with
the plane wave, the matrix element corresponding to the inelastic collision is obtained in an analytic form and calculation

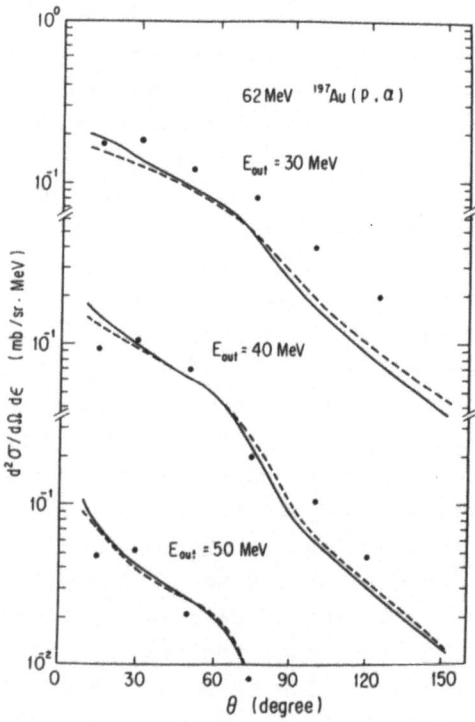

FIG. 5. 62 MeV ^{197}Au(p,α) angular distribution for three
outgoing energies.

is rather easily done. An example of this calculation is shown
in Fig. 6 for 62 MeV ^{120}Sn(p,p´) reaction for three outgoing
energies. As seen in this figure, the angular distribution is
greatly improved compared with the original treatment. Calculation of several other reactions were done and qualitatively
similar fitting to the data were obtained. From this, we can
say that the exciton model modified in this manner is very promising to easily calculate the angular distribution including
the backward angles.

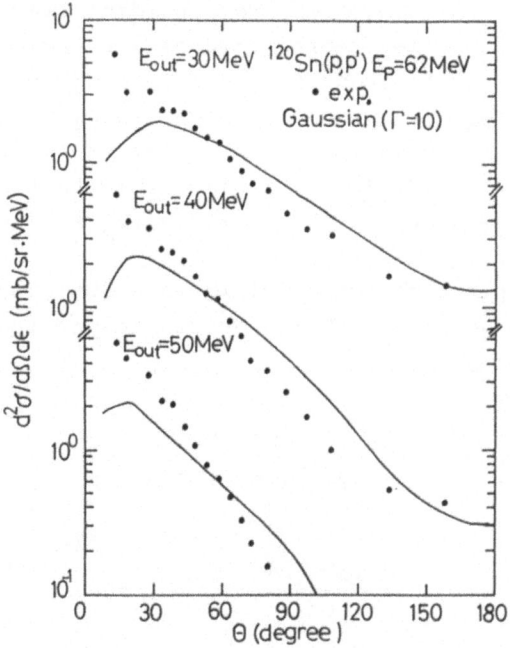

FIG. 6. 62 MeV ^{120}Sn(p,p´) angular distribution for three
outgoing energies.

3. APPLICATION TO HEAVY-ION REACTION

To calculate the fast particle emission in heavy-ion reactions,
several models were developed /10/. Now I present one another
based on exciton model which is developed for the fusion-like
reactions. From the incomplete fusion experiment, it is suggested /11/ that the emission of fast particles is strongly

correlated to the velocity of nucleons measured in the centre-of-mass frame. Also it is widely known that angular distribution of fast particle is forward peaked. Fast particles are thought to be emitted in the early stage of fusion reaction and in this stage, there are two flows of flux with opposite directions, one coming from the projectile and the other from the target. Considering the Fermi motions of projectile, target and compound nucleus, the momentum space for nucleons is represented as given in Fig. 7. This is an example for the $^{16}O+^{40}Ca \rightarrow {}^{58}Ni$ reaction of 157 MeV incident energy at the position of Coulomb barrier. Fermi momenta are taken to be the same for all nuclei and their centres are denoted by C_p, C_t and C_c. The difference $C_p - C_t$ is the centre-of-mass momentum per nucleon of the projectile, $C_c - C_t$, that of the target. When the mass of the projectile is lighter than that of the target, as is the case of Fig. 1, the relation $|C_p - C_c| > |C_c - C_t|$ holds.

Basic assumption of our model is to assume two sources of particle emission, one with its origin coming from projectile

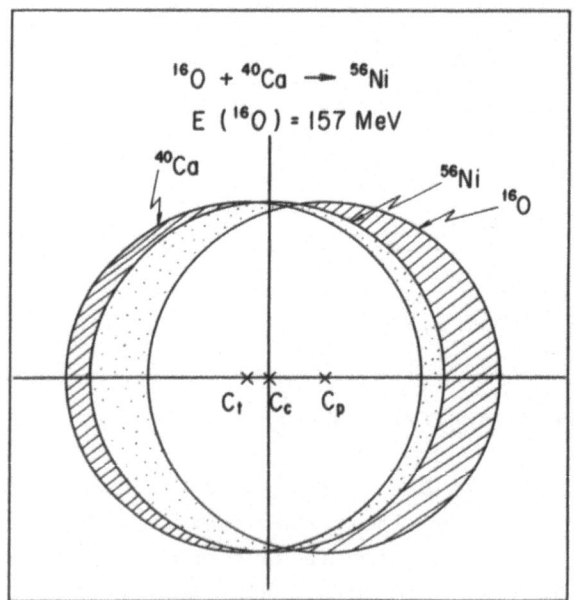

FIG. 7. Momentum space relation for the reaction $^{16}O+^{40}Ca$ of energy 157 MeV at the Coulomb barrier.

particles and the other from target particles. In Fig. 7 hatched region on the right is the particle state of projectile particles and hatched region on the left, that of the target ones. The dotted regions shown in both right and left are the hole states. The linear momentum of right hatched and left dotted regions are directed to the right and left hatched and right dotted regions are directed to the left. We identify the former regions as the source 1 (projectile source) and the latter as the source 2 (target source). These two are assumed to emit independently fast particles. The numbers of particles and holes in source 1 are the same and are denoted by N_1 and number of particles and holes in source 2 are denoted by N_2. We calculate the total linear momentum of source 1 and denote it by P_1 and that of source 2 by P_2. The total excitation energy of source 1 is E_1 and that of source 2 is E_2. These two sets of parameters, $\{N_i, P_i, E_i\}$ (i = 1, 2) are used as the initial values for the exciton model calculation.

For the momentum dependence of the exciton model, we start from the model of Mädler and Reif /12/. Important quantity is the level density $\omega(p,h,E,P)$ of the system, where in addition to the particle and hole numbers and energy, we specify the total linear momentum $P = |\vec{P}|$ of the excitons. I skip to give the formula to obtain this quantity but only say that starting from the Laplace transformation of the level density, it is calculated by using the saddle point method /13/. Main approximations are to neglect the Pauli effect and to replace the single particle momentum by its average value to calculate the momentum-dependent part of the level density. The cross-section for the nucleon emission is written as

$$\frac{d^2\sigma}{d\varepsilon\,d\Omega} = \sigma_{abs} \sum_n \left[\left\{\tau_n W_n(\varepsilon,\Omega)\right\}_{proj} + \left\{\tau_n W_n(\varepsilon,\Omega)\right\}_{targ}\right],$$

(6)

$$W_n(\varepsilon,\Omega) = \frac{2s+1}{\pi^2\hbar^3}\mu\varepsilon\frac{\sigma_{inv}}{4}\frac{\omega_r(p-1,h,U,|\vec{P}-\vec{p}|)}{\omega_c(p,h,E,P)},$$

where τ_n is the duration time for exciton state n obtained from the master equation and $\vec{p}$ is the momentum of emitted particle.

25

Detailed balance argument was assumed to obtain this relation. For the alpha emission, we introduce the pick-up type components together with the formation factor $F_{1,m}$ as given in eq. (2). For the calculation of momentum-dependent part, we also included the momentum of holes caused by the pick-up type reaction.

We show some results of the calculation of this model. Since this model is specially developed from the model of incomplete fusion reaction, it is desirable to fit the data of fast particle in coincidence with the evaporation residue. Examples of this kind are shown in Figs. 8 and 9 for the reaction of $^{16}O + ^{40}Ca \rightarrow ^{56}Ni$ at incident energy of 157 MeV /20/, which corresponds to Fig. 7. In these figures, the evaporation residues

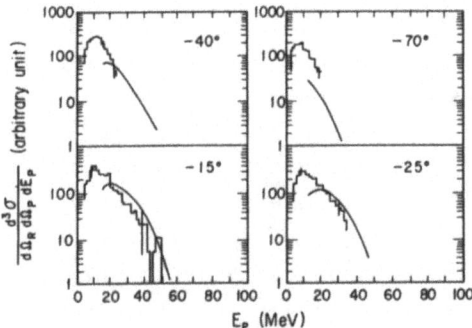

FIG. 8. Energy spectra of proton at four detection angles from the reaction 157 MeV ^{16}O on ^{40}Ca. Evaporation residue is observed at 9° in the experiment /20/.

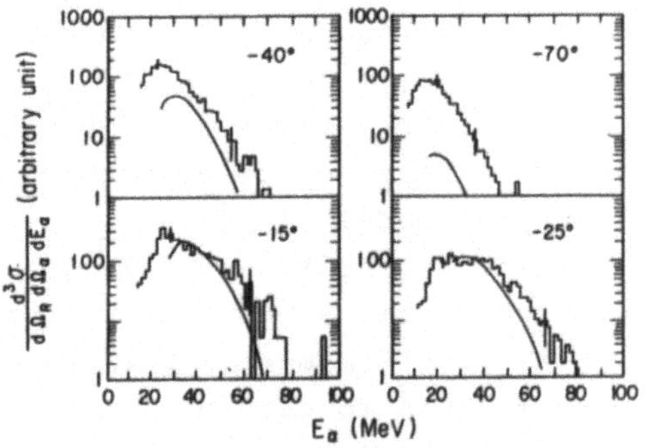

FIG. 9. Energy spectra of alpha at four detection angles from the reaction 157 MeV ^{16}O on ^{40}Ca. Evaporation residue is observed at 9° in the experiment /20/.

(mass > 40) are detected at $+9^o$ and Fig. 8 is the proton spectra at four angles and Fig. 9, that of alpha particle. Data are represented by triple differential cross-sections with arbitrary unit. The calculation, however, is only possible for the double differential cross-section and thus in these figures, we show by smooth solid lines the calculated results with an arbitrary normalization. Important thing is that both in experimental data and in calculations, the relative values at different angles and relative values of proton and alpha are fixed, that is, there is only one overall normalization. From Fig. 9, we find that our model fits the data rather well. Especially the tendency of the change of spectra shape as the angle changes from backward to forward is reproduced nicely. High-energy component is prominent at forward angles and decreases as the angle changes to backward direction. The lack of low-energy component in the calculation comes from the fact that in our calculation, only first 5 steps of exciton states are included and thus the evaporation part is not contained. In Fig. 9, the alpha spectra are shown together with the calculation. As is the case of proton, the general tendency of the spectra shape are reproduced nicely. At angles of -70°, the cross-section is too small to fit the data. Though not shown in the figure, however, the calculation based on evaporation LILITA code reproduces the backward data, which means that the pre-equilibrium component is not necessary in the backward direction. What I wish to stress is that the relative cross-section of proton and alpha is reproduced fairly well. The spectra shape is explainable from the coalescence model but the enhancement of the alpha strength compared to those of other composite particles is impossible to deduce. In our present calculation, this hard problem is solved at least qualitatively. In the calculation of Figs. 8 and 9, the initial exciton number is 4-particle 4-hole state (target source emission is negligible as far as forward emission of fast particle is concerned). The excitation energy is about 65 MeV and exciton state till the 8-particle 8-hole state is included in the calculation. The most part of the fast particle is coming from the first 2 steps, as was the case of light ion reaction both for nucleon and alpha emission. The main term of alpha emission is two-particle pick-up type reaction and the next term is the three-particle pick-up term and these two terms dominate the cross-section. Thus the pick-up type was found to be very important also in the heavy ion reaction. In fact more than one order enhancement occurs by the inclusion of the pick-up term. One another reason to the enhancement comes from the momentum-dependent part of the level density. The initial linear momentum of the excitons is rather high as is seen in Fig. 7 and reduction of the level density occurs because of the small phase space due to the small number of combination with respect to the momentum. After the emission of fast particle in the forward direction, the momentum of residual nucleus is reduced depending on the energy and angle of emitted particle. As a result, the momentum-dependent factor of residual nucleus becomes much larger than the pre-emission system. From this factor, the enhancement of the cross-section occurs for the fast particle in the forward direction. This factor causes the reduction for the backward emission of fast particles because in this case, the linear momentum of the residual nucleus becomes larget than the pre-emission system and more severe restriction on the momentum space causes the reduction of the factor in the level density. What happens when we compare the nucleon and alpha emission? For the nucleon and alpha with the same energy, the momentum of alpha is twice that of the nucleon. Thus the forward emitted alpha is enhanced more than that of the proton from the momentum-dependent part of the level density. From this factor, the alpha particle is enhanced over the proton by factor 5 or so. This is the second and completely new reason of the alpha enhancement.

Calculation was performed for the $^{16}O+^{27}Al$ system for three incident energies and also for $^{16}O+^{90}Zr$. In these cases the proton spectra calculated are sometimes overestimated up to one order of the data. The alpha spectra are reproduced very nicely without any tuning of parameter values. From the above comparison of data and calculation, it follows that our model is promising to be used for the heavy-ion reaction of these energy regions.

REFERENCES

1. J. J. Griffin, Phys. Rev. Lett. 17 (1966) 478.
2. I. Ribanský and P. Obložinský, Phys. Lett. 45B (1973) 318.
3. E. Gadioli, E. Gadioli-Erba and J. J. Hogan, Phys. Rev. C16 (1977) 1404.
4. A. Iwamoto and K. Harada, Phys. Rev. C26 (1982) 1821. K. Sato, A. Iwamoto and K. Harada, Phys. Rev. C28 (1983) 1527.
5. A. Iwamoto and K. Harada, Nucl. Phys. A419 (1984) 472.
6. G. Mantzouranis, D. Agassi and H. A. Weidenmüller, Phys. Lett. 57B (1975) 220.

7. J. M. Akkermans, Phys. Lett. 82B (1977) 20.
8. S. Ziyang et al., Z. Phys. A305 (1982) 61.
9. K. Sato, Phys, Rev. C, to be published.
10. M. Blann, Phys. Rev. C23 (1981) 205.
    S. Yoshida, Z. Phys. A308 (1982) 133.
    T. Otsuka and K. Harada, Phys. Lett. 121B (1983) 106.
    H. Machner, Phys. Rev. C28 (1983) 2173.
    K. Niita, Z. Phys. A316 (1984) 309.
11. H. Morgenstern et al., Phys. Rev. Lett. 52 (1984) 1104.
    C. S. F. Stephans et al., ANL preprint.
12. Mädler and R. Reif, Nucl. Phys. A337 (1980) 445.
13. A. Iwamoto, 4th Int. Conf. on Nuclear Reaction Mechanisms,
    Varenna, 1985.

MULTISTEP DIRECT REACTION ANALYSES OF CONTINUUM REACTIONS[*]

T. Tamura, T. Udagawa and M. Benhamou

Department of Physics, University of Texas,
Austin, Texas, USA

ABSTRACT. Essence of the concept of the direct reaction, and
of its corresponding theory is first discussed. Then the
reason why this theory may be used to describe continuum cross
sections is discussed, it being supported by results of a few
realistic calculations. It is also shown, somewhat surprisingly, that the direct reaction technique can be used to describe
even the fusion between heavy ions.

1. INTRODUCTION

A reaction is called direct, if it proceeds from a state to
another, without going through a formation of a compound system.
In most of the applications of the direct reaction (DR) theory,
such states are of two-body channels. Given a pair of nuclei,
or of ions, we are interested in describing their relative
motion. If it is assumed that both nuclei stay in their ground
states, what can happen is only the elastic scattering. And
our experience shows that we have to use optical model (OM)
to describe it, the channel wave function becoming what is
called distorted wave (DW). Therefore, the DR theory may be
regarded as a theory to describe the transition from a state
with a DW into another.

If more than one-step transformation must be made, for one
reason or another, in order to arrive at the final state, then
the theory to be used becomes the MSDR, i.e. multistep DR

---

[*]This work was supported in part by the U.S. Department
of Energy.

theory. For simplicity, however, we include the single-step DR
(SSDR) as a part of MSDR, in the following. Since the construction of DW, and hence the use of OM, are so unseparately tied
with any DR theory, we may even go one step further, so as to
consider the elastic scattering as an extreme of MSDR.

The OMP, i.e. the OM potential, that we use in practice is
complex. It thus describes an absorption, and a large fraction
of the absorbed amplitude is consumed to form a compound system.
Thus, somewhat contradicting to what was said above, the DR
theory does describe the formation of the compound system as
well. An important use of this possibility will be discussed
later in section 4.

Until a few years ago, the use of the MSDR theory was done
almost exclusively, so as to analyze data to reactions that
resulted in discrete final states. In such cases, it is rather
natural that one is most interested in transitions that have
large cross sections. This then means that the final channel
is strongly coupled to the initial channel, and this is why
the coupled-channels (CC), and the coupled-reaction-channels
(CRC) calculations were done quite often. For the case of
weak coupling, however, it is permissible to treat the transition perturbatively, i.e., under the Born approximation (BA).
Then the theory becomes the well known DWBA.

In the present paper, we apply the MSDR theory to describe
continuum cross sections. As is expected, and as will be explained later, it then means that a large number of final
states are involved. In other words, the dominance of selected
final states can normally be forgotten. Then the use of CC or
CRC theory becomes unnecessary. Thus, we basically understand
that the MSDR theory means a DWBA theory, including, however,
the first, second, etc., order versions of it.

In section 2, we discuss continuum reactions to which the
MSDR theory can be used in its most conventional form. In
section 3, we then comment very briefly on breakup reactions,
and in section 4, we discuss the possible use of the DR concept
in describing the fusion of two heavy ions. This paper will
then be summarized in section 5.

2. MSDR THEORY AND CONTINUUM CROSS SECTIONS

Let us consider first the simplest possible DR, i.e., the (n,n′) reaction, taking a doubly closed shell nucleus, e.g., ^{208}Pb as target. The low-lying states may be of the particle-hole (ph) nature, and the excitation of these states can be described rather easily in terms of DWBA. (For simplicity, we do not consider the formation of any collective state.) Actually, these low-lying states are discrete states, and are outside the interest of the present paper, as we have stated above.

Let us next consider a state that is still of a ph nature, but is lying rather high. Such a state can be formed, e.g., by promoting a nucleon in the target from a deep lying orbit into a vacant orbit. If this state is left alone, the inelastic excitation of this state can still be treated in terms of DWBA, because this state is nothing but a discrete state.

When excitation energy is high, as we have assumed, however, such a ph state cannot be left alone. In the same region of the excitation energy E_x, there are a number of 2-ph, 3-ph, ..., states, that are formed by promoting nucleons from high-lying occupied orbits into vacant orbits. These states are coupled to the above 1-ph state, via the residual interactions, and thus the amplitude of the original 1-ph state is distributed over a large number of complicated eigenstates in the neighborhood of E_x. To make the matter worse, there can be in fact a number of 1-ph states, also in the same neighborhood of E_x.

There are thus a very large number of very complicated eigenstates, and to perform DWBA calculations to obtain cross sections for exciting all these eigenstates is impracticable. It may in fact be awkward, even to attempt this, because we do not know the detail of the structure of these eigenstates, in the first place.

The matter is drastically simplified, however, if we reconsider the problem in the following way. We have already remarked that the amplitude of the starting 1-ph state is distributed over many eigenstates. We then consider whether it might be justified to forget about this distribution first, and then carry out the DWBA calculation to excite this 1-ph state. Then, instead of distributing the amplitude, we may simply distribute this cross section over the eigenstates. If this procedure is justified, it is clear that we do not have much difficulty in carrying out actual calculations.

A possible danger involved in the simplified treatment, described in the preceding paragraph, is the following. To distribute the cross section means to distribute the probability. It may thus fail to take into account the contributions of the interference terms, which may play an important role in a better theory, in which the amplitude is distributed first, and then the cross sections are calculated later. However, if we can show that these interference contributions cancel among themselves, and thus do not contribute to the final form of the continuum cross sections, then the above simplified treatment is justified. We shall now show that it is possible to demonstrate the cancelleration of the interference contributions.

We may write the wave function of an eigenstate as

$$|N\rangle = \sum_B a_B^{(N)} |B\rangle + \delta|N\rangle , \tag{1}$$

where $|B\rangle$ stands for a 1-ph state, it being mixed in $|N\rangle$ with an amplitude denoted by $a_B^{(N)}$, while $\delta|N\rangle$ describes the contributions to $|N\rangle$, e.g., of k-ph states with $k \geq 2$. It is legitimate to assume that the magnitude and sign of $a_B^{(N)}$ vary randomly, when it is seen as a function of N.

The (one-step) DWBA amplitude $T_N^{(1)}$ to excite this state $|N\rangle$ may be written as

$$T_N^{(1)} = \langle x_b|\langle N|x|0\rangle|x_a\rangle = \sum_B a_B^{(N)} t_{ba}^B \tag{2}$$

with

$$t_{ba}^B = \langle x_b|v_{B0}|x_a\rangle ; \quad v_{B0} = \langle B|v|0\rangle . \tag{3}$$

Here, $|0\rangle$ is the target ground state, while $|x_a\rangle$ and $|x_b\rangle$ are respectively, DWs in the incident and exit channels. Note that $\delta|N\rangle$ of (1) does not contribute to (2), because a one-step DR can create only 1-ph states.

By using (2) and (3), we can write the energy averaged one-step cross section as

$$d^2\sigma^{(1)}(E_n')/dE_n' \, d\Omega = \sum_N |T_N^{(1)}|^2 \, \delta(E_N - E_x) =$$

$$= \sum_N \delta(E_N - E_x) \sum_{BB'} a_B^{(N)} a_{B'}^{(N)} t_{ba}^B t_{ba}^{B'*} | =$$

(4)

$$= \sum_B c_B(E_x) \, |t_{ba}^B|^2 =$$

$$= \sum_B c_B(E_x) \, [d\sigma_B^{(1)}(E_n')/d\Omega] \; .$$

In (4), $E_x = E_n - E_n'$ is the excitation énergy due to the inelastic scattering, and $\sum_N \delta(E_N - E_x)$ means to take sum over states $|N\rangle$, whose energies come within an energy cell around E_x. In obtaining the third equality, we used a relation that

$$\sum_N a_B^{(N)} a_{B'}^{(N)} \, \delta(E_N - E_x) = \delta_{BB'} \, c_B(E_x). \tag{5}$$

A key concept used to obtain (5) is the statistical assumption that $a_B^{(N)}$ is random, and what the quality in (5) means is that the interference terms do cancell out. This is what we wished to happen.

The last line of (4) shows that we have succeeded in obtaining the continuum cross section in a desired form. There, $d\sigma_B^{(1)}(E_n')/d\Omega$ is nothing but the DWBA cross section for exciting a simple ph state $|B\rangle$, and it appears in (4) being multiplied by $c_B(E_x)$. This $c_B(E_x)$ describes the distribution of the unit probability, with which $|B\rangle$ is created at a ph energy E_B, over a range of E_x. This $c_B(E_x)$ should thus be constructed in practice, e.g., as a Gaussian distribution, centered at E_B, and very conveniently replaces the distribution of the probability over states $|N\rangle$, which was mentioned above.

The above is a review of the derivation of the MSDR formula, given in ref. /1/, which we shall call TUL henceforth. We did this only for the case of SSDR, but TUL also discussed it for two-step cases. Our practical use of the MSDR formula was first reported in ref. /2/. Prior to our work, however, Reif /3/ and

Lewis /4/ had reported on their use of the SSDR version of MSDR. See a long list of other related works given as ref. /46/ in TUL.

In TUL, results of applying the MSDR method for a variety of reactions were reported. These include (p,p'), (p,n) and (p,α) reactions, and, since the proton energy E_p used in the experiment was fairly high (45—65 MeV) we always considered one- and two-step contributions. As seen in a few figures given in TUL, our results fit data rather nicely, making us convinced that it is indeed senseful to use the MSDR method to describe continuum cross sections. An interesting application of this method to a heavy-ion induced reaction is found in Lenske et al. /5/.

More recently, a few analyses were made /6/ of (n,n') data /7/ taken with E_n = 25.7 MeV, and the results with ^{56}Fe and ^{209}Bi targets are given, respectively, in Figs. 1 and 2. Since the incident energy is now rather low, we expect that the two-step contributions are small, and we confirmed this numerically. The results shown in Figs. 1 and 2, which are thus essentially the results of SSDR calculations, agree rather nicely with data. Nevertheless, the theory has a tendency to underestimate some-what low E_n' (high E_x) data.

Analysis of (n,p) data /8/ was also done /6/, and the results

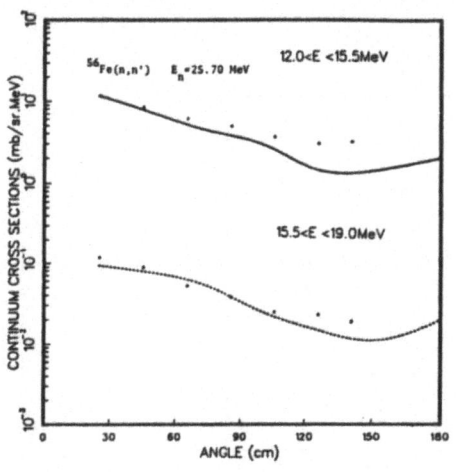

FIG. 1.   FIG. 2.

are shown in Fig. 3, for the case with ^{93}Nb target. The E_n used
this time is rather low; 14.1 MeV, and the out going proton
energy gets as low as 8 MeV. With these low energies, the validity of using DR theory gets rather tricky. We are also
suffered from a lack of knowledge of dependable optical model
parameters. Nevertheless, Fig. 3 shows that we were able to
fit the data rather nicely.

There are given two theoretical results in Fig. 3. In obtain-
ing the BG results (solid lines), we used the optical model
parameters (OMP) as given by Becchetti and Greenlees /9/,

Nb(n,p) E = 14 MeV

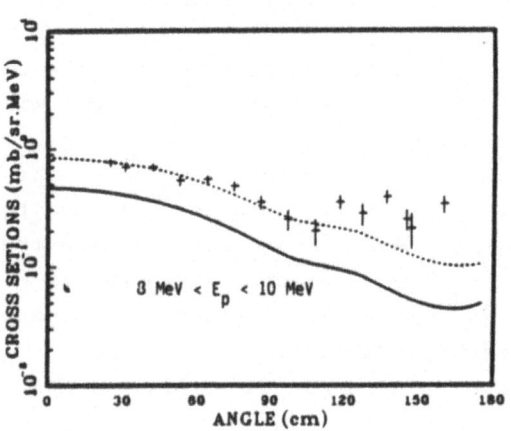

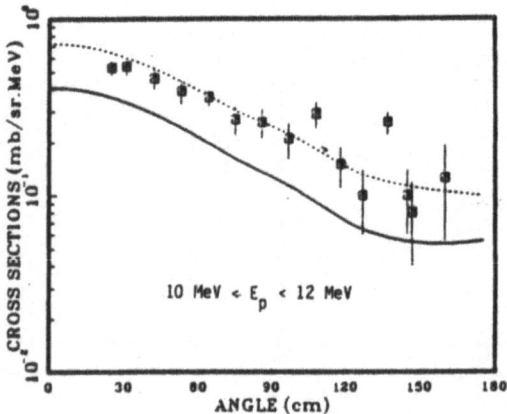

FIG. 3.

while the MBG (dotted lines) results were obtained by decreasing the imaginary potential so that lower energy protons became less absorptive. This helped to increase the theoretical cross sections somewhat, particularly for lower E_p, and thus to improve the fit.

We also present here the result of our recent analysis of ^{65}Cu(p,n) reaction data /10/, taken at E_p = 26 MeV. Again good fits to data were achieved as shown in Fig. 4. The same figure also shows a set of theoretical curves, obtained /10/ by using the pre-equilibrium decay model, recently modified by Blann et al. /11/, so as to obtain angle-dependent spectra. It is seen, however, that this model systematically underestimate the large angle cross sections.

The reason why we do not have this trouble is, we believe, that it is quantum-mechanical, in that we use DWBA. There the

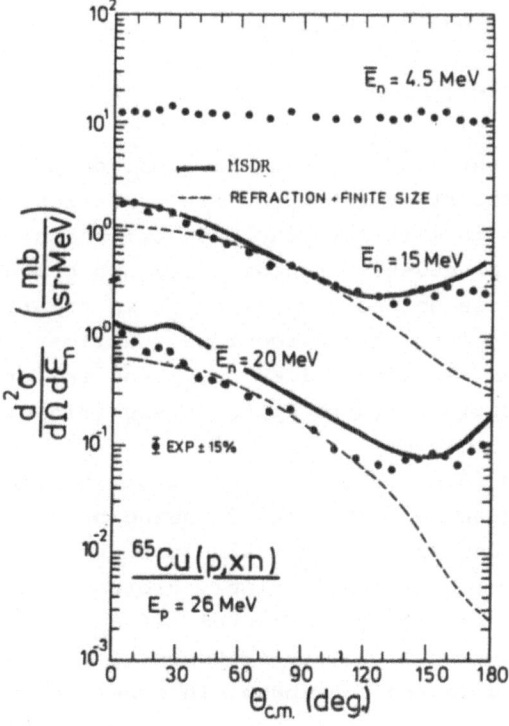

FIG. 4.

important diffraction effect is properly taken into account.
In other words, we retain the quantum-mechanical interference
effect. This should not be mixed up with the disappearance of
the interference term, which we emphasized above in relation
to eq. (4).

We want to remark here that a computer program (ORION-
TRISTAR-1) for the SSDR calculations has been published /12/,
and we hope that the reader will find this program useful to
analyze data like those seen in Figs. 1—4. Note that we mentioned above the need of modifying OMP, in fitting the low-
energy data of Fig. 3. It may be that a systematic analysis
of this type of data provide a new way in improving our knowledge of OMP. It would not be necessary to stick to the conventional method to analyze just the elastic scattering data,
in order to obtain OMP. We also note here that another program
(ORION-TRISTAR-2), which calculates also the two-step contributions, will soon be published.

3. BREAKUP REACTIONS

The reader might wonder why we have not discussed above any
reaction of the stripping type, i.e. reactions in which the
ejectile mass is less than the projectile mass, as in (d,p),
(α,d), etc., processes. When discrete state transition is concerned, there is no basic differences in treating e.g., in
terms of DWBA, these stripping type reactions on the one hand,
and pickup type reactions, like (p,α), on the other. In the
continuum reactions, there appear a significant difference
between them, however.

Consider a (d,p) type reaction, i.e., a reaction induced by
deuteron, and one observes the outgoing proton of the continuum nature. One may then naively consider that the neutron
in the incident deuteron is simply captured in one of the vacant orbits in the target. However, if $E_d - E_p$ exceeds 10 MeV
or so, the neutron cannot be accomodated in a bound orbit; it
must stay also in the continuum. In other words, the final
state of the reaction under consideration is of a three-body
nature. Such a process is normally called a breakup process,

and it does not fall under the DR process we defined in the introduction. In order to avoid causing any unnecessary confusion, we defined DR as a transition from a two-body channel into another.

We shall not go into any further discussion of the breakup reaction here, in spite of its interest and of the fact that it is one of the major current subjects in the nuclear reaction studies. The reason of this choice is the fact that this symposium is for neutron induced reactions, and that neutron never induces a breakup process. We may simply note that the interested reader can find a few comprehensive review articles in ref. /13/.

4. FUSION OF HEAVY IONS

Just above, at the end of last section, we stated that we would not discuss breakup reactions, because it is unrelated to the "neutron induced reactions". Thus the reader would wonder, by seeing the title of this section, how the fusion of two heavy ions can ever be related to neutron induced reactions.

There are two reasons why we decided to discuss this subject in the present paper. One is that we thought that a large fraction of the participants of this symposium would be interested in the problem of fission. And, in a rather broad sense of the word, fusion is a process that is inverse to fission. The second reason of the above decision is that we found recently that the DR reaction concept can very conveniently and powerfully be used, in understanding a large body of heavy-ion fusion data. More precisely, we found that a proper use of the imaginary part of the optical potential was the key to achieve this understanding.

We have stressed in the introduction that the use of the optical potential is the very starting point of any practical DR theory. It would thus be desirable to expose ourselves as much as possible to problems, the study of which allows us to see the physics played by the optical potential, in particular its imaginary part. This is why we decided to discuss fusion here.

As we remarked above, it is not unnatural that one considers fusion as a process inverse to fission. Since one knows that the concept of barrier, and hence that of barrier penetration play important roles in the understanding of fission, one is not surprised to find that the so-called barrier penetration mode (BPM) was considered first, in attempting to fit the fusion data.

This model has, however, a few problems. Unless one assumes an unreasonably thin barrier, BPM predicts a too fast decrease of the fusion cross section, as the incident energy is decreased. This is shown in Fig. 5 by a dashed line. Also, the potential used in BPM is purely real outside a radius, at which the incoming boundary condition is imposed. With such a potential, however, the elastic scattering data cannot be fitted.

As we emphasized in the introduction, to have an optical

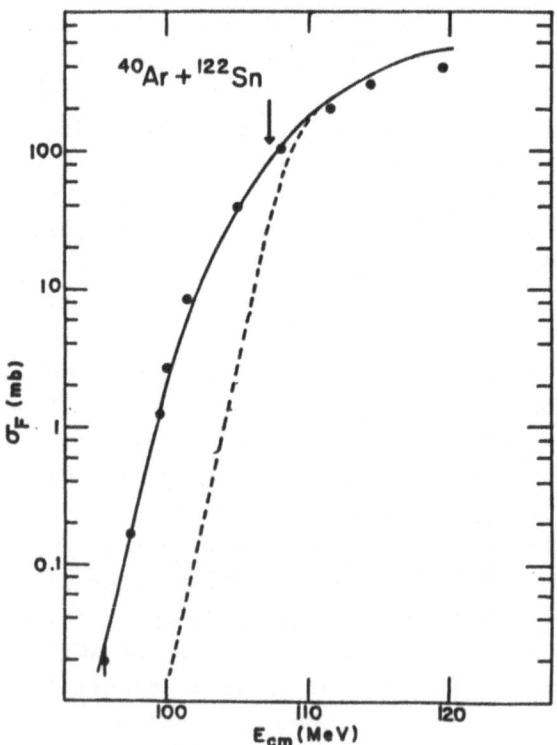

FIG. 5.

model, and thus to be able to fit the elastic scattering data, was what made the whole DR theory meaningful. We actually believe that, not only DR but also any reaction theory may loose its validity, if it cannot be consistent with the very simple but very basic elastic scattering. And we may say that the optical model is the only known practical tool to fit the elastic scattering data. This means that it is more or less mandatory to put the understanding of fusion under the category of DR, in its very broad sense.

What are measured in the fusion reaction are not, however, the products of DR reactions. Experiment measures the formation cross sections of the evaporation residues, and of the fusion-fission products. Therefore, we are not going to perform the DR calculation in its usual sense. Instead, we are going to calculate its counter part, i.e., the absorption cross section. This is what was alluded to in the introduction.

As is well known, this absorption cross section, which is normally called the total reaction cross section, and is denoted by σ_R, is quite often calculated in conjunction with the elastic scattering analysis. Therefore, it is not actually an object that is entirely foreign to the concept of the DR theory.

This σ_R can be expressed as

$$\sigma_R = (8\pi/\hbar v k^2) \sum_l (2l+1) \int_0^{R_o} |\chi_l(r)|^2 \, W(r) \, dr. \qquad (6)$$

In (6), v and k are, respectively the velocity and the wave number in the incident channel, while $\chi_l(r)$ is the radial part of the DW corresponding to the partial wave with angular momentum l. The $W(r)$ is the imaginary part of the optical model potential, and the integral in (6) is just an expectation value of this imaginary part with respect to $\chi_l(r)$. The upper limit R_o denotes a radius at which $W(r)$ de facto vanishes.

Our idea is to obtain the fusion cross section as

$$\sigma_F = (8\pi/\hbar v k^2) \sum_l (2l+1) \int_0^{R_f} |\chi_l(r)|^2 \, W(r) \, dr. \qquad (7)$$

This σ_F differs from σ_R only in the choice of the upper limit
of the integral. This means that, with the choice of $R_f \prec R_o$,
we are saying that σ_f is given just as a part of σ_R. More precisely, we are claiming that σ_R can be divided into two parts.
The one contributed by r, with $R_f \prec r \prec R_o$, is the total DR
cross section, while the other contributed by r, with $r \prec R_f$,
is nothing but the fusion cross section.

In both (6) and (7), we chose the lower limit of the integ-
ration to be r = 0. In practice, however, $|\chi_1(r)|^2$ is very
small for small r, due of course to the strong absorption.
Thus, this lower limit can in fact be replaced by r = R_i, say.
Note that R_i and R_o are calculated quantities; they are known
once optical parameters are known. With the reasonable parameters, we find that R_i is smaller than R_o only by 2 fm, or
less. Thus the total absorption takes place within an amazing-
ly narrow range of r.

We are going to express R_f as $R_f = r_f(A_1^{1/3}+A_2^{1/3})$, and see
what should be the value of r_f which we choose, in order to be
able to fit the data. In other words, we have formulated the
theory as a one-parameter theory.

By analyzing a large amount of fusion cross-section data,
we found that a fixed choice of r_f = 1.45 fm can be used in
most of the cases. Note that a fixed r_f means that it can be
taken independent of E_{cm}, as well as independent of the choice
of the pair A_1 and A_2 that are going to be fused. Our approach
is certainly a phenomenology. However, once this amazing con-
stancy of r_f is found, we begin to feel that our approach
acquires some physical significance, the value of r_f = 1.45 fm
describing an important aspect of the fusion reaction.

In order to show to what extent our method works, we present two figures here. Figure 5 is for the ^{40}Ar+^{122}Sn case
/14/. As noted above, the BPM result represented by a dashed
line exemplifies the trouble of this model. On the other hand,
our result represented by a solid line agrees almost perfectly
with the data. The capability of our method is clearly seen.

In Fig. 6, we show two cases: ^{58}Ni+^{64}Ni and ^{58}Ni+^{58}Ni cases
/15/. Out of these two, the former has a rather unusual property, in that it has two-neutron transfer channels that are

formed with "positive" Q-values. This means that, in these new channels, the two ions become easier to come closer together, and thus there will be a significant contribution to the fusion cross section from these channels.

Faced with this specific situation, we see that our work here begins to take back the feature of the DR theory, in its ordinary sense. We should first carry out the DR calculation so as to construct (distorted) waves in the above two-neutron transfer channels, and then only afterwards we calculate absorptive cross sections based on these waves. We in fact found that the use of the Born approximation was too crude to describe this DR process. The CRC calculations had to be done.

The result of this rather sophisticated calculation is shown by a solid line in Fig. 6, and it is seen that it fits data rather nicely. On the other hand, a very straightforward use of eq. (7) gives the result shown by a dot-dash line,

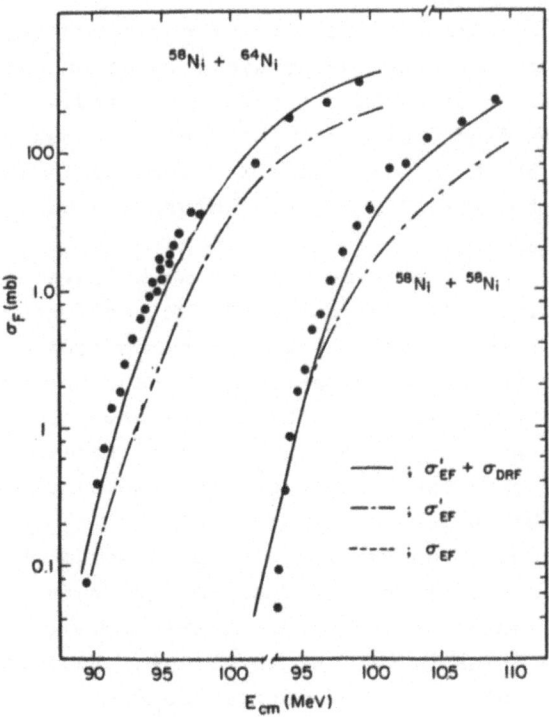

FIG. 6.

which, as expected, is too small compared with experiment. In Fig. 6, the ^{58}Ni+^{58}Ni case is also shown, which does not have any reaction channel with positive Q-values. In this case we thus expect that the use of CRC results in essentially the same result, as the use of eq. (7) would do. As seen in Fig. 6, this expectation is indeed fulfilled. We have thus shown that our method explains on one footing two cases, which are significantly different from one to another, in their structural aspects.

We may remark here that the use of the BPM was considered for the so-called subbarrier fusion, and some other theories had been conceived of to describe above barrier fusion. Our method, on the other hand, treats these two cases on one footing, as seen, e.g., in Fig. 5. The arrow shown there indicates the position of the (s-wave) barrier. It is seen that our theoretical curve (solid line) does reproduce data on both sides of this arrow.

Based on the successful results of our approach, as shown above, we may now discuss the possible relation between fission and fusion. Note that we have introduced a radius called R_i. Numerically, we find that this R_i is rather close of R_b, which is the radius at which the s-wave barrier top is located. This then means that $|x_1(r)|^2 \simeq 0$, even before the barrier top is reached.

We thus see that our theory, which we believe is basically correct, says that all the reactions, DR and fusion, take place outside the barrier. Thus, there remains no room to introduce the concept of barrier penetration. On the other hand, we remarked above that fission and fusion are inversely related, and for the former concept of barrier penetration is known to be vital. Does this mean that there is something wrong in our understanding of fusion?

There is actually no problem. There is not much room to discuss this in detail here, and we want to refer the reader to our recent paper /16/. As shown there, only one needs to note that the above (s-wave) barrier is defined based on the two-body A_1+A_2 channel, while the barrier considered in fission is a many-body barrier. In other words, these two barriers are completely different physical entities. Therefore a theory which predicts different behaviors of the two does not have any self-contradiction in it.

5. SUMMARY

In the present paper, two major subjects were discussed. One is the description of continuum cross sections, in terms of the MSDR method. The other is the description of fusion, also within the concept of the DR theory. We hope that the reader now shares with us the view, that these two seemingly different reaction processes are in fact rather deeply related. The very concept of the DR theory is what to connect them.

REFERENCES

1. T. Tamura, T. Udagawa and H. Lenske, Phys. Rev. C26 (1982) 109.
2. T. Tamura, T. Udagawa, D. H. Feng and K. K. Kan, Phys. Lett. 66B (1977) 109.
3. R. Reif, Acta Phys. Slov. 25 (1975) 208.
4. M. B. Lewis, Phys. Rev. C11 (1975) 145.
5. H. Lenske et al., Phys. Lett. 122B (1983) 66.
6. M. Benhamou, Ph.D. Thesis, University of Texas, 1984.
7. A. Marcinkowski et al., Nucl. Sci. Eng. 83 (1983) 13.
8. M. Uhl, personal communication.
9. F. D. Becchetti and G. W. Greenlees, Phys. Rev. 182 (1969) 1190.
10. W. Scobel, personal communication.
11. M. Blann, W. Scobel and E. Plechaty, Phys. Rev. C30 (1984) 1493.
12. T. Tamura, T. Udagawa and M. Benhamou, Comp. Phys. Comm. 29 (1983) 391.
13. R. J. de Meijer and R. Kamermans, Rev. Mod. Phys. 57 (1985) 147.
    X. H. Li, T. Udagawa and T. Tamura, Phys. Rev. C, to be published.
14. W. Reissdorf et al., Phys. Rev. Lett. 49 (1982) 1811.
15. M. Beckerman et al., Phys. Rev. Lett. 45 (1980) 1472.
    M. Beckerman et al., Phys. Rev. C23 (1981) 1581; 25 (1982) 837.
16. T. Udagawa, B. T. Kim and T. Tamura, Phys. Rev. C, to be published.

NOVEL APPROACH TO THE STATISTICAL THEORY
OF NUCLEAR REACTIONS

H. A. Weidenmüller

Max Planck Institute for Nuclear Physics,
Heidelberg, FRG

ABSTRACT. I describe the statistical input, and some of the
results obtained recently, in the statistical theory of nuclear reactions with the help of a novel theoretical approach.
This approach makes use of and generalizes methods developed
by Wegner, Schäfer, and Efetov in the theory of disordered
solids.

1. INTRODUCTION

Statistical nuclear physics encompasses both statistical
spectroscopy, and the statistical theory of nuclear reactions.
In the latter domain, we deal with such diverse problems as
compound-nucleus reactions, pre-equilibrium processes, isospin
mixing in the compound nucleus, and the entire field of non-
equilibrium processes in heavy-ion reactions. In recent years,
the physical foundations of this field have been much clarified. It is now widely believed that the statistical properties of nuclei are manifestations of a generic property of
the nuclear Hamiltonian. In the classical limit, this property would cause the trajectories to undergo chaotic motion in
phase space. This property is believed to be generic because
Hamiltonians describing real physical systems are likely to
be non-integrable , and therefore would give rise to chaotic
motion in the classical limit. For simplicity, I refer to such
quantum systems as chaotic quantum systems. There exists a
growing body of evidence /1/ that the fluctuations which occur
in chaotic quantum systems can be quantitatively and generi-
cally simulated by the fluctuation properties of an ensemble
of random matrices - the Gaussian Orthogonal Ensemble (GOE)

- introduced roughly thirty years ago by Wigner. The combination of these facts leads to an entirely new approach to the quantum many-body problem: For chaotic quantum systems, it may be both impossible and uninteresting to calculate the spectrum (or other observables) exactly. Instead, a division of labor seems to be called for. Average and gross properties of such systems must be worked out in terms of the dynamical features of the underlying Hamiltonian, while the effect of fluctuations on any observable is best simulated in terms of a GOE.

The implementation of this kind of programme requires methods for calculating average properties, and other methods for evaluating the fluctuations simulated by the GOE. Average properties can frequently be well computed by semiclassical methods (including the mean-field approach). This is not what we are concerned with here. Rather, we address the second point: The calculation of fluctuation properties of observables as simulated by a GOE. As examples, we discuss the cases of compound-nucleus and pre-equilibrium reactions. From the remarks made above it should be clear, however, that the method is not restricted to these problems, not even to the domain of nuclear physics, but should apply in general to the evaluation of noise generated by quantum chaos.

2. THE PROBLEM

We use the GOE to generate the fluctuations associated with quantum chaos. This ensemble is defined as follows. We consider a Hamiltonian matrix $H_{\mu\nu}$ ($\mu, \nu = 1, \ldots, N$) with dimension $N \gg 1$. (We eventually take the limit $N \to \infty$.) Hermitecity and time-reversal invariance allow us to choose H real and symmetric, $H = H^+ = H^T$. The independent elements in $H_{\mu\nu}$ are those with indices $\mu \leq \nu$. We consider these elements as uncorrelated random variables with a Gaussian probability distribution. The distribution is defined in terms of the first two moments of H,

$$\overline{H_{\mu\nu}} = 0, \tag{1a}$$

$$\overline{H_{\mu\nu}H_{\mu'\nu'}} = \frac{\lambda^2}{N} (\delta_{\mu\mu'}\, \delta_{\nu\nu'} + \delta_{\mu\nu'}\, \delta_{\nu\mu'}). \tag{1b}$$

Here, λ is a strength parameter, and the bar denotes the ensemble average. It is easy to check that eqs. (1) are invariant under orthogonal transformations of H: No basis in the N-dimensional space plays a preferred role. (This explains the name Gaussian orthogonal ensemble.) The eqs. (1) together with the definition of the volume element, $dV = \prod_i dH_{ii} \prod_{i<j} dH_{ij}$ (which is likewise orthogonally invariant) complete the definition of the distribution function and thus of the GOE.

In the first example - compound-nucleus scattering - we assume that the compound-nucleus resonances are coupled to each other by a GOE Hamiltonian. The compound nucleus is unstable against decay into the open channels labelled a, b, c, ... The scattering matrix pertaining to fixed spin and parity as a function of energy E has a set of poles, each pole being associated with one of the resonances. We consider both cases of isolated and overlapping resonances. A suitable model for the elements $S_{ab}(E)$ of the S-matrix is given by

$$S_{ab}(E) = \delta_{ab} - 2i\pi \sum_{\mu,\nu}^{N} W_{\mu a}(D^{-1})_{\mu\nu} W_{\nu b}, \tag{2a}$$

$$D_{\mu\nu} = E \delta_{\mu\nu} - H_{\mu\nu} + i\pi \sum_c W_{\mu c}W_{\nu c}, \tag{2b}$$

with

$$\sum_\mu W_{\mu a}W_{\mu b} = \delta_{ab} \sum_\mu W_{\mu a}^2. \tag{2c}$$

Eqs. (2a),(2b), define the S-matrix in a form which may be viewed as the extension of the single-level Breit-Wigner formula to N resonances. It is easy to check that S is unitary, $SS^+ = 1$. The matrix elements $W_{\mu a}$ are real and independent of energy. They are subject to the condition (2c). This condition together with the Kronecker symbol in eq. (2a) quaranted that the average S-matrix is diagonal, $\overline{S_{ab}} = \delta_{ab} \overline{S_{aa}}$. This is tantamount to the neglect of direct reactions. If present, such reactions can be included in the formalism; the inclusion requires no more than notational complications. These are here omitted for clarity. The strength of the matrix elements $W_{\mu a}$ determines together with the number of channels whether the resonances do overlap or not. If H stands for the GOE of eqs. (1), then eqs. (2) define an ensemble of S-matrices. The task consists in calculating $\overline{S_{ab}(E)}$, $\overline{S_{ab}(E)S^*_{cd}(E)}$, and $\overline{S_{ab}(E_1)S_{cd}(E_2)S^*_{ab}(E_1)S^*_{cd}(E_2)}$. The average S-matrix $\overline{S}$ is needed to connect the parameters of the model (2) with physical input: Optical-model (or coupled-channels) calculations yield a scattering matrix which we identify with $\overline{S}$. For the special choice (2c) made above, the independent parameters of the model – the strength λ appearing in eq. (1b) and the quantities $\sum_{\mu} W^2_{\mu a}$ – are equal in number to the input, i.e. the specification of $\overline{S_{aa}}$ and of the average level spacing d. (Orthogonal invariance of H implies that only the bilinear forms $\sum_{\mu} W_{\mu a} W_{\mu b}$ and no other combination of matrix elements occur in the ensemble-average of any observable.) Once the model is specified in this way, the calculation of $\overline{S_{ab}(E)S^*_{cd}(E)}$ yields average cross-sections and other observables, while the fourth moment of S indicated above yields a measure of cross-section fluctuations.

We note that the model embodied in eqs. (2) is a special case of the general scheme outlined in section 1. The average S-matrix $\overline{S_{aa}}$ is derived from non-statistical considerations which utilize the specific dynamical features of the problem under consideration. The fluctuations of S about this average are generated by the GOE.

The problem of calculating the second and fourth moments of S has a long history in nuclear physics. It was implicitly formulated by N. Bohr, partially treated by Bethe, and, in more detail, by Hauser and Feshbach. An important contribution was made by T. Ericson. The problem was at the centre of P. Moldauer's work throughout most of his life, and it has continued to receive attention in recent years. Reviews may be found in refs. /2, 3/. The formulation /4/ adopted in eqs. (2), although equivalent to that of most earlier approaches, is

different in that it emphasizes the orthogonal invariance of
the GOE. This is one of the reasons why it has led to a solution of the problem.

A model for pre-equilibrium - or precompound - processes
can be formulated in a similar vein. To be more specific, I
consider the cases of isospin mixing in compound-nucleus reactions and of multistep-compound reactions in the terminology
of ref. /5/, as opposed to the case of multistep-direct reactions. In the case of isospin mixing, we consider two classes
of compound-nucleus resonances with isospins T_1 and $T_2 = T_1 + 1$,
respectively. These classes are labelled 1 and 2. In the case
of multistep compound-nucleus reactions, we consider m classes of states, each class being specified by a fixed number
of particle-hole excitations. (In this way, we generate a
hierarchy of states of growing complexity.) To simplify the
presentation, I consider only the case with m = 2 in the following. Thus, isospin mixing and precompound reactions are
formally described by the same model.

The formulation of the model follows lines very similar to
eqs. (2). Let us suppress direct reactions, and let $|n, \mu\rangle$ be
the states in class n (n = 1, 2) labelled by a running index
μ, with a range extending from 1 to $\dot{N}_n$ (n = 1, 2), $N \gg 1$. We
introduce the matrix elements $W_{n\mu, a}$ in an obvious fashion,
and write the Hamiltonian connecting the compound-nucleus
resonances $|n, \mu\rangle$ in the form $H_{n\mu, n'\mu'}$. Then eqs. (2) take
the form

$$S_{ab}(E) = \delta_{ab} - 2i\pi \sum_{n,n'=1}^{2} \sum_{\mu,\mu'} W_{n\mu,a} (D^{-1})_{n\mu,n'\mu'} W_{n'\mu',b}, \quad (3a)$$

$$D_{n\mu,n'\mu'} = \delta_{nn'} \delta_{\mu\mu'} E - H_{n\mu,n'\mu'} + i\pi \sum_{c} W_{n\mu,c} W_{n'\mu',c} \quad (3b)$$

and

$$\sum_{\mu} W_{n\mu,a} W_{n\mu,b} = \delta_{ab} \sum_{\mu} w^2_{n\mu,n} . \quad (3c)$$

To specify the model completely, we have to specify the distribution of the elements of $H_{n\mu,n'\mu'}$. For n = n', we take H

to be a GOE. For $n \neq n'$, we require H to be real and symmetric with elements that are not correlated with those of $H_{n,n}$, and that have a Gaussian distribution centred at zero. Thus,

$$\overline{H_{n\mu,n'\mu'}} = 0 \quad \text{for all} \quad n,n', \mu, \mu',$$

$$\overline{H_{n_1\mu_1,n_1\mu_1'} H_{n_2\mu_2,n_2\mu_2'}} = \delta_{n_1 n_2} \frac{\lambda_n^2}{N_n} (\delta_{\mu_1\mu_2} \delta_{\mu_1'\mu_2'} + \delta_{\mu_1\mu_2'} \delta_{\mu_2\mu_1'}),$$

$$H_{1\mu,2\nu} = H_{2\nu,1\mu}.$$

$$\overline{H_{1\mu,2\nu} H_{1\mu',2\nu'}} = \rho^2 \delta_{\mu\mu'} \delta_{\nu\nu'}.$$

The extension of this model to $m > 2$ classes is straihtforward.

3. RESULTS

Only the first model described in section 2 has been fully solved; work on the second problem is in progress. The method of solution employed in both cases is similar. It is an extension of methods devised by Wegner and Schäfer /6, 7/ and Efetov /8/ for the solution of the Anderson model with N orbitals per site in the limit $N \to \infty$. In the Anderson model one studies whether a disordered solid possesses a finite conductivity. One considers a single electron moving in a regular lattice consisting of an infinite number of sites. In the extension of the model due to Wegner, the electron may occupy any one of N orbitals with associated single-particle energies at each site. The electron may hop from any site to any of the neighbouring sites. To simulate the disorder, caused by the fact that chemically different ions occupy the sites or in some other way, one describes the single-particle energies associated with the N orbitals at each site as well as the hopping matrix elements in terms of a random-matrix model. In this way, the Anderson model becomes similar in formal structure to the models formulated in section 2, and methods of solution developed by the Anderson model become applicable to the calculation of fluctuation properties caused by quantum chaos.

Since the compound-nucleus problem formulated in eqs. (2) is formally similar to an Anderson model with a single site or, put differently, to an Anderson model on a lattice with zero dimensions, the solution of the compound-nucleus problem is actually simpler than that of the Anderson model for a realistic three-dimensional lattice. It turns out, in fact, that the model formulated in eqs. (2) can be solved exactly in the limit $N \rightarrow \infty$, i.e. under omission of terms proportional to N^{-k} with $k > 0$ and k integer. This in turn is of interest for the understanding of theoretical approaches to the Anderson model: An exact solution is always useful for the understanding of assumptions and/or approximations introduced to handle more complex situations /9/.

Limitations of space and the complexity of the mathematical technique make it impossible for me to describe the method used to solve the model of eqs. (2). A detailed account is given in ref. /10/. I likewise abstain from reproducing here the exact expression obtained for $\overline{S_{ab}(E_1)S^*_{cd}(E_2)}$, see refs. /10, 11/, because of its length and lack of transparency. I prefer to give instead a number of comments relating to this result, and to the way it compares with previous approaches /12/.

(i) The average $\overline{S_{ab}(E_1)S^*_{cd}(E_2)}$ depends, aside from trivial multiplicative factors, only on the transmission coefficients $T_a = 1 - |\overline{S_{aa}}|^2$ and not on the phases of the average S-matrix elements. This result confirms earlier findings /13/ and justifies earlier efforts to parametrize the average compound-nucleus cross-sections in terms of the T_a. The dependence on the transmission coefficients is the same over the entire spectrum of the GOE - it is stationary. This stationarity property enhances the belief that the result may claim some universal validity, and that it might also be obtained from a different set of assumptions as furnished, for instance, by the maximum entropy principle /14/. Numerical evidence /12/ suggests that this principle when applied to fluctuations of the S-matrix yields the same result as derived from the model of eqs. (2) for $\overline{S_{ab}(E_1)S_{cd}^*(E_2)}$.

(ii) In the case of many open channels or strongly overlapping resonances, more precisely: In the case when $\sum_a T_a \gg 1$, the exact result can be used to generate /12/ an asymptotic expansion in inverse powers of $(\sum_a T_a)$. This expansion agrees with the result obtained from the replica trick /4/, and earlier /15/ from a partial summation (using diagram methods) of the double Born series. The asymptotic expansion has the form

$$\overline{S_{ab}(E_1)S_{cd}^*(E_2)} = \overline{S_{aa}} \; \overline{S_{cc}^*} \; \delta_{ab} \delta_{cd} + (\delta_{ac}\delta_{bd} + \delta_{ad}\delta_{bc}) \cdot$$

$$\cdot \frac{T_a T_c}{\sum_f T_f + \frac{2i\pi}{d}(E_2 - E_1)} + \dots , \tag{5}$$

where the dots indicate terms of higher order in $(\sum_f T_f)^{-1}$. To first order in $(\sum_f T_f)^{-1}$, eq. (5) yields for $E_1 = E_2$, the Hauser-Feshbach formula with a factorization of the compound-nucleus cross-section and an elastic enhancement factor of two. Factorization is equivalent to independence of formation and decay (the Bohr hypothesis); this in turn is often identified with internal equilibration of the system before it undergoes decay. We now see that factorization only applies in the limit $\sum_a T_a \gg 1$. This is physically reasonable because this limit must be identified with the thermodynamic limit, and equilibration can only be expected in the thermodynamic limit. Indeed, the thermodynamic limit corresponds to infinite volume or, in our case, infinitely small level spacing d. The denominator of the second term on the r.h.s. of eq. (5) shows that fluctuations of the S-matrix element have a correlation width Γ given by

$$\Gamma = \frac{d}{2\pi} \sum_a T_a . \tag{6}$$

(The quantity $\hbar/\Gamma$ can be shown to be the average life time of the compound nucleus.) The expansion in powers of $(\sum_a T_a)^{-1}$ is therefore also an expansion in powers of $d/(2\pi\Gamma)$.

This expansion is good if d $\ll 2\pi\Gamma$, i.e. in the thermodynamic limit.

(iii) In the case of many open channels, $\sum_a \gg 1$, but weak absorption in each channel $T_a \ll 1$ so that $\sum_a T_a = t$ finite for $\sum_a \to \infty$, an asymptotic expansion in inverse powers of the number of channels can be introduced, which includes the case of non-overlapping resonances. In this case, the elastic enhancement factor W is given by /12/

$$W = 3 - \frac{1}{2} t + \ldots \tag{7}$$

which agrees with earlier results by Moldauer /16/.

(iv) The time-development function needed to evaluate experiments on compound-nucleus life times performed with the blocking technique was evaluated exactly /17/, and was compared with approximate formulas derived earlier /18/. Excellent agreement was observed.

(v) The expression for $S_{ab}(E_1)S^*_{cd}(E_2)$ was rewritten in such a way that it could easily be evaluated numerically /12/. The numerical results showed good agreement with earlier Monte Carlo calculations /13/.

(vi) The inclusion of direct reactions in compound-nucleus theory is possible by means of a unitary transformation /19/. In the present context, this transformation becomes particularly transparent /20/.

In summary, we see that the exact solution to the model formulated in eqs. (2) gives a complete answer to many questions of compound-nucleus theory including direct reactions as well as fluctuation properties of the S-matrix. The one exception is the evaluation of the four-point function $\overline{S_{a_1 b_1}(E_1)S_{a_2 b_2}(E_2)S^*_{a_3 b_3}(E_3)S^*_{a_4 b_4}(E_4)}$.
This quantity contains complete information on the fluctuation properties of cross-sections and other observables. The evaluation of this quantity would require more work.

The treatment of precompound reactions along similar lines but starting from the model formulated in eqs. (3) is in prog-

ress. It is expected that this will provide a mathematically
satisfactory justification of the results derived in ref. /15/.

The author would like to thank his collaborators on these
problems, especially J. J. M. Verbaarschot and M. R. Zirnbauer,
for their enthusiasm, stimulation and cooperation.

REFERENCES

1. O.Bohigas and M.-J. Giannoni, in "Lecture Notes in Physics",
   Vol. 209, p. 1, Springer-Verlag, Berlin - Heidelberg - New
   York - Tokyo, 1983.
2. C. Mahaux and H. A. Weidenmüller, Ann. Rev. Nucl. Part. Sci.
   29 (1979) 1.
3. T. A. Brody, J. Flores, J. B. French, P. A. Mello,
   A. Pandey and S. S. M. Wong, Rev. Mod. Phys. 53 (1981) 385.
4. H. A. Weidenmüller, Ann. Phys. (N.Y.) 158 (1984) 120.
5. H. Feshbach, A. K. Kerman and S. Koonin, Ann. Phys. (N.Y.)
   125 (1980) 429.
6. F. Wegner, Phys. Rev. B19 (1979) 783.
7. L. Schäfer and F. Wegner, Z. Phys. B38 (1980) 113.
8. K. B. Efetov, Adv. Phys. 32 (1983) 53.
9. J. J. M. Verbaarschot and M. R. Zirnbauer, J. Phys. A17
   (1985) 1093.
10. J. J. M. Verbaarschot, H. A. Weidenmüller and M. R.
    Zirnbauer, Phys. Rep., in press.
11. J. J. M. Verbaarschot, H. A. Weidenmüller and M. R.
    Zirnbauer, Phys. Lett. B149 (1984) 263.
12. J. J. M. Verbaarschot, Ann. Phys. (N.Y.), in press.
13. H. M.Hofmann,J. Richert and J. W. Tepel, Ann. Phys. (N.Y.)
    90 (1975) 391.
    H. M. Hofmann, J. Richert, J. W. Tepel and H. A. Weiden-
    müller, Ann. Phys. (N.Y.) 90 (1975) 403.
    H. M. Hofmann, T. Mertelmeier, M. Herman and J. W. Tepel,
    Z. Phys. A297 (1980) 153.
    P. A. Moldauer, Phys. Rev. C11 (1975) 426.
14. P. A. Mello, P. Pereya and T. H. Seligman, Ann. Phys. (N.Y.)
    in press.
15. D. Agassi, H. A. Weidenmüller and G. Mantzouranis, Phys.
    Rev. C22 (1975) 145.
16. P. A. Moldauer, Phys. Rev. 123 (1961) 968.
17. J. J. M. Verbaarschot and S. Yoshida, Preprint, 1985.
18. K. Yazaki and S. Yoshida, Nucl. Phys. A232 (1974) 249.
    S. Yoshida and K. Yazaki, Nucl. Phys. A255 (1975) 173.
    H. Matsuzaki, Nucl. Phys. A308 (1978) 95.
19. C. A. Engelbrecht and H. A. Weidenmüller, Phys. Rev. C8
    (1973) 859 and the second of ref. /13/.
20. H. Nishioka and H. A. Weidenmüller, Phys. Lett. 157B
    (1985) 101.

INELASTIC SCATTERING CROSS-SECTION OF ^{160,162,164}Dy USING THE TIME-OF-FLIGHT TECHNIQUE

T. J. Al-Janabi, A. B. Kadhim, S. J. Hasan,
K. M. Mahmood and A. J. Al-Azawi

Nuclear Research Centre, Iraqi Atomic Energy Commission,
Baghdad, Iraq

ABSTRACT. Reactor fast neutrons were used to measure the inelastic scattering cross-section of ten low-lying states in ^{160,162,164}Dy isotopes using the time-of-flight technique in the energy range of 0.4—4 MeV. Charge comparison method has been employed for (n,γ) discrimination during the off-line analysis. The integral cross-section values relative to a secondary standard value of ^{56}Fe were obtained. The results are compared with the deduced values based on the statistical Hauser—Feshbach model.

1. INTRODUCTION

In this paper the inelastic scattering cross-section of (n,n') reaction to separate levels in three even-mass Dy isotopes is presented. The measurements were performed using reactor fast neutrons with a TOF facility constructed at the IRT-5000 reactor in Baghdad. Different target elements were studied during the course of the measurements, some of these studies were devoted to basic understanding of the reaction mechanism and model testing, while the rest of the measurements were carried out for applied purposes. Examples of the later type were ^{93}Nb, ^{27}Al, ^{58}Ni, ^{56}Fe, etc., whereas the former measurements were conducted on some highly enriched rare earth isotopes, such as ^{172,174}Yb and the subject of the present contribution on ^{160,162,164}Dy.

As a matter of fact the cross-section data on Dy isotopes are not available in the literature yet, also the ENDF evaluations are incomplete regarding these isotopes. Therefore, and in order to test the reliability of our method, the data on

^{93}Nb were analysed at the start and we were able to reproduce the published results quite well.

2. EXPERIMENTAL METHOD

In the present measurements use was made of the reactor fast neutrons and of the TOF system installed on a horizontal channel of the IRT-5000 reactor. Full descriptions of the system used and of the experimental procedure were given previously /1, 2/. The targets of the present study were prepared from oxide powder, Dy_2O_3, enriched to 91%, 94% and 96% in ^{160}Dy, ^{162}Dy and ^{164}Dy, respectively. Five grams of each sample were packed in a polyethylene bag of suitable dimensions and positioned at an angle of 45° relative to the incident neutron beam. The scattered neutrons were detected by 8 large NE213 detectors (1 m long x 9 cm in diameters) arranged into two groups and situated at the forward and the backward angles relative to the incident beam at one-meter-distance from the target position. Gamma-ray spectra were measured by two co-axial Ge(Li) detectors having 20% efficiency and 1.9 keV resolution at 1.332 MeV. These detectors provided the start signals for the TOF spectra in two channels, namely γ_1n and γ_2n. The electronic logic of the system was designed to select coincidences between any of the two gamma-ray detectors and any of the 8 neutron detectors. Coincidence between the two gamma events observed by these two detectors was also provided.

The (n,γ) discrimination was performed using the charge comparison method /3/, in which, two gates (50 ns short and 250 ns long) were generated from the leading edge of the anode signal. The charge ratio (within the short and the long gates) has been plotted for each detector, hence we were able to determine the best (n,γ) separation conditions. These conditions were applied in the TOF spectra, resulting in a good reduction for gamma-ray events associated with neutrons. The TOF spectra were then converted into energy spectra using a standard un-folding technique, while the variation in the flight path for each individual detector and the interaction position within the detector itself have been taken into account.

3. RESULTS AND DISCUSSION

The level scheme information of the samples studied was taken over from refs. /4, 5/. Gamma transitions de-exciting well-known levels have been selected during the on-line data collection. Excitation functions determined were relative to the 847(2^+) keV state in ^{56}Fe. The cross-section ratios have been determined per 200 keV energy interval.

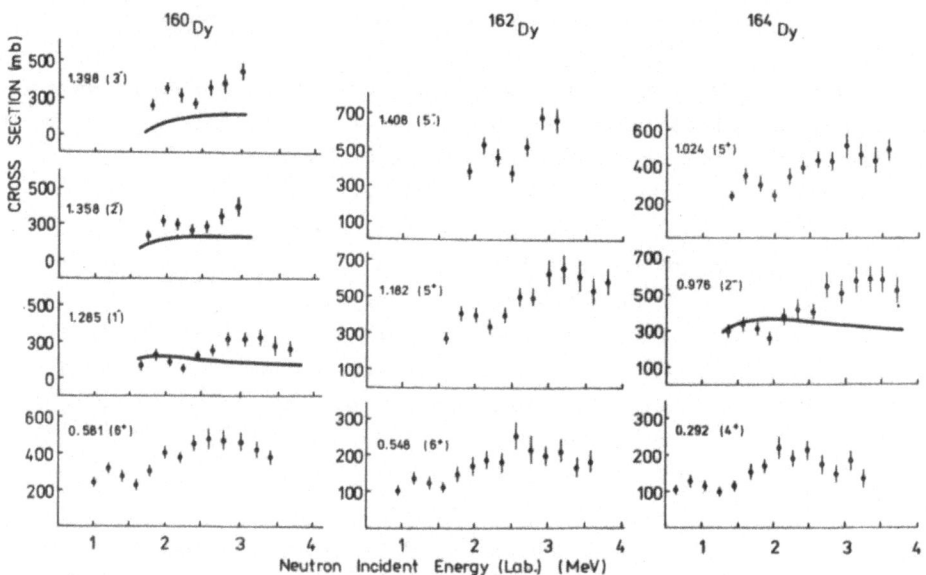

FIG. 1. Cross-section for levels in ^{160,162,164}Dy. Solid lines represent compound nucleus calculations.

The cross-section results plotted in Fig. 1, were found relative to the published data in ref. /6/. It should be noted, however, that the error bars shown in the figure, represent statistical as well as quoted errors of the reference values. The cross-section values were not corrected for multiple scattering and attenuation of neutrons within the target, which did not exceed ~ 4%. The contribution of direct reaction observed as pronounced forward peak associated with ^{56}Fe data has been corrected in the normalization procedure.

The solid curves shown in Fig. 1, represent the calculated

values based on a compound nucleus statistical model implemented
in the CINDY programme /7, 8/. The calculations are in reasonable agreement for some states (e.g. 1.285, 1.358 MeV levels in
^{160}Dy), while for other states the discrepancies in the magnitude are evident. It should be noted, however, that discrepancies
become serious at incident neutron energy exceeding 2.0 MeV,
and apart from the poor statistics at higher energies, the present measurements indicate the presence of other effects, such
as precompound and direct interaction contributions to the
cross-section as one might expect with increasing bombarding
energy.

REFERENCES

1. J. D. Jafar et al., Nucl. Inst. Meth. (1985), to be published.
2. T. J. Al-Janabi et al., Paper submitted to the International
   Conference Nuclear Data for Basic and Applied Science,
   Santa Fé, May 13—17, 1985.
3. C. L. Morris et al., Nucl. Inst. Meth. 137 (1976) 397.
4. C. M. Lederer et al., "Table of Isotopes", 7th ed., John
   Wiley and Sons, New York, 1978.
5. U. Abbondanno et al., Nuovo Cimento 29 (1980) 8.
6. D. L. Smith, Report ANL/NDM-20, 1976.
7. E. Sheldon et al., Comp. Phys. Comm. 6 (1973) 99.
8. W. B. Gilboy et al., Nucl. Phys. A42 (1963) 86.

MICROSCOPIC DESCRIPTION OF DIRECT CONTRIBUTION TO NEUTRON INELASTIC SCATTERING

R. Antalík

Institute of Physics, Electro-Physical Research Centre of
the Slovak Academy of Sciences, Bratislava, Czechoslovakia

ABSTRACT. The DWBA theory with the RPA multipole response functions to an external isovector neutron field are used to calculate the direct contributions to 14 MeV neutron inelastic
scattering of ^{56}Fe, ^{58}Ni and ^{52}Cr. Summing these microscopic
direct spectra with the parametrized evaporation contributions
we obtained a quite good agreement with the experimental double-differential cross-sections.

The direct contribution to nuclear reactions should be done in
a frame of the quantum mechanical collision theory, what in
low energy region means that we have used the DWBA theory /1/.
This theory is well understood and very well working. However,
besides the determination of the optical model potential the
one, but crucial problem of determination of energies and matrix elements of excited residual levels still remains to be
solved. This last problem has been solved here in a frame of
the microscopic nuclear structure theory /2/. For this purpose
we have calculated RPA multipole response functions to the external isovector neutron field, for multipolarities λ from 1
to 6. Typical example of the response functions for the neutron and the proton systems of ^{56}Fe of multipolarity $\lambda = 2$
(for the 2^+ states) is given in Fig. 1. It should be stressed,
however, that our response functions include not only the collective, but also weakly collective and two-quasiparticle states.

Because in DWBA calculations we use a collective model form
factor, for simplicity, we must express the RPA reduced matrix

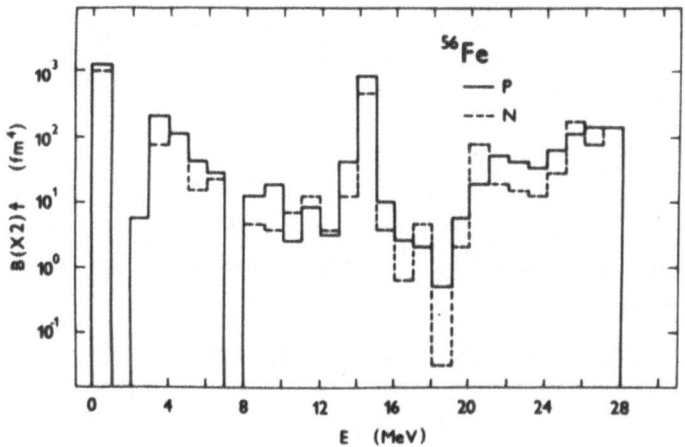

FIG. 1. Quadrupole response functions of neutron (N) and proton (P) fields of ^{56}Fe.

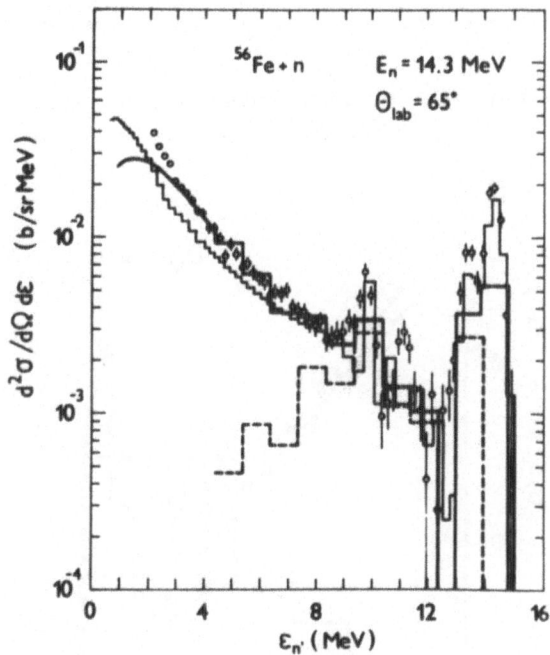

FIG. 2. Double-differential cross-sections for 14 MeV scattered neutrons in ^{56}Fe to the laboratory angle 65°. The dashed histogram represents the direct contribution, the thick solid histogram, the total theoretical cross-section, and the thin histogram, the evaluation from ENDF/B-IV. The points are experimental data of Takahashi et al. /7/.

element for a particular state B(Xλ) (where X is N for the neutron and P for the proton systems) in terms of effective dynamical deformations /3/

$$\beta^2(X\lambda) = B(X\lambda)(\frac{3}{4\pi} N_X R^\lambda)^{-2}$$

where N_X is the number of particles of the type X. For the effective dynamical deformation of inelastically scattered neutrons we have

$$\beta_{n,n'}(\lambda) = \left| \frac{V_{np} N_p \beta(P\lambda) + I V_{nn} N_n \beta(N\lambda)}{V_{np} N_p + V_{nn} N_n} \right|$$

where N_p, N_n are numbers of protons and neutrons participating in building of the level under consideration, V_{np}, V_{nn} are n-p and n-n potentials, respectively, and I is equal to (+1) or (-1) for the isoscalar or isovector excited levels, respectively.

To complete our DWBA requirements optical model potentials

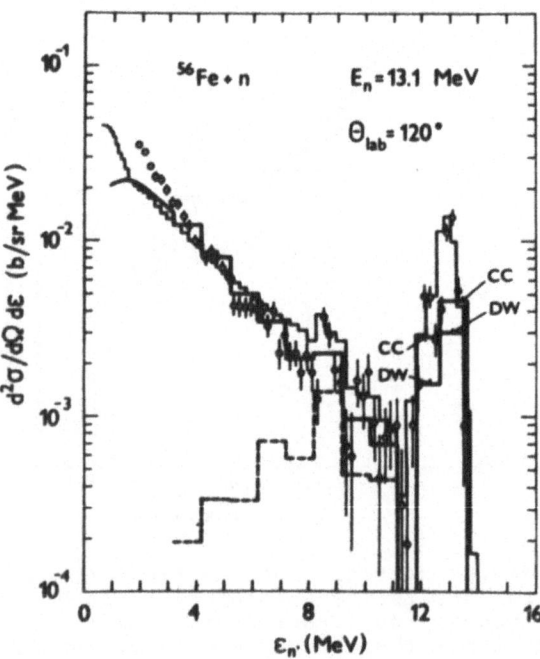

FIG. 3. The same as in Fig. 2, except for $\theta_{lab} = 120^0$.

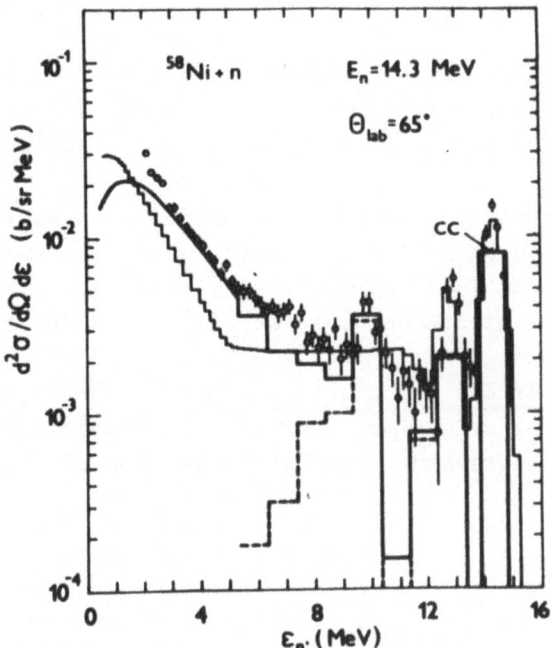

FIG. 4. The same as in Fig. 2, except for ^{58}Ni.

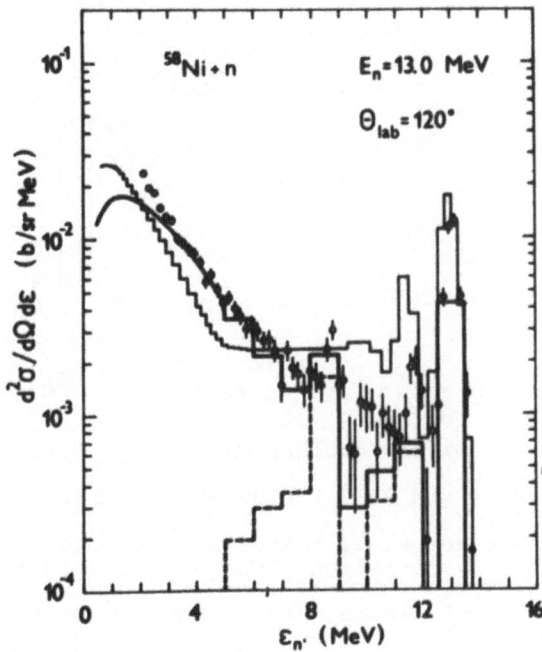

FIG. 5. The same as in Fig. 2, except for ^{58}Ni, $\theta_{lab} = 120°$.

have been chosen as follows: for the target nucleus ^{56}Fe we have used Arthur and Young potential /4/, and for the targets ^{58}Ni and ^{52}Cr it has been Becchetti and Greenlees potential /5/.

To describe the compound contribution to neutron inelastic scattering spectra a parametrized Maxwellian evaporation shape has been used. This choice is very convenient for our purpose and its feasibility has been attested for many times, see e.g. /6/.

Results of our calculations are given in Figs. 2-6. From these figures it can be seen that the theoretical predictions are in good agreement with both experimental and empirical data.

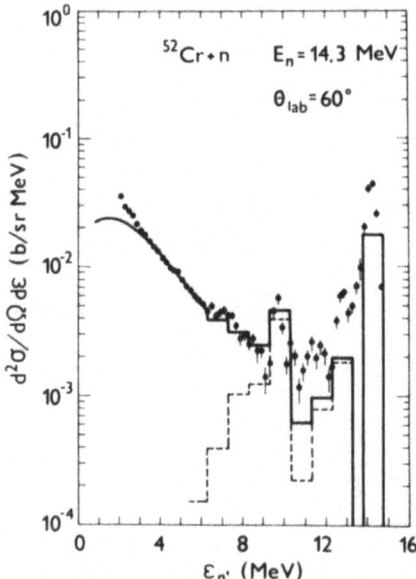

FIG. 6. The same as in Fig. 2, except for ^{52}Cr, θ_{lab} = 60°.

REFERENCES
1. N. Austern, "Direct Nuclear Reactions", Wiley, New York, 1970.
2. V. V. Voronov and V. G. Soloviev, Particles and Nuclei 14 (1983) 1380.
   A. I. Vdovin and V. G. Soloviev, ibid. 14 (1983) 237.
3. A. Bohr and B. Mottelson, "Nuclear Structure II", Benjamin, New York, 1974.
4. E. Arthur and P. Young, Report BNL-NCS-51245, p. 731, Upton, 1980.
5. F. Becchetti and G. Greenlees, Phys. Rev. 182 (1969) 1190.
6. A. Lukyanov et al., Sov. J. Nucl. Phys. 21 (1985).
7. A. Takahashi et al., OKTAVIAN Report A-1983-1, Osaka, 1983.

WALKERS, SLAVES, OR BOTH?

E. Běták

Institute of Physics, Electro-Physical Research Centre
of the Slovak Academy of Sciences, Bratislava,
Czechoslovakia

ABSTRACT. The correspondence of the random walks and Markovian
master equations for the two-dimensional problems of nuclear
physics (namely the nucleon transfer in heavy-ion reactions and
the two-component exciton model) is briefly sketched. The preference of the mathematical description for the real calculations should be connected to the physical conditions of the
experiment.

The statistical description of many features of nuclear reactions, like the equilibration process in the exciton model or
the nucleon transfer in heavy-ion reactions (and many others)
is usually based on the corresponding master equation (or some
approximation to it, e.g. the Fokker-Planck equation). In parallel to that, the same processes can be treated as random walk
problems. Both the approaches are intuitively expected to yield
similar results. Unfortunately, the comparison of both of them
is by no means straightforward, mainly because of the different
bases used in the description: the evolution of a process is
expressed as a function of time t in the master equation, whereas it is a function of the number of steps q in the random walk.
Though there is some correlation between both t and q, no unique
relation that couples them together can be given. Therefore,
a comparison can be done only assuming some average correspondence between t and q to be valid /1, 2/.

We follow here the ideas of Akkermans´ paper /1/. Assuming
that the state space is discrete (which is really the case) and
one-dimensional (true for some processes), the master equation
can be written in its matrix form

$$\dot{P}(t) = \underline{\underline{A}} \, \underline{P}(t). \tag{1}$$

Similarly, random walk equation reads

$$\underline{P}(q+1) = \underline{\underline{T}} \, \underline{P}(q). \tag{2}$$

The processes described here need not to be those of nearest neighbour transitions (i.e. $A_{ij} = 0$ or $T_{ij} = 0$ for $|i-j| \geq 2$), and can describe also open systems (i.e. $\sum_i A_{ij} < 0$ or $\sum_i P_{ij} < 1$). We suppose that the walker takes his steps at random times $t_1, t_2, \ldots, t_q$ governed by (a set of) probability density(ies) $\psi_i(T) \equiv \psi_i(t_i - t_{i-1})$ (waiting time distribution(s)). Obviously, normalization of the type

$$\int_0^\infty dT \, \psi_i(T) = 1 \tag{3}$$

must be valid. The solutions of (1) and (2) are coupled via /3/

$$\underline{P}(t) = \sum_{q=0}^\infty \psi(T) \, \underline{P}(q) \tag{4}$$

for the case of position-independent waiting time distribution. Bedeaux et al. /4/ have shown that the solutions $\underline{P}(t)$ and $\underline{P}(q)$ satisfy simultaneously eqs. (1) and (2), if the waiting time distribution takes the form

$$\psi(T) = \Lambda \exp(-\Lambda T). \tag{5a}$$

Unfortunately, the condition (5a) is rather restrictive for the real physical situation in nuclear reaction. Generally, one should have /5, 6/

$$\psi_i(T) = \Lambda_i \exp(-\Lambda_i T). \tag{5b}$$

The two approaches are shown to be equivalent if and only if /1, 6/

$$\underline{\underline{A}} = (\underline{\underline{T}} - \underline{\underline{1}})\underline{\underline{\Lambda}}, \tag{6a}$$

where

$$\Lambda_{ij} = \delta_{ij}\Lambda_i. \tag{6b}$$

Till now we have just repeated the main ideas of Akkermans thesis /1/, given for the one-dimensional case therein. Note, that none of the considerations required any special form of $\underline{\underline{A}}$ or $\underline{\underline{T}}$ (it is also the case of all the derivation done by Akkermans /1/, but skipped for brevity from the present paper). In the two-dimensional cases of interest for nuclear physics, either in the two-component exciton model or in the nucleon transfer in heavy-ion reactions, one is dealing only with finite set of states (e.g. because of finite number of nucleons in nuclei). Such a set can be re-ordered as to yield the one-dimensional case. Of course, the great advantage of restriction to the nearest neighbour transitions has to be released here, but all the derivation can be repeated straightforward-ly. So we arrive also in the two-dimensional (finite) case to the equivalence of both approaches. The vanishing N, Z correlation reported for master equation approach (in the contrary to the random walk) by Hiller et al. /2/ seems, therefore, to be caused by assuming the oversimplified form (5a) instead of the more realistic one (5b). Nevertheless, due to the (generalization of the) relation (4), the solutions of both the approaches are not strictly identical, and some minor differences might emerge. The preference of the mathematical formalism for the process studied should be closely related to its physical background, namely, which of the two independent variables is the one more closely related to the measured quantities, as pointed in ref. /2/.

The author is grateful to Prof. Griffin for valuable comments.

REFERENCES

1. J. M. Akkermans, Ph.D. Thesis (Report ECN-121), Petten, 1982.
2. B. Hiller et al., Nucl. Phys. A424 (1984) 335.
3. E. W. Montroll and G. H. Weiss, J. Math. Phys. 6 (1965) 167.
4. D. Bedeaux et al., J. Math. Phys. 12 (1971) 2116.
5. W. J. Shugard and H. Reiss, J. Chem. Phys. 65 (1976) 2827.
6. J. M. Akkermans and H. Gruppelaar, Z. Phys. A300 (1981) 345.

THE NEUTRON EMISSION FROM LEAD BOMBARDED WITH 14 MeV NEUTRONS

T. Elfruth, D. Hermsdorf, H. Kalka, D. Seeliger,
K. Seidel and S. Unholzer

Technical University Dresden, GDR

ABSTRACT. Neutron emission spectra obtained with time-of-flight technique are compared with calculations which take into account equilibrium and pre-equilibrium emissions as well as direct collective excitations.

1. INTRODUCTION

In the last years pre-equilibrium processes have been studied in many experiments. After satisfying descriptions of the angle-integrated spectra the angular distributions calculated for many nuclei also show a reasonable agreement. In the present paper the neutron emission from lead is investigated. Lead has been chosen, on the one side, because it is an important material of future fusion reactor blankets, and on the other, because the influence of closed-shell effects on the neutron emission is of special interest.

2. EXPERIMENTS

Double-differential neutron emission cross-section for lead are measured at the TUD DT generator using the pulsed-beam time-of-flight spectroscopy with a flight path of about 5 m. The cross-sections are determined in the emission energy range between 2 MeV and the incident neutron energy, and within an angular range from 15° to 165° in steps of 15° by means of the ring geometry. To get a comparatively weak angular dependence of the incident neutron energy, the scattering sample is arranged perpendicular to the deuteron axis. Thus the experimental arrangement is especially suitable to study the angular distribution. The experimental arrangement and the data analysis have been presented in detail in Gaussig /1/ and elsewhere /2, 3/.

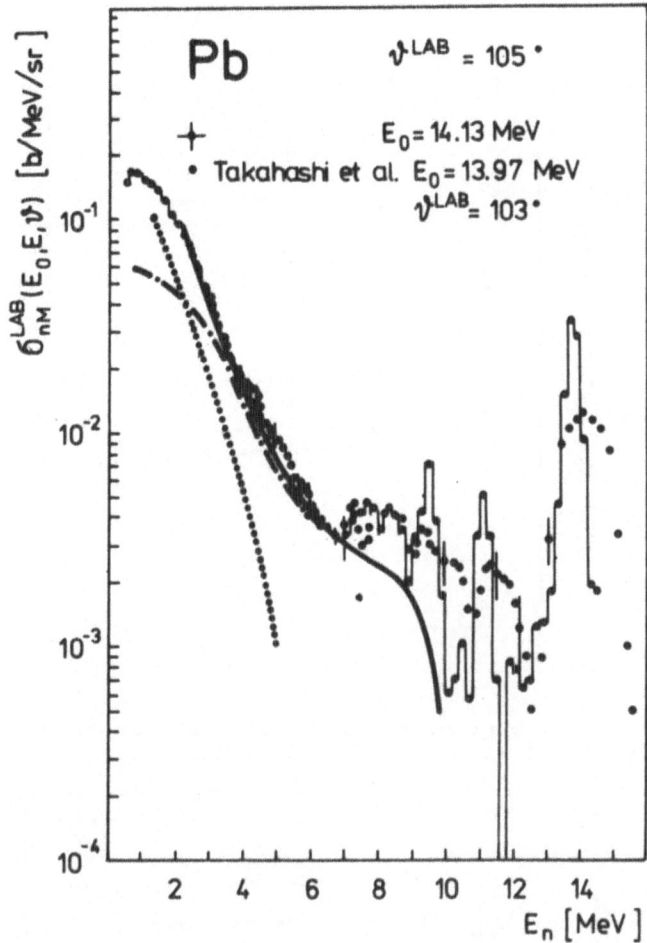

FIG. 1. Neutron emission spectrum of lead. Experiments of TUD and Osaka University /4/ are compared with calculations. Contributions from secondary neutron emission (n,2n) have been predicted by the STAPRE code and the AMAPRE code to get the total one. ... STAPRE (n,2n), -·- AMAPRE, —— sum.

An example of the derived double-differential neutron emission cross-section compared with Takahashi's data /4/ is given in Fig. 1. Double-differential cross-section at the emission

energies of 3.1, 7.0 and 11.5 MeV are also compared with the
recent measurements made at the Osaka University as well as
with the former TUD experiment /5/ in Figs. 2–4. A general consistency between the three experiments is obvious at higher

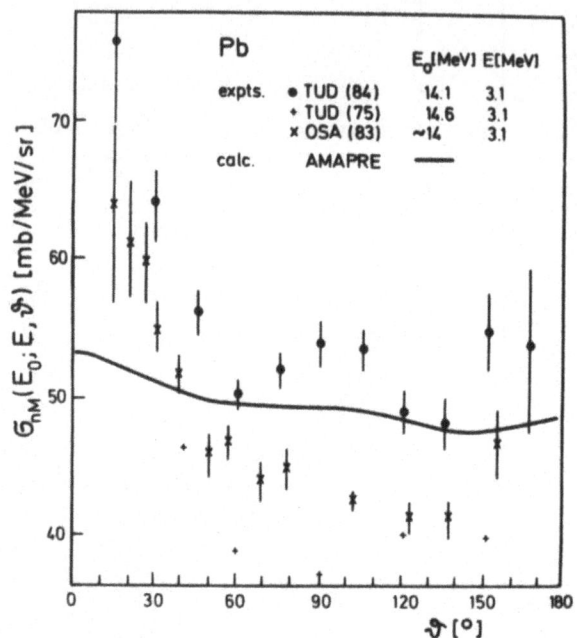

FIG. 2. Angular distribution at 3 MeV emission energy. Experiments are compared with data from AMAPRE.

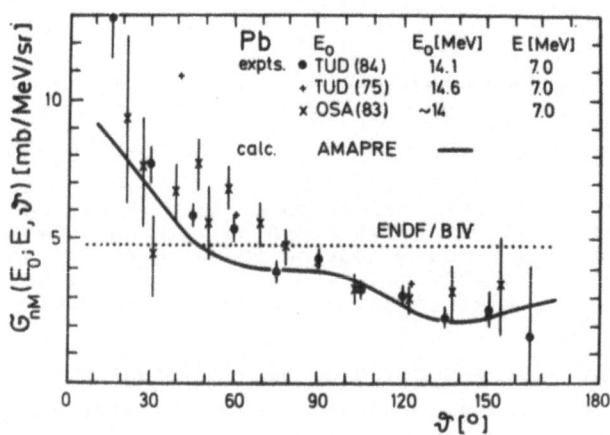

FIG. 3. Same as in Fig. 2 at an emission energy of 7 MeV.

emission energies, whereas remarkable differences occur at lower
emission energies. With the exception of the data point at 15°
the new TUD data show an almost symmetric angular distribution
relative to 90°, which is in agreement with the physical point
of view, whereas the Osaka data show a forward-peaking usually
expected for higher emission energies only. This may be caused
by the strong angular dependence of the incident neutron energy
in this experiment.

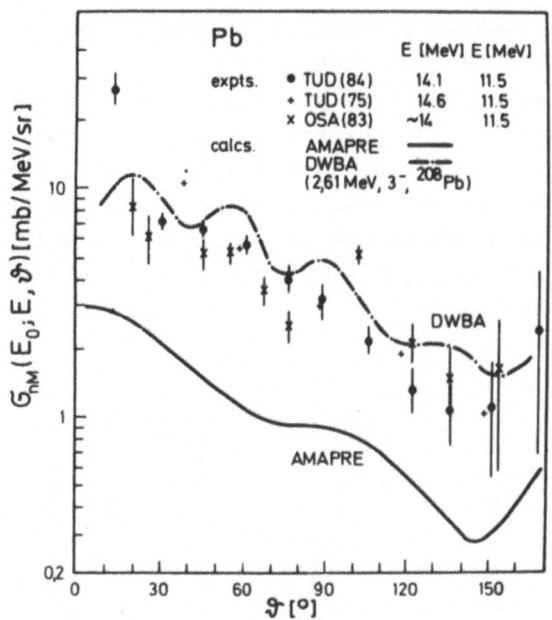

FIG. 4. Same as in Fig. 2 at an emission energy of 11.5 MeV as
well as a comparison with data from /7/.

3. CALCULATIONS

Model calculations have been carried out taking into account
pre-equilibrium and equilibrium contributions by AMAPRE code
/6/ and collective direct contributions by DWBA code /7/.

The AMAPRE code is based on the Generalized Exciton Model
/8/. Due to the solution of the master equations up to infinite
time both pre-equilibrium and equilibrium particle emissions
are obtained within the same concept. To predict the angular

dependence the Correlated Emission Model /9/ has been adopted. The Fermi motion of target nucleons and Pauli's principle are taken into account. The level densities for lead are taken from Ignatyuk's /10/ phenomenological formula, taking into consideration shell-structure effects. The emission spectra of lead cannot be reproduced with level density parameters from the low energy resonance counting or from the Fermi gas model. A random walk-like emission of secondary neutrons due to (n,2n) processes is assumed to be valid for energies above 20 MeV only. Therefore, the spectra of secondary neutrons calculated with the Hauser-Feshbach STAPRE code /11/ must be added.

At low emission energies the calculation reproduces the right order of magnitude of the experimentally obtained data as well as slight asymmetry of the angular distribution (Fig. 2). In the intermediate region, in which there prevails the pre-equilibrium emission, the experiments are reproduced very well (Fig. 3). Strong collective enhancement of the 3^- octupole vibration in ^{208}Pb becomes evident at high emission energies (Fig. 4).

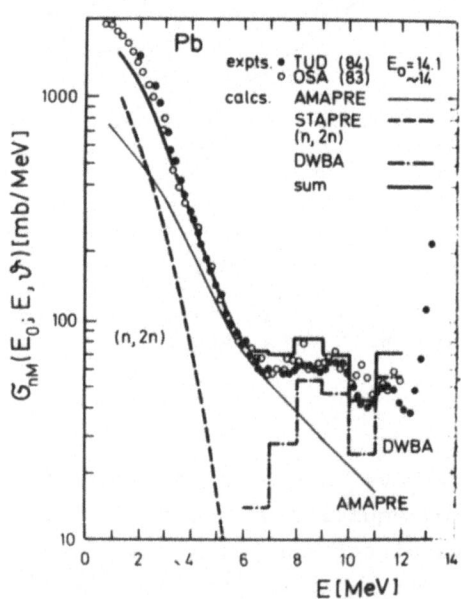

FIG. 5. Angle-integrated neutron emission spectrum. Experiments of the TUD and of /4/ are compared with calculations of /6/, /7/ and /11/.

Figure 1 gives an example of energy spectrum at a fixed angle. Besides the strong collective excitations the influence of the closed shells, especially of the level densities, on the determination of the pre-equilibrium part of the spectrum is also obvious. Finally in Fig. 5 the angle-integrated data are compared with calculations. Again the dominating influence of the collective excitations at emission energies above 8 MeV is apparent.

4. CONCLUSIONS

The experimentally and theoretically obtained data are in good agreement. The spectrum of the pre-equilibrium emission becomes more "similar" to an equilibrium emission spectrum due to the influence of the closed shells. Collective excitations dominate in the high-energy part of the emission spectrum. If these contributions are adequately taken into account, a reasonable description of the whole emission spectrum is obtained.

REFERENCES

1. T. Elfruth et al., Proc. of the 1st 14 MeV Coordinated Research Programme Meeting, Gaussig, 1983 (to be published by IAEA).
2. T. Elfruth et al., Report ZfK-530, June 1984, p. 13.
   H. Helfer et al., Report ZfK-530, June 1984, p. 126.
3. T. Elfruth et al., Proc. XIVth Int. Symp. on Nuclear Physics, Gaussig, 1984 (to be published).
4. A. Takahashi et al., OKTAVIAN Report A-83-01, June 1983.
5. D. Hermsdorf et al., Report ZfK-277(Ü), 1975.
6. H. Kalka, Dissertation, Technical University Dresden, 1983.
7. A. I. Ignatyuk and V. P. Lunev, private communication, 1984.
8. J. M. Akkermans et al., Phys. Rev. C22 (1980) 73.
   J. M. Akkermans, Report ECN-121, 1982, p. 51.
   H. Gruppelaar et al., Report ECN-114, 1982.
9. C. Costa et al., Report ECN-172, 1982.
10. A. I. Ignatyuk et al., Yad. Fiz. 21 (1975) 485.
11. M. Uhl and B. Strohmaier, Report IRK-76/1, 1975.

THE (n,α) REACTIONS ON THE ^{95}Mo AND ^{143}Nd ISOTOPES WITH 3 MeV NEUTRONS

M. Florek, I. Szarka, J. Oravec, K. Holý, E. Mišianiková,
*H. Helfer, *U. Jahn, +Yu. P. Popov

Department of Nuclear Physics, Comenius University,
Bratislava, Czechoslovakia
*Physics Section, Technical University Dresden, GDR
+Joint Institute for Nuclear Research, Dubna, USSR

ABSTRACT. Partial cross-sections of (n,α) reactions induced by 3 MeV neutrons on ^{95}Mo and ^{143}Nd have been measured. Cross-sections have been deduced from alpha particles spectra measured by the gas proportional telescope.

1. INTRODUCTION

The (n,α) reactions in an energy region of $\sim$ 3 MeV (the neutrons obtained in neutron generators on the basis of the D(d,n)^{3}He reaction) on middle nuclei deeply under the Coulomb barrier of nuclei occur with a very low probability (rare processes $\sigma_\alpha \sim 10^{-32}$ m^2). The products of the reactions can affect the quality of constructional materials of a power breeder and thermonuclear reactors /1/, besides the values of the (n,α) reaction cross-sections supply the theory with a valuable experimental material.

The (n,α) reaction cross-sections on the ^{55}Mn and ^{69}Ga isotopes were measured using the activation analysis method /2/. In the present paper the direct measurement of (n,α) reaction cross-sections on the neutron beam are described and the first results on the ^{95}Mo and ^{143}Nd nuclei are obtained.

2. EXPERIMENT

The gas proportional telescope was used for measuring the alpha particles' spectra. The telescope consisted of the gas proportional dE/dx chamber equivalent to $\sim$ 1 μm thick Si and

of the surface barrier semiconductor Si-Au detector having an
effective surface of 200 mm^2. The chamber was filled with a
Ar + 6% CO_2 gas mixture at 46 kPa pressure. The arrangement
of the apparatus is shown in Fig. 1. Output signal from the
proportional chamber is driven on the input of single channel
analyser, where the energetic window has been adjusted for the
specific ionization losses of the alpha particles /3/. The
pulses from the semiconductor detector are driven on the multi-
channel amplitude analyser in coincidence with the pulses from
the proportional chamber. This configuration causes background
lowering and eliminates the possibility of alpha particles
detection originating in the semiconductor detector.

The targets were made of MoO_3 (Mo enriched to 95% with
^{95}Mo) and Nd_2O_3 (Nd enriched to 83.2% with ^{143}Nd) and were
deposited onto a Pb-foil backing with a total area of 200 mm^2.
The thickness of Mo and Nd was equal to 18 gm^{-2} and 22.7 gm^{-2},
respectively. The targets were placed 34 mm from the deuterium
target of the 400 keV neutron generator /4/ perpendicular to
the deuteron flux axis. The flux density of neutrons and their
energy (averaged over the whole area of samples accounting
neutron angular distribution) was 6 x 10^{10} m^{-2} s^{-1} and

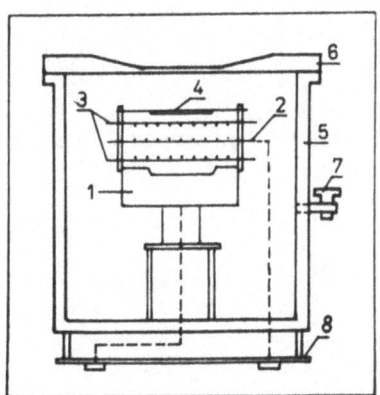

FIG. 1. Scheme of the gas proportional telescope. 1 - surface
barrier Si-Au detector, 2, 3 - anode and cathode wires of the
dE/dx chamber, 4 - target, 5 - vacuum vessel, 6 - copper
cover, 7 - vacuum pumpe flange, 8 - isolator.

3.0±0.2 MeV, respectively. The neutron flux was calibrated by the standard method of the In-foil /5/.

The ^{6}Li(n,α)T reaction with the well known cross-section (150±20) x 10^{-31} m^2 /6/ has been used for the calibration of the gas proportional telescope and for the measured results. The enriched isotope ^{6}Li, in form of LiF, has been deposited onto a Pb-foil with a thickness of 13 gm^{-2}.

3. RESULTS AND DISCUSSION

The alpha particle energy spectrum of the ^{95}Mo(n,α)^{92}Zr reaction during 65 hours is shown in Fig. 2, where the peak corresponds to alpha transition to the ground state of ^{92}Zr with

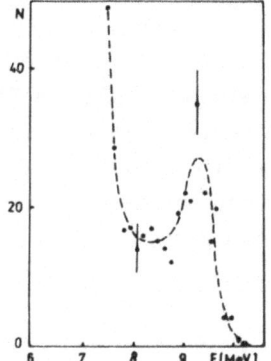

FIG. 2. Alpha particle energy spectrum of the ^{95}Mo(n,α)^{92}Zr reaction.

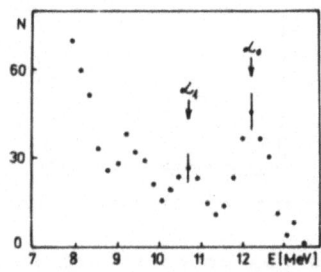

FIG. 3. Alpha particle energy spectrum of the ^{143}Nd(n,α)^{140}Ce reaction.

76

an energy of 9.4 MeV (reaction energy $Q = 6.67$ MeV). The counts in the peak are equal to 140 ± 12.

The results of the ^{143}Nd$(n,\alpha)^{140}$Ce reaction measurement are shown in Fig. 3. From the figure we can identify two peaks, which correspond to alpha transitions to the ground state ($E_{\alpha_0} = 12.4$ MeV) and to the first excitation state of ^{140}Ce with an energy of 10.8 MeV. The counts in the relevant peaks are equal to 165 ± 13 and 42 ± 7, respectively.

Using the calibration ^{6}Li(n,α)T reaction (from this reaction tritons were detected because of the absorption effect of alpha particles in the sample LiF), we obtained the partial cross-sections of the (n,α) reactions induced by 3 MeV neutrons of the following values:

$$^{95}\text{Mo} \quad \sigma_{\alpha_0} = (90\pm40) \times 10^{-34} \text{ m}^2,$$

$$^{143}\text{Nd} \quad \sigma_{\alpha_0} = (130\pm60) \times 10^{-34} \text{ m}^2,$$

$$\sigma_{\alpha_1} = (30\pm20) \times 10^{-34} \text{ m}^2.$$

REFERENCES

1. N. P. Balabanov et al., Proc. Conf. on Neutron Physics, Obninsk, 1974, p. 126.
2. M. Florek et al., Czech. J. Phys. B34 (1984) 30.
3. I. Szarka et al., Acta Phys. Univ. Comen. 25 (1985)
4. A. Meister et al., Techn. Univ. Dresden Information 05-12, 1980.
5. G. S. Nargolwalla et al., J. Radioanal. Chem. 5 (1970) 403.
6. A. I. Aliev et al., "Nuclear Data for Neutron Activation Analysis", Atomizdat, Moscow, 1969.

EXCITON MODEL AND MULTISTEP COMPOUND REACTIONS

H. Kalka, D. Hermsdorf and D. Seeliger

Technical University Dresden, GDR

ABSTRACT. The exciton model (EM) contains both statistical multi-step compound (SMC) and direct (SMD) contributions. The SMC part of the EM can be derived exactly from the microscopic random-matrix model (RMM) of Agassi et al. /1/ without use of the detailed balance argument and the Golden Rule.

We consider the dynamical matrix Γ_{nm} of the RMM which enters the master equation /1, 2/

$$\hbar \frac{d}{dt} P_{km}(t) \equiv - \sum_n P_{kn}(t) \Gamma_{nm}. \tag{1}$$

$P_{nm}(t)$ is the probability that the system finds itself in class m, if at t = 0 it was in class n. Using the notation of /1, 2/, Γ_{nm} can be written in an EM-like form /3/

$$\Gamma_{nm}/\hbar \equiv [W_n + \sum_k \lambda_{nk}]\delta_{nm} - \lambda_{nm}, \tag{2}$$

where the total emission rate is defined by the escape width

$$W_n \equiv \Gamma_n^{\uparrow}/\hbar = \frac{1}{2\pi\hbar g_n^b} \sum_c T_n^c \equiv \int_0^{E-B} W_n(\mathcal{E}) \, d\mathcal{E} \tag{3}$$

and the transition rates

$$\lambda_{nm} \equiv (2\pi/\hbar)[\overline{(V_{nm})^2} + \sum_c \Gamma_n^c \Gamma_m^c] g_m^b \tag{4}$$

consist of an internal and external part. g_n^b symbolizes the exciton density of bound configurations. $\overline{(V_{nm})^2}$ is the mean square bound-bound residual interaction, whereas the transmission coefficients T_n^c and the Γ_n^c are proportional to the mean square bound-continuum interaction $\overline{(v_n^c)^2}$.

1. PARTICLE EMISSION RATES

Changing the sum over channels in (2) into an integration, introducing the residual state density $\varphi_{n-1}(U)$, and replacing the transmission coefficients by the "absorption cross-section in bound configurations of class n" σ_n^b we obtain

$$W_n(\varepsilon) \, d\varepsilon = \frac{2s+1}{\pi^2 \hbar^3} \varepsilon \sigma_n^b(\varepsilon) \frac{\varphi_{n-1}(E-B-\varepsilon)}{\varphi_n^b(E)} . \tag{5}$$

Here $\varphi_{n-1}(U)$ means the "full" exciton state density /3, 4/ taking into account that the residual nucleus is left after emission in either a bound or unbound configuration.

2. TWO-BODY TRANSITION RATES

A substitution of the many-body interaction V by a two-body one causes: (i) the vanishing of $\overline{(V_{nm})^2}$ unless $|n-m| = 2$ and $\overline{(v_n^c)^2}$ unless $|n-m| \leq 2$ (the exit mode definition /5/ for the channel c is used); (ii) the replacing of all state densities adjoined to matrix elements by partial (final) state densities /3/, proposed for the first time by Williams /6/; (iii) the interaction of the incident channel only with the first exciton class n_o (1 – entrance doorway model).

Neglecting all terms of the second order in $\overline{(v_n^c)^2}$ and applying (i) and (iii) we obtain the transition rates from (4) as

$$\lambda_n^+ = \frac{2\pi}{\hbar} \overline{(V_{nn+2})^2} \varphi_n^{(bb)+}, \quad \lambda_n^- = \frac{2\pi}{\hbar} \overline{(V_{nn-2})^2} \varphi_n^{(bb)-} \tag{6}$$

which obey the steady-state equilibrium condition

$$\lambda_n^{(bb)+} \varphi_n^b = \lambda_{n+2}^{(bb)-} \varphi_{n+2}^b .$$

3. EMISSION CROSS-SECTION

The equivalence between the time-integrated master eq. (1) and
the fluctuation cross-section

$$\langle \sigma_{cc'}^{fl} \rangle = \pi \lambda^2 \sum_{nm} T_n^c (\Gamma^{-1})_{nm} T_m^{c'} / 2\pi \rho_m^b \tag{7}$$

is shown in ref. /1/. Using the continuum approach, eq. (5) and
(iii) we arrive at the emission cross-section formula /3/

$$\sigma(\varepsilon) \, d\varepsilon = \sigma_{n_o}^b \sum_m \tau_m W_m(\varepsilon) \, d\varepsilon \tag{8}$$

with the mean life time

$$\tau_m \equiv \tau_{n_o m} = \delta_{nn_o} \int_o^\infty P_{nm}(t) \, dt. \tag{9}$$

To make the EM practicable for calculations ad hoc assumptions
must be made. At first $\overline{(V_{nm})^2}$ and $\overline{(V_n^c)^2}$ are assumed to be independent of m and n. Then $\overline{M^2} \equiv \overline{(V_{nm})^2}$ will be treated as a free
(fit) parameter, and $\sigma^b \sim \overline{(V_n^c)^2}$ can be replaced by the optical
absorption cross-section diminished by the absorption in open
configurations $\sigma^b = \sigma^{abs} - \sigma^u$. However, the fraction σ^u is unknown.

4. CONCLUSIONS

The EM (which contains pre-equilibrium and equilibrium contributions, see eq. (9)) is a RMM. The neglection of the continuum-
continuum interaction in the shell-model approach /1, 7/ reduces
the EM to an SMC model where the exciton state densities ρ_n are
replaced by the "bound" ones and the absorption cross-section
σ^{abs} by the absorption in bound configurations σ^b.

Deviations from a symmetric angular distribution about 90°
belong to the SMD-part of the EM. The necessary correlations between S-matrix elements pertaining to different values of J and
π arise from: (i) violation of the statistical assumptions in
the first stages of the reaction /8/ and/or (ii) inclusion of
the continuum-continuum interaction.

REFERENCES

1. D. Agassi, et al., Phys. Rep. 22C (1975) 145.
2. K. W. McVoy and X. T. Tang, Phys. Rep. 94 (1983) 139.
3. H. Kalka et al., to be published.
4. G. Schütte, Nucl. Phys. A126 (1969) 513.
5. H. Feshbach et al., Ann. Phys. (N.Y.) 125 (1980) 429.
6. F. C. Williams Jr., Phys. Lett. 31B (1970) 184.
7. G. Mahaux and H. A. Weidenmüller, "Shell-model Approach to Nuclear Reactions", North-Holland, Amsterdam, 1969.
8. G. Mantzouranis et al., Phys. Lett. 56B (1975) 220.

NUCLEAR LEVEL DENSITY DETERMINED FROM THE ANALYSIS OF SPECTRA OF (n,n´) AND (p,n) REACTIONS

G. N. Lovtsikova, O. A. Salnikov, S. P. Simakov
and A. M. Trufanov

Physical and Power Institute, Obninsk, USSR

ABSTRACT. Nuclear level density is determined in a wide range of energies, starting from low-lying states through the region at the neutron binding energy up to continuum. The method is applied for ^{113}In, Mo, ^{59}Co (p,n) and (n,n´) reactions.

The nuclear level density is one of the fundamental characteristics of atomic nuclei. To a considerable degree, it determines the cross-section of any nuclear reaction. Many experimental and theoretical papers have been devoted to this problem.

It is believed, that the most straightforward information on the nuclear level density can be obtained from the experiments on total cross-sections in the resonance region of the energy spectrum, where we can get their number by a simple counting of the number of resonances (levels) per unit energy interval.

However, there are several disadvantages of this method:

(1) It can be applied for compound nuclei only. Target nuclei, however, are of the same interest.

(2) The data obtained by this method are limited in energy by one point - by the binding energy of neutron in the nucleus.

(3) For the purpose of theoretical interpretation (extrapolation with respect to the excitation energy) it is necessary to know the type of resonance concerned (s,p,d, etc.). In some cases, however, it cannot be identified unambiguously.

That is why there is much effort exerted to determine the nuclear level density based on other reactions (cross-sections) in a wider range of energies. Such a possibility is offered by the analysis of the form of the studied equilibrium neutron spectra from the (n,n′) and (p,n) reactions measured at various energies of incident nucleons. When analysing the experimental spectra at the determination of the nuclear level density values, it is necessary to identify that part of the spectrum which is influenced by non-statistical processes, i.e. to find the mechanism of emission of neutrons.

If, for the neutrons coming from the (p,n) reaction, the contribution from non-statistical processes can be neglected over the whole spectrum up to the energy of protons of 9 to 10 MeV, then for the neutrons coming from the (n,n′) reaction this contribution is to be taken into account even in the case that the initial energy of neutrons is of about 5 MeV. In this case the following method of constructing the equilibrium part of the spectrum′ is used. In spectra of the neutrons coming from the (n,n′) reaction there occurs an interval in which the contribution from the non-statistical processes can still be neglected (criterion is the angular and energy distribution of the emitted neutrons). From the analysis it follows that it is the energy interval of the emitted neutrons ranging from 0.6 to 2.5 MeV that is concerned. When changing the incident neutron energy, various excitation energies of the residual nucleus still staying in the same energy interval of the emitted neutrons and the equilibrium region of the spectra in a wide range of excitation energies can be obtained.

In Fig. 1 an example of the construction of such a spectrum is given for the case of ^{113}In, Mo and ^{59}Co. A more detailed description of the determination of the equilibrium part of the spectrum in the (n,n′) and (p,n) reactions was described in /1-3/.

The presented method of the determination of nuclear level density from the equilibrium neutron spectrum is based on the following principle: The precise Hauser-Feshbach expression is used for the description of the equilibrium part of the spectrum, which for the case of the (p,n) reaction assumes the form:

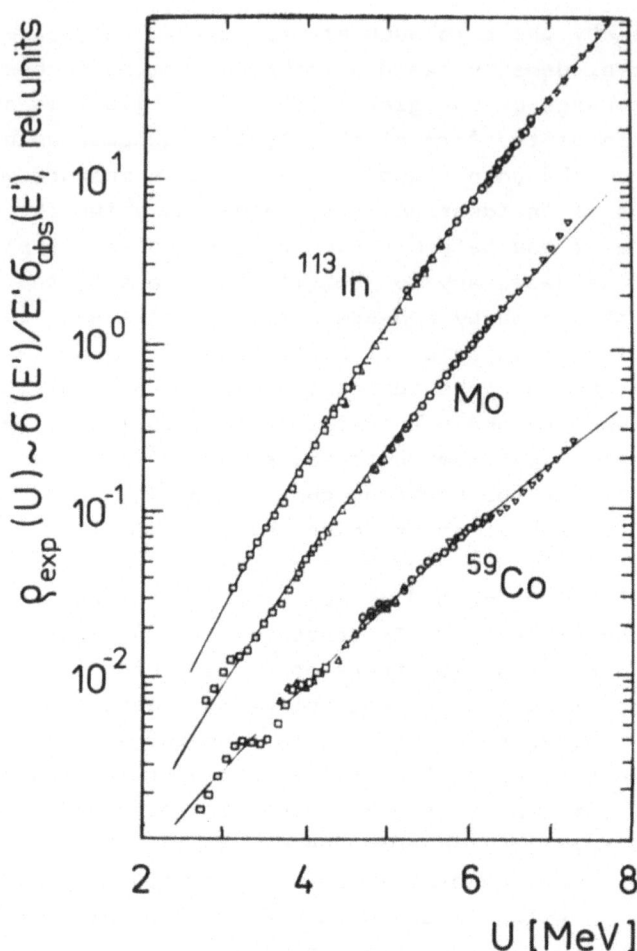

FIG. 1. Energy dependence of the nuclear level density. The symbols □, △, O, ▽ stand for the experimental data obtained for various incident energy values. The curves are the result of calculations.

$$\mathfrak{G}(E_p,E_n) = \frac{\pi \lambda_p^2}{2(2I_0+1)} \sum_{lj} T_{lj}^p \sum_J \frac{2J+1}{D_{J\pi}} \sum_{I'\pi'} \sum_{l'j'} T_{l'j'}^n \cdot$$

$$\cdot (E_p+Q_{p,n}-U') \mathcal{G}(U',I',\pi') \ . \tag{1}$$

Here the usual notation is used, whereby the p and n indexes denote that the symbol is related to the proton or neutron, respectively. The nuclear level density is denoted by $\varrho(U',I',\pi')$

$$\varrho(U',I',\pi') = \frac{1}{2}\frac{\sqrt{\pi}}{12}\frac{\exp(2\sqrt{a\ U^*})}{a^{1/4}\ U^{*5/4}} \ .$$

$$\cdot\ \frac{(2I'+1)\exp\left[-(J+\frac{1}{2})^2/2\ \sigma^2\right]}{2\ \sqrt{2\pi}\ \sigma^3} \ , \tag{2}$$

where $U' = U^* = U - \Delta$, Δ is the back shift, $\sigma^2 = (6/\pi^2)\ a$. $\langle m^2\rangle\ \sqrt{U^*/a}$ is the spin cut-off parameter, and $\langle m^2\rangle = 0.24\ A^{2/3}$ is determined according to the Fermi gas model using the a and Δ parameters yielding the best description of the spectrum. In this approach all possible effects upon the level density which are relevant to the real nucleus (the shell structure in the one-particle state spectra, pair interaction, collective effects) and which cause the difference from the foreseen Fermi gas model are simultaneously taken into account. The determination of the absolute value of the level density is done by extrapolation of the calculated spectrum to the low-energy region of the excited nuclei, where the energy level scheme along with their parameters are known

$$\sigma(E_p,E_n) = \frac{\pi\lambda^2}{2(2I_0+1)}\sum_{l,j} T^p_{l,j}\sum_J \frac{2J+1}{D_{j,\pi}}\sum_{l',j'}$$

$$T^n_{l',j'}(E_p+Q_{p,n}-U') \tag{3}$$

and the calculated spectrum can be compared with the real one.

In Figs. 2 and 3 the experimental and the calculated neutron spectra for $^{57}Fe(p,n)^{57}Co$ and $^{115}In(p,n)^{115}Sn$ reactions are shown. The calculations have shown that the parameters "a" and "Δ" are bound with each other, each of them showing a strong influence upon the form of the neutron spectrum; the parameter a principally determines the form of its continuous part, while Δ exerts the influence upon the discrete part. In

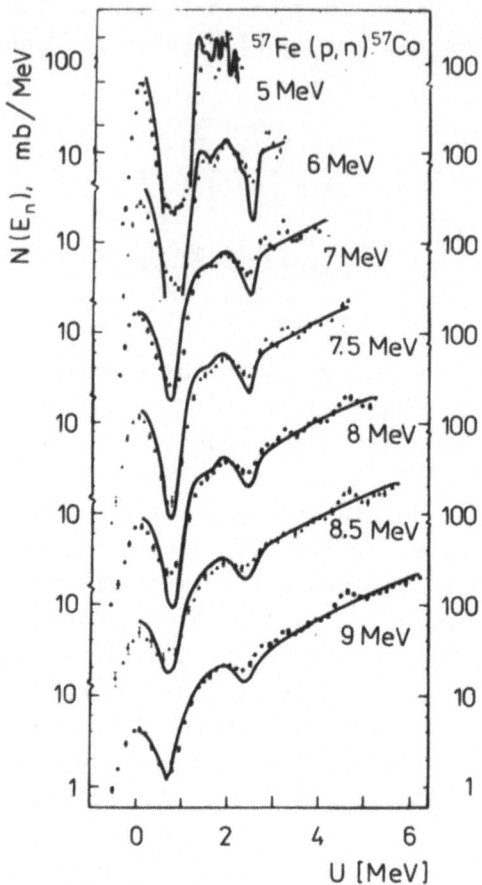

FIG. 2. Experimental and calculated spectra of the neutrons from the reaction $^{57}Fe(p,n)^{57}Co$.

the calculations of the spectrum the important role is played by their combination.

The parameters of the level density formula obtained by means of the above-mentioned many-nuclei procedures are given in Table 1.

Attention should be paid to the good agreement of the mean distance between the levels obtained by the present method and from the resonance values. However, in the case of the isotopes measured, there are only few resonance values that have been obtained. The agreement between the density parameters of the nuclear levels for ^{113}In, obtained from the analysis of

TABLE 1

Nucleus	Reaction	a (MeV^{-1})	Δ (MeV)	D_0 (eV) This work	D_0 (eV) Ref. /4/
^{57}Co	(p,n)	7	- 0.25	281.90	
^{59}Co	(n,n´)	7.9	0.6		
^{65}Zn	(p,n)	10.7	- 0.40	563.06	
^{89}Y	(n,n´)	8.9	1.8		
^{91}Nb	(p,n)	10.5	0.25	9.66	
^{93}Nb	(n,n´)	11.2	- 0.20		
^{94}Nb	(p,n)	11.39	- 0.70	110.63	90 /5/
Mo	(n,n´)	13.3	0.80	Data on Mo have been averaged over all the isotopes	
^{95}Tc	(p,n)	11.75	- 0.90	6.03	
^{98}Tc	(p,n)	11.9	- 0.75	62.30	
^{107}Cd	(p,n)	15.25	- 0.18	43.09	
^{109}Cd	(p,n)	15.10	- 0.20	93.68	
^{113}In	(p,n)	18.25	0.70	1.13	
^{113}In	(n,n´)	17.9	0.80		
^{115}In	(n,n´)	18.0	0.80		
^{115}Sn	(p,n)	14.0	0.47	355.1	320±80
^{117}Sb	(p,n)	20.0	0.50	0.103	
^{119}Sb	(p,n)	18.6	0.36	0.504	
^{120}Sb	(p,n)	16.8	- 0.50		
^{122}Sb	(p,n)	16.3	- 0.54	11.96	10.3±1
^{181}Ta	(n,n´)	16.9	0.8		
^{181}W	(p,n)	20.3	- 0.25	17.75	18±4
^{209}Bi	(n,n´)	12.7	1.6		

spectra from the two reactions, (p,n) and (n,n´), is a very interesting point, too. All what has been mentioned above gives evidence to the reliability of the nuclear level density data obtained from the analysis of the spectra from the (p,n) and (n,n´) reactions, and it allows us to overcome the vacancy in the level density data yet existing between the resonance data (binding energy of neutrons) and the data from the low-lying nuclear states.

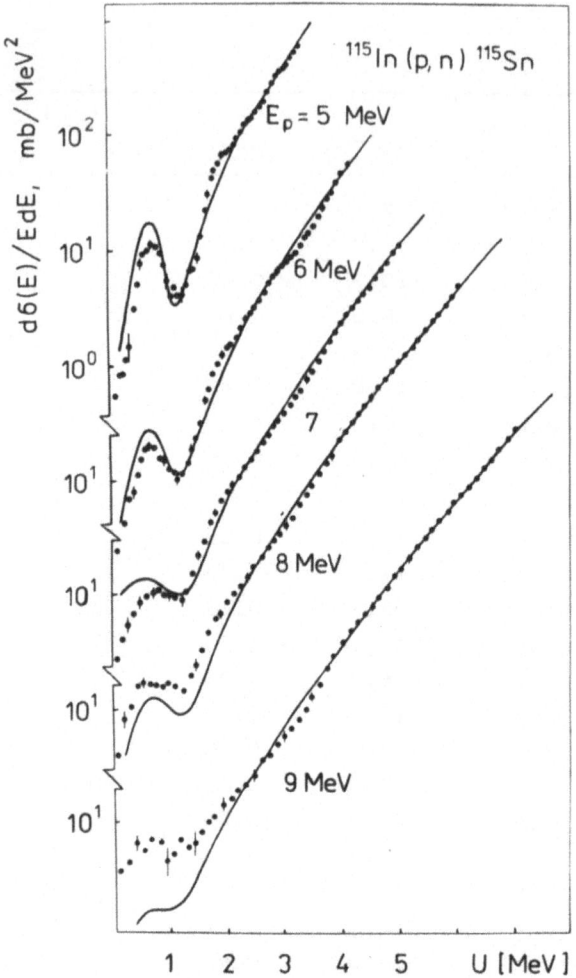

FIG. 3. Experimental and theoretical spectra of the neutrons from the reaction ^{115}In(p,n)^{115}Sn.

REFERENCES

1. S. P. Simakov, G. N. Lovtsikova, O. A. Salnikov and N. N. Titarenko, Nucl. Phys. 38 (1983) 3.
2. V. G. Pronjaev, G. V. Kotelnikova, G. N. Lovtsikova and O. A. Salnikov, Nucl. Phys. 30 (1979) 604.
3. G. N. Lovtsikova, A. M. Trufanov and O. A. Salnikov, Nucl. Phys. 33 (1981) 41.
4. A. R. Musgrove, B. S. Allen et al., Proc. Int. Conference, Harwell, 1978, p. 449.
5. A. A. Litshagin, V. A. Vinogradov and O. T. Grudzevich, Preprint FEI-1385, Obninsk, 1983.

NEUTRON EMISSION IN LARGE AMPLITUDE COLLECTIVE MOTIONS

B. Milek and R. Reif

Technical University Dresden, GDR

ABSTRACT. Neutron emission during large amplitude collective
motions is treated (i) in terms of coupled adiabatic single-
particle states of a two-centre shell model with finite-depth
potentials and (ii) within a three-body model for the reaction
dynamics.

Some aspects of a heavy-ion reaction can be modelled crudely
as a single nucleon in the field of two potentials in a distance R(t), the time dependence of which is determined by the
relative motion of the centres along given classical trajectories. Within such a picture, nucleon emission proceeds by
coupling the bound adiabatic two-centre states, occupied initially or populated during the course of the reaction, to
states in the single-particle continuum. This dynamical coupling due to non-adiabatic effects in the relative motion is
dominated by its radial component, and the main contribution
to the emission probability should be grasped by restricting
the continuum to single-particle resonances, for which a boundary condition corresponding to the radioactive decay (Gamow
states) seems to be suitable. Then the nucleon emission
spectrum is determined by the occupation probability n and the
decay width Γ of the Gamow state populated, both quantities
being dependent on time.

The position and decay width of the Gamow states of a two-
centre shell model with two Woods—Saxon potentials have been
calculated as a function of R by extending the method described
in ref. /1/. For the system $^{16}O + ^{16}O$ (compare Fig. 1) it follows

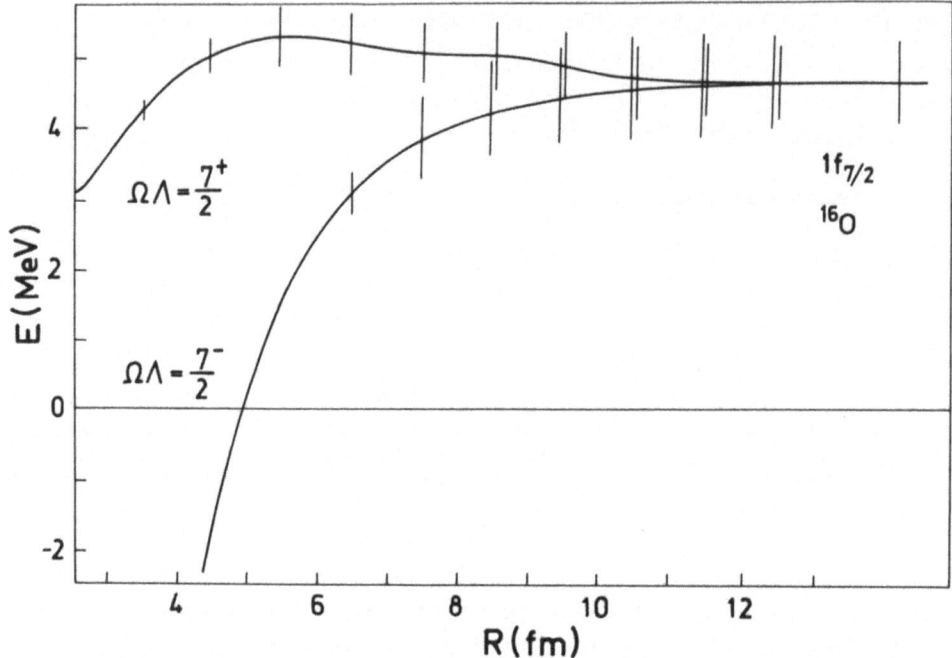

FIG. 1. Single-particle states in the $^{16}O+^{16}O$ potential as
a function of the relative distance R of the centres. The
bars are marking the width of the Gamov states.

that the width exhibits pronounced bumps for those regions, in
which the wave function of the Gamow state contains highly
excited components of the oscillator basis, which has been used
to represent the mean potentials. Calculations of the differential emission probability

$$\frac{d\,\omega(b)}{d\varepsilon} = \frac{2}{\pi} \sum_{\alpha} \int dt' \; \frac{\Gamma_{\alpha}^{2}(t')n_{\alpha}(t')}{(\varepsilon - E_{\alpha}(t'))^{2} + \Gamma_{\alpha}^{2}(t')}$$

have been performed for the reaction $^{16}O+^{17}O$, $E_{c.m.}$ = 25 MeV,
within a truncated set of adiabatic states α. The levels resulting from the following asymptotic states are taken into account
by setting up the coupling scheme: $1d_{5/2}$ - $1s_{1/2}$ - $1d_{3/2}$;
$1d_{5/2}$ - $1f_{7/2}$. After integrating over the impact parameter b,
the peaks in the emission spectrum (Fig. 2) at ≈ 1 MeV and
≈ 6 MeV reflect the shell structure in the continuum. The spec-
90

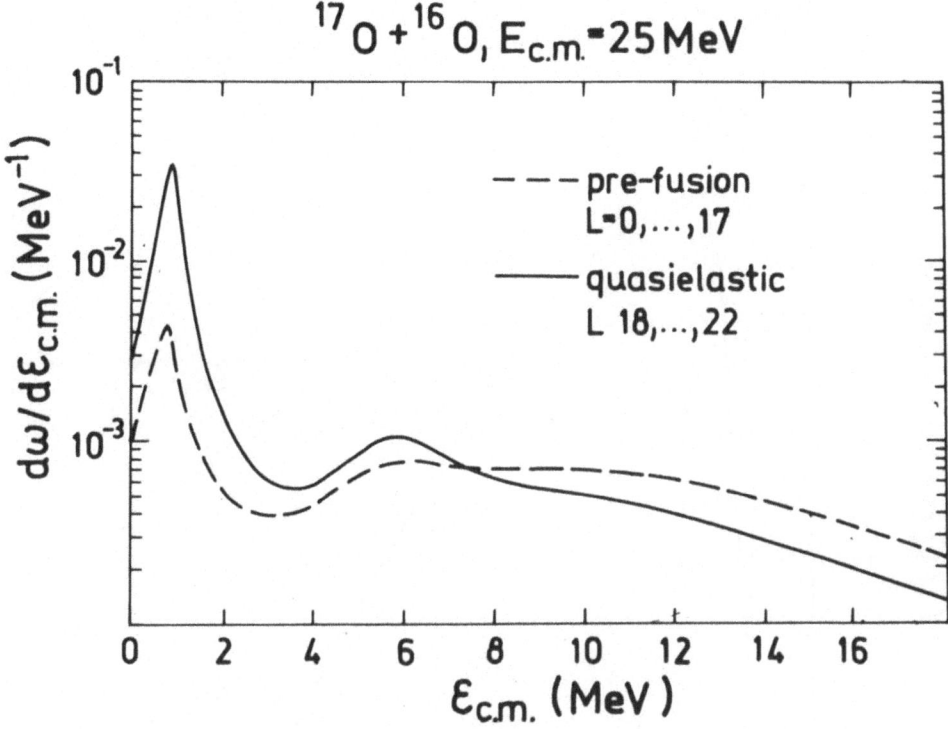

FIG. 2. Neutron emission spectrum for the reaction $^{17}O+^{16}O$, $E_{c.m.}$ = 25 MeV.

trum resulting from close trajectories leading to fusion is similar in shape, but somewhat harder than that from more peripheral trajectories.

In a Faddeev-like formulation /2/ the nucleon emission probability has been investigated in ref. /3/ in dependence on bombarding energy $E_{c.m.}$, impact parameter b, and initial binding energy E_B of the nucleon in an 1s state using single-term separable potentials characterized by the range parameter r_o of the form factor as well as a uniform relative motion of both potentials along straight-line trajectories. It was found that for a given reaction the emission probability exhibits a monotonic increase with decreasing impact parameter, with an emission probability for central collision being rather independent of E_B. The calculated double differential emission

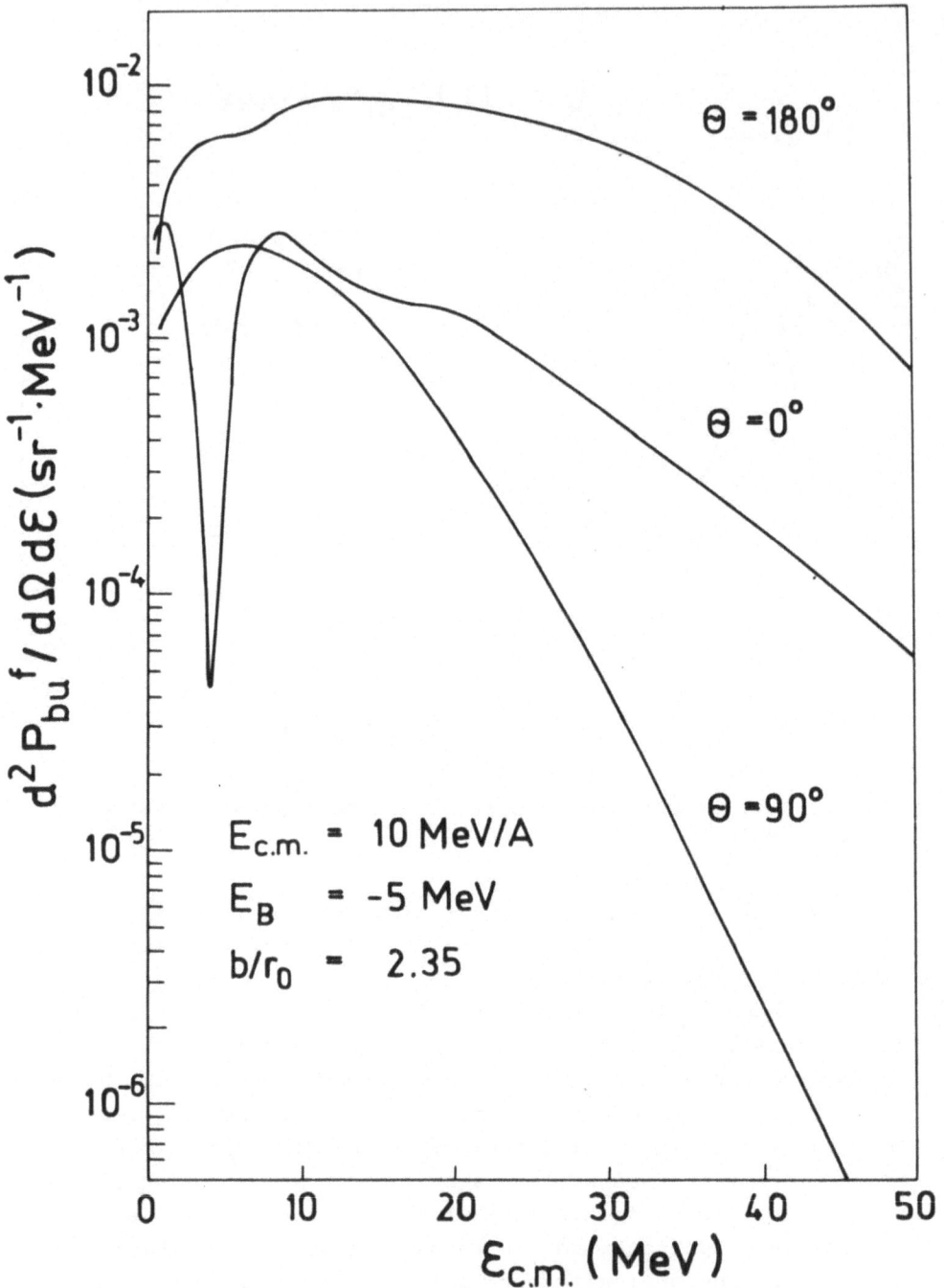

FIG. 3. Double-differential emission probability calculated within a three-body model ($E_{c.m.}$ = 10 MeV/A, E_B = 5 MeV, b/r_0 = 2.35).

probability as a function of the emission energy $\mathcal{E}$ is shown in Fig. 3 for various emission angles θ, with $\theta = 0^\circ$ specifying the direction of flight of the potential V_1, to which the particle is bound initially. The dominating mechanism is a knock-out of the particle by the potential V_2. For higher emission energies the spectrum falls down rapidly, and in all cases the emission perpendicular to the beam axis is suppressed by orders of magnitude.

REFERENCES

1. F. A. Gareev, M. Ch. Gizzatkulov and J. Revai, Nucl. Phys. A286 (1977) 512.
2. J. Revai, Report KFKI-1984-28, 1984.
3. M. Milek, R. Reif and J. Revai, Phys. Lett. 50B (1985) 65.

PARTICLE-HOLE LEVEL DENSITIES IN THE APPROXIMATION
OF HOMOGENEOUS SPECTRUM OF SINGLE-PARTICLE LEVELS

P. Obložinský

Institute of Physics, Electro-Physical Research Centre of
the Slovak Academy of Sciences, Bratislava, Czechoslovakia

ABSTRACT. An analytical relation for particle-hole level densities has been derived, assuming the homogeneous spectrum of
single-particle levels. The constraints on bound configurations,
finite potential well and the Pauli blocking have been taken
into account.

1. INTRODUCTION

Incorporation of angular momentum coupling into pre-equilibrium
models of nuclear reactions requires two artefacts. Namely,
particle-hole level densities which call for a specific model
/1/ and the spin-dependent part of matrix elements for two-body
interactions where correct averaging procedure is essential
/2/.

The so-called approximation of homogeneous spectrum of single-particle levels extends usual equidistant-spacing model by
considering each level twice degenerated according to the sign
of a <u>constant</u> magnetic quantum number /1/. Our goal is to derive a relation for the corresponding particle-hole level densities considering the Pauli blocking, finite potential well
and bound configurations. In this communication we present essential points postponing all details to more complete report.

2. METHOD

We follow the Darwin—Fowler statistical method, by starting from the generating functions for particle and hole level densities, expressing these densities via contour integrals, performing their folding and by evaluating the integrals via the residue theorem /3, 4/ rather than via the saddle point. In this way, using two approximations in performing the final integration, we end up at the level density expression of the Ericson-like form. The density of total angular momentum J can be obtained from the density of its projection M by a standard procedure.

We define the Fermi level midway the last filled and the first vacant level and consider the system of equispaced levels $1/g$, $3/g$, ..., $B - 1/g$ for particles and $1/g$, $3/g$, ..., $F - 1/g$ for holes, where B is the binding and F is the Fermi energy. Each level is twice degenerated according to magnetic quantum number $+m$, $-m$.

The generating functions for particle and hole level densities are

$$Z_p = \prod_{l=1}^{\frac{1}{2}gB} \left(1 + xy^{\frac{2}{g}(l-\frac{1}{2})}u^m\right) \prod_{\lambda=1}^{\frac{1}{2}gB} \left(1 + xy^{\frac{2}{g}(\lambda-\frac{1}{2})}u^{-m}\right),$$

$$\tag{1}$$

$$Z_h = \prod_{k=1}^{\frac{1}{2}gF} \left(1 + vy^{\frac{2}{g}(k-\frac{1}{2})}u^m\right) \prod_{\varkappa=1}^{\frac{1}{2}gF} \left(1 + vy^{\frac{2}{g}(\varkappa-\frac{1}{2})}u^{-m}\right).$$

The partial densities are

$$\omega(p,E_p,M_p) = \frac{1}{(2\pi i)^3} \oint\oint\oint \frac{Z_p}{x^{p+1}y^{E_p+1}u^{M_p+1}} \, dx \, dy \, du \,,$$

$$\tag{2}$$

$$\omega(h,E_h,M_h) = \frac{1}{(2\pi i)^3} \oint\oint\oint \frac{Z_h}{v^{h+1}y^{E_h+1}u^{M_h+1}} \, dv \, dy \, du \,,$$

then the particle-hole level density

$$\omega(p,h,E,M) = \sum_\mu \int_0^E \omega(p,E-\varepsilon,M-\mu)\omega(h,\varepsilon,\mu) \, d\varepsilon \,. \tag{3}$$

The products in eqs. (1) can be rearranged and integrals (2) over x,u and v,u, respectively, evaluated exactly. Using substitution $y = \exp(-\beta)$ and the convolution theorem for Laplace transforms, and defining

$$P_{M-\mu} = \frac{1}{2}\left(p - \frac{M-\mu}{m}\right), \quad h_\mu = \frac{1}{2}\left(h - \frac{\mu}{m}\right)$$

eq. (3) reads

$$\omega(p,h,E,M) = \sum_\mu \frac{1}{2\pi i} \int_{c-i\infty}^{c+i\infty} d\beta\, e^{-\beta(E-\alpha_{ph}^{M\mu})} \cdot$$

$$\cdot \frac{\displaystyle\prod_{l=1}^{p-P_{M-\mu}}\left(1 - e^{-\beta[B-\frac{2}{g}(p-P_{M-\mu}-1)]}\right)}{\displaystyle\prod_{l=1}^{p-P_{M-\mu}}\left(1 - e^{-\beta\frac{2}{g}l}\right)} \cdot$$

$$\cdot \frac{\displaystyle\prod_{\lambda=1}^{P_{M-\mu}}\left(1-e^{-\beta[B-\frac{2}{g}(P_{M-\mu}-\lambda)]}\right)\prod_{k=1}^{h-h_\mu}\left(1-e^{-\beta[F-\frac{2}{g}(h-h_\mu-k)]}\right)}{\displaystyle\prod_{\lambda=1}^{P_{M-\mu}}\left(1-e^{-\beta\frac{2}{g}\lambda}\right)\prod_{k=1}^{h-h_\mu}\left(1-e^{-\beta\frac{2}{g}k}\right)} \cdot$$

$$\cdot \frac{\displaystyle\prod_{\varkappa=1}^{h_\mu}\left(1-e^{-\beta[F-\frac{2}{g}(h_\mu-\varkappa)]}\right)}{\displaystyle\prod_{\varkappa=1}^{h_\mu}\left(1-e^{-\beta\frac{2}{g}\varkappa}\right)}, \tag{4}$$

where

$$\alpha_{ph}^{M\mu} = \frac{(p-P_{M-\mu})^2 + P_{M-\mu}^2}{g} + \frac{(h-h_\mu)^2 + h_\mu^2}{g} =$$

$$= \frac{1}{2g}\left[p^2 + \left(\frac{M-\mu}{m}\right)^2 + h^2 + \left(\frac{\mu}{m}\right)^2\right] \tag{5}$$

96

is the minimum energy required by the Pauli principle for the configuration with $p-p_{M-\mu}$ particles having spin projection $+m$, h_μ holes also $+m$ and the rest $-m$. Note that $\omega(p,h,E,M) = 0$ if $E < \alpha_{ph}$.

We evaluate the integral in eq. (4) using two approximations. (i) Since $gB \gg p$ and $gF \gg h$, we neglect terms of the type $2(p-p_{M-\mu}-1)/g$ and expand $(1-e^{-\beta B})^{p-p}M-\mu$ using the binomial theorem. (ii) We make use of $1-e^{-t} = 2e^{-1/2t}\sinh(1/2t) \approx te^{-1/2t}$. Then, the residue theorem yields

$$\omega(p,h,E,M) = \frac{(\tfrac{1}{2}g)^{p+h}}{p!\,h!\,(p+h-1)!} \sum_\mu \left(\frac{1}{2}\left(\begin{array}{c} p \\ p-\frac{M-\mu}{m} \end{array} \right) \right)\left(\frac{1}{2}\left(\begin{array}{c} h \\ h-\frac{\mu}{m} \end{array} \right) \right)$$

$$\cdot \sum_{i=0}^{P} \sum_{j=0}^{h} (-1)^{i+j}\binom{P}{i}\binom{h}{j} \theta(E-\alpha_{ph}^{M\mu}-iB-jF) \cdot$$

$$\cdot (E-A_{ph}^{M\mu}-iB-jF)^{p+h-1}, \tag{6}$$

where θ is the unit step function and $A_{ph}^{M\mu}$ is the Pauli correction function

$$A_{ph}^{M\mu} = \frac{1}{4g}\left[p^2 - 2p + \left(\frac{M-\mu}{m}\right)^2 + h^2 - 2h + \left(\frac{\mu}{m}\right)^2 \right]. \tag{7}$$

3. DISCUSSION

Eq. (6) can be simplified by performing summation over μ. This is achieved via approximately replacing $A_{ph}^{M\mu}$ by appropriate average value A_{ph}^{M}.

By further summing of eq. (6) over M we get the particle-hole state density $\omega(p,h,E)$ in accord with ref. /5/ which considers a somewhat different definition of the Fermi level.

The obtained density which refers to angular momentum J is $\omega(p,h,E,J) = \omega(p,h,E,M=J) - \omega(p,h,E,M=J+2m)$ giving approximately

$$\omega(p,h,E,J) \approx \frac{\frac{J}{m}+1}{2^{p+h-1}(p+h+\frac{J}{m}+2)} \begin{pmatrix} p+h \\ \frac{1}{2}(p+h-\frac{J}{m}) \end{pmatrix} \cdot$$

$$\cdot \frac{g^{p+h}}{p!h!(p+h-1)!} \sum_{i=0}^{p} \sum_{j=0}^{h} (-1)^{i+j} \begin{pmatrix} p \\ i \end{pmatrix} \begin{pmatrix} h \\ j \end{pmatrix} \theta(E-\alpha_{ph}-iB-jF) \cdot$$

$$\cdot (E-A_{ph}-iB-jF)^{p+h-1} , \tag{8}$$

where

$$\alpha_{ph} = \frac{p^2}{2g} + \frac{h^2}{2g} \quad \text{and} \quad A_{ph} = \frac{p^2-p}{4g} + \frac{h^2-h}{4g} . \tag{9}$$

Eq. (8) can be seen as a product of two terms. One term keeps all dependence on J and the other one is the state density $\omega(p,h,E)$.

The value of m can be estimated consistently within our approach by means of the maximum spin of the configuration. It leads to

$$m^2 \approx \frac{2J_{rig}}{g} \approx 0.18 \, A^{\frac{2}{3}} , \tag{10}$$

where J_{rig} is the moment of inertia and A is the atomic number.

REFERENCES

1. A. V. Ignatjuk, "Statistical Properties of Excited Atomic Nuclei", Energoatomizdat, Moscow, 1983, 34 p. (in Russian).
2. H. Feshbach et al., Ann. Phys. (N.Y.) 125 (1980) 429.
3. F. C. Williams, Nucl. Phys. A166 (1971) 231.
4. E. Běták and J. Dobeš, Z. Phys. A279 (1976) 319.
5. P. Obložinský, this volume.

PARTICLE-HOLE STATE DENSITIES FOR STATISTICAL MULTISTEP COMPOUND REACTIONS

P. Obložinský

Institute of Physics, Electro-Physical Research Centre of
the Slovak Academy of Sciences, Bratislava, Czechoslovakia

ABSTRACT. An analytical relation is derived for particle-hole
state densities with energy constraints. The constraints, following the concept of statistical multistep compound reactions,
mean that the particles are bound and the holes respect the bot-
tom of the potential well. The corresponding densities of accessible final states for escape and damping are suggested.

1. INTRODUCTION

Particle-hole state densities are commonly calculated within
the equidistant-spacing approximation. Density of states with
all particles bound was treated approximately by Kalbach /1/;
a simple recurrent derivation was given by Stankiewicz /2/.

   We aim to obtain a justified relation for state densities
where energy constraints are applied to both particle and hole
excitations. Further, we aim to derive the multistep compound-
like /3/ densities of accessible final states.

   In the following we give a very brief account of the work
which should be published in full elsewhere /4/.

2. PARTICLE-HOLE STATE DENSITIES WITH ENERGY CONSTRAINTS

Our approach is based on the Darwin—Fowler statistical method
which employs complex generating functions. The mathematical
techniques used are similar to those applied in refs. /5, 6/.

   Single-particle excitations are given by a series of energies $1/g$, $2/g$, ..., B; the hole excitations by O, $1/g$, ...,

$F-1/g$. Generating functions for particle and hole state densities are

$$Z_p = \prod_{l=1}^{gB} \left(1 + xy^{\frac{l}{g}}\right),$$

$$Z_h = \prod_{k=0}^{gF-1} \left(1 + vy^{\frac{k}{g}}\right).$$

(1)

The residue theorem yields

$$\omega(p,E_p) = \frac{1}{(2\pi i)^2} \oint\oint \frac{Z_p}{x^{p+1} y^{E_p+1}} \, dx \, dy ,$$

$$\omega(h,E_h) = \frac{1}{(2\pi i)^2} \oint\oint \frac{Z_h}{v^{h+1} y^{E_h+1}} \, dv \, dy$$

(2)

and $\omega(p,h,E)$ is obtained by folding the above partial densities. After algebraic transformations, i.e. integrating over x, v, substituting $y = \exp(-\beta)$ and using the convolution theorem for Laplace transforms we get

$$\omega(p,h,E) = \frac{1}{2\pi i} \int_{c-i\infty}^{c+i\infty} e^{\beta(E-\alpha_{ph})} \prod_{l=1}^{p} \left(\frac{1 - e^{-\beta(B-\frac{p-1}{g})}}{1 - e^{-\beta\frac{l}{g}}} \right) \cdot$$

$$\cdot \prod_{k=1}^{h} \left(\frac{1 - e^{-\beta(F-\frac{h-k}{g})}}{1 - e^{-\beta\frac{k}{g}}} \right) d\beta .$$

(3)

In evaluating the integral (3) we apply two approximations. (i) Consider $gB \gg p$, $gF \gg h$, neglect the terms $(p-1)/g$, $(h-k)/g$ and expand binomially the terms $(1-e^{-\beta B})^p$ and $(1-e^{-\beta F})^h$. (ii) Make use of $1-e^{-t} = 2e^{-1/2t}\sinh(1/2t) \approx te^{-1/2t}$. Using these approximations and the residue theorem we arrive at

$$\omega(p,h,E) = \frac{g^{p+h}}{p!h!(p+h-1)!} \sum_{i=0}^{p} \sum_{j=0}^{h} (-1)^{i+j} \binom{p}{i} \binom{h}{j} .$$

$$\cdot \; \theta(E - \alpha_{ph} - iB - jF)(E - A_{ph} - iB - jF)^{p+h-1}, \qquad (4)$$

where $\alpha_{ph} = p(p+1)/2g + h(h-1)/2g$ is the minimum energy of the (p,h) state due to the Pauli principle and $A_{ph} = p(p+1)/4g + h(h-3)/4g$ is the Pauli correction function.

A few examples of $\omega(p,h,E)$ can be seen in Fig. 1. The density reaches ist maximum at about $E \approx (pB+hF)/2$ and drops back to zero at $E = pB + hF - \alpha_{ph}$.

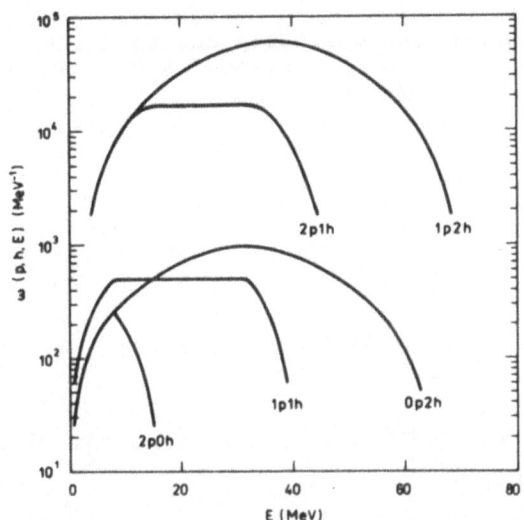

FIG. 1. Particle-hole state densities with energy constraints as a function of the excitation energy (g = 8 MeV, B = 8 MeV, F = 32 MeV).

3. DENSITY OF ACCESSIBLE FINAL STATES

Applying the notation by Feshbach /3/ one should calculate the Y-functions for escape and damping widths. To perform this we adopt the technique of ref. /7/ which unlike ref. /2/ avoids superfluous normalization.

The diagrams to be evaluated are shown in Fig. 2. Escape of a nucleon is due to three processes, namely $\Delta n = -1, 0, +1$; damping is due to $\Delta n = +1$. The evaluation is rather straightforward. As an example we show here the latter case only.

Damping refers to the last two diagrams

101

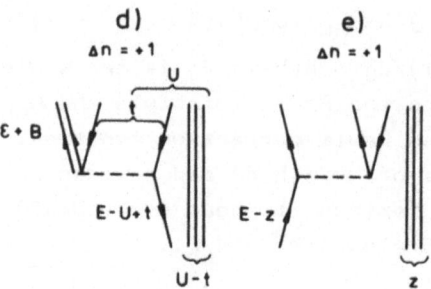

FIG. 2. Diagrams of the multistep compound transition modes.
Diagram d displays damping, provided the escaping particle is
replaced by a bound particle.

$$Y_n^{n+1} \downarrow = {_d}Y_n^{n+1} \downarrow + {_e}Y_n^{n+1} \downarrow \ , \tag{5}$$

where

$$_e Y_n^{n+1} \downarrow =, \int_0^E \frac{\omega(p-1,h,z)}{\omega(p,h,E)} \ \omega(1, \ 0,E-z) \ \omega(2,1,E-z) \ dz \tag{6}$$

and similarly for ${_d}Y_n^{n+1}\downarrow$. We insert from eq. (4), integrate
and get, $N = p + h$,

$$_e Y_n^{n+1} \downarrow = \frac{\frac{1}{2} g^4}{\tilde{\omega}(p,h,E)} \left[\frac{\omega(p-1,h,E^{N+1}) - \omega(p-1,h,(E-B)^{N+1})}{(N-1)N(N+1)} - \right.$$

$$\left. - B \frac{\omega(p-1,h,(E-B)^N)}{(N-1)N} - B^2 \frac{\omega(p-1,h,(E-B)^{N-1})}{2(N-1)} \right] \tag{7}$$

and similarly

$$_d Y_n^{n+1} \downarrow = \frac{\frac{1}{2} g^4}{\omega(p,h,E)} \left[\frac{\omega(p,h-1,E^{N+1}) - \omega(p,h-1,(E-B)^{N+1})}{(N-1)N(N+1)} + \right.$$

$$+ B \frac{\omega(p,h-1,(E-F)^N)}{N} +$$

102

$$+ \frac{\omega(p,h-1,E^2(E-F)^{N-1}) - \omega(p,h-1,(E-B)^2(E-F)^{N-1})}{2(N-1)} \Bigg] . \qquad (8)$$

Here, we have used a short-hand notation

$$\omega(p,h,U^{N+\nu}) = \begin{cases} \dfrac{q^N}{p!h!(N-1)!} \displaystyle\sum_{i=0}^{p} \sum_{j=0}^{h} (-1)^{i+j} \binom{p}{i}\binom{h}{j} \cdot \\[2em] 0 \quad \text{for } U < 0. \end{cases}$$

$$\cdot (U-iB-jF)(U-iB-jF)^{N+\nu} \quad \text{for } U \geq 0. \qquad (9)$$

The above result given by eqs. (5, 7, 8) can be compared with ref. /2/, provided E < F. We find that the last four terms of their eq. (17) should be multiplied by a factor of 1/2 and their Bg^2 should be read as B^2g^2.

REFERENCES

1. C. Kalbach, Phys. Rev C23 (1981) 124; C24 (1981) 819.
2. K. Stankiewicz et al., Nucl. Phys. A435 (1985) 67.
3. H. Feshbach et al., Ann. Phys. (N.Y.) 125 (1980) 429.
4. P. Obložinský, Nucl. Phys., submitted.
5. F. C. Williams, Nucl. Phys. A166 (1971) 231.
6. E. Běták and J. Dobeš, Z. Phys. A279 (1976) 319.
7. P. Obložinský et al., Nucl. Phys. A226 (1974) 347.

DERIVING EXCITATION FUNCTION OF ^{27}Al[(n,t)+(n,nt)] REACTION
FROM MEASUREMENT PERFORMED IN DIVERSE NEUTRON FIELDS

S. Sudár, *R. Wölfle and *S. M. Qaim

Institute of Experimental Physics, Kossuth University,
Debrecen, Hungary
*Institute of Chemistry 1 (Nuclear Chemistry) Jülich GmbH,
Jülich, FRG

ABSTRACT. Aluminium samples together with sets of 12 flux monitor foils were irradiated in six different d/Be neutron
fields (E_d = 17.5–30 MeV). The shapes of the neutron spectra
were determined by spectrum unfolding. In the second calculation step the excitation function for ^{27}Al[(n,t)+(n,tn)] process was obtained from the neutron flux distribution and the
activities measured.

1. INTRODUCTION

The determination of excitation function for neutron reaction
can be performed relatively easily only in those energy regions
in which there exist monoenergetic neutron sources (e.g. 3 to
10 MeV and 13.5 to 20 MeV). The consequences of these limitations are apparent in compilation of experimental differential
cross-section data. Therefore there arose the question whether
the irradiation with monoenergetic neutrons inducing a particular nuclear reaction could be replaced by a number of irra-
diations in different neutron fields exhibiting extended spectra, followed by the application of mathematical unfolding
procedures to obtain the desired excitation function. A theoretical approach to this goal, illustrated by several hypo-
thetical numerical examples, was made by Smith [1].

2. SPECTRUM UNFOLDING CODES

Unfolding codes for the calculation of the differential neutron
flux from measurements of a number of activities induced by
nuclear reactions having different but known excitation func-

tions have already been used for many years /2–4/. The starting
equation of "spectrum unfolding" is

$$A_i = \int_{E_o}^{E_m} \phi(E) \mathbf{G}_i(E) \, dE \qquad (i = 1, \ldots, n), \qquad (1)$$

where A_i represents the specific reaction rate of the i-th flux
monitor foil, $\mathbf{G}_i(E)$ is the excitation function of the i-th reaction and $\phi(E)$ is the unknown differential neutron flux. In
the case of the "excitation function unfolding" we have n measured activities for the same reaction in different neutron fields

$$A_i = \int_{E_o}^{E_m} \phi_i(E) \mathbf{G}(E) \, dE \qquad (i = 1, \ldots, n) \qquad (2)$$

with $\mathbf{G}(E)$ as the unknown function. Mathematically, both the
cases represent the same problem, only $\mathbf{G}(E)$ and $\phi(E)$ are ex-
changed. This integral equation can be approximated as

$$A_i = \sum_{k=1}^{N} \phi(E_k) \mathbf{G}_i(E_k) \Delta E_k = \sum_{k=1}^{N} \phi_k \mathbf{G}_{ik}, \qquad (3)$$

where $\mathbf{G}_{ik} = \mathbf{G}_i(E_k) \Delta E_k$ and N is the number of energy channels.
In the usual case n is less than N, therefore this set of
equations mathematically represents an indeterminate system.
It can be solved only if some additional information is given
to the experimental data. This additional information usually
involves the initial differential neutron flux (and error correlation information if the least squares unfolding code is
used). For details concerning the solution see refs. /1, 3, 4/.

3. EXPERIMENTAL

Composite samples consisting of two parts of aluminium and
a set of 12 flux monitor foils between them were irradiated
using a Jülich Isochronous Cyclotron (JÜLIC) at 17.5, 20.0,
22.5, 25.0, 27.5 and 30.0 MeV deuteron energies. The activities
of foils were determined by means of a coaxial Ge(Li) detector.

Corrections for pile-up effect, coincidence losses, self-absorption, etc. were applied. Tritium activities induced in the Al
samples were separated after the addition of a H_2 carrier by
vacuum extraction and counted in a proportional gas counter
in anticoincidence arrangement.

4. RESULTS AND DISCUSSION

The unfolding codes (SAND II and two versions of the least
squares method) were applied in two steps: in the first step
- for the determination of the needed set of differential
neutron fluxes and in the second - for the determination of
the [(n,t)+(n,tn)] excitation function on ^{27}Al. The calculated
differential neutron fluxes at different deuteron energies are
shown in Fig. 1. All differential neutron fluxes have been
normalized to give the same area.

The excitation function of the ^{27}Al[(n,t)+(n,tn)] calculated by the least squares method is given in Fig. 2. Earlier
experimental data points obtained using monoenergetic neutrons
as well as results of statistical model calculation using the
Hauser—Feshbach method are also given. The agreement is satis-

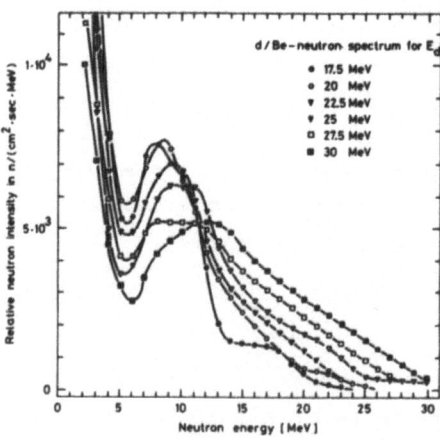

FIG. 1. d/Be neutron flux
distribution with SAND II for
deuteron energies from 17.5
to 30 MeV.

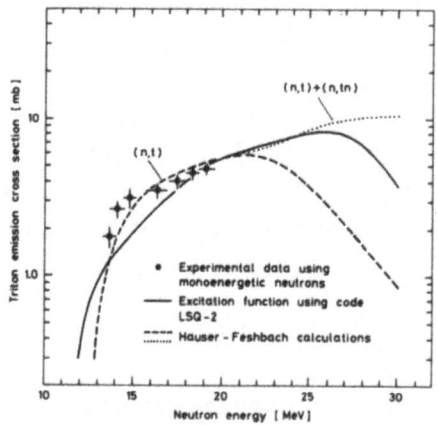

FIG. 2. Excitation function
of the [(n,t)+(n,tn)] reaction on ^{27}Al.

factory from 16 MeV up to about 25 MeV, beyond which there is probably not enough information in the measurements to give real excitation function. The details of the results are given elsewhere /5/.

5. CONCLUSION

Our results have shown that determination of the excitation function by the application of unfolding codes to activities obtained experimentally in diverse neutron fields, even with intervening calculation of fluxes, is feasible.

REFERENCES

1. D. L. Smith, Report ANL/NDM-77 Argonne National Laboratory, 1982.
2. C. A. Oster, Proc. 2nd ASTM-EURATOM Symp. Reactor Dosimetry, Vol. 3, p. 1365, Palo Alto, 1977.
3. W. N. McElroy, S. Berg, T. Croskett and R. G. Hawkins, Report AFW-TR-67-41, Vol. 1, Atomic International, 1967.
4. F. G. Perey, Report ORNL/TM-6062 Oak Ridge National Laboratory, 1977.
5. R. Wölfle, S. Sudár and S. M. Qaim, Nucl. Sci. Eng., in press.

ON MECHANISMS OF REACTION ^{9}Be(n,2n)

Du Xiang Wan and *Zhang Ben Ai

Institute of Atomic Energy, Academia Sinica,
Beijing, China

*Institute of Applied Physics and Computational Mathematics,
Beijing, China

ABSTRACT. The theoretical calculations for different direct processes of reaction ^{9}Be(n,2n) have been performed. The results
indicate that the two-step process – direct inelastic scattering and decay with the second neutron emission is the main mechanism of this reaction. The one-step 3-body break-up process
gives important contribution, too.

1. INTRODUCTION

It is of interest to study the mechanisms of the ^{9}Be(n,2n) reaction which is quite complex. The possible processes are as
follows:

A. Two-step processes:

a) Compound nucleus cascade

$$n + {}^9\text{Be} \rightarrow {}^{10}\text{Be}^* \rightarrow {}^9\text{Be}^{*K} + n_1$$
$$\hookrightarrow {}^8\text{Be}^{*M} + n_2 \ (\text{or } 2\,{}^4\text{He} + n_2);$$

b) Direct inelastic scattering and statistical decay with
the second neutron emission

$$n + {}^9\text{Be} \xrightarrow{\text{direct}} {}^9\text{Be}^{*K} + n_1$$
$$\hookrightarrow {}^8\text{Be}^{*M} + n_2 \ (\text{or } 2\,{}^4\text{He} + n_2).$$

B. One-step direct processes:

c) Direct 3-body break-up

$$n + {}^9\text{Be} \rightarrow {}^8\text{Be} + n_1 + n_2 \ (\text{or } {}^4\text{He} + {}^5\text{He} + n_1$$
$$\hookrightarrow {}^4\text{He} + n_2);$$

d) Direct 4-body break-up

$$n + {}^9Be \longrightarrow 2\,{}^4He + 2n.$$

The ref. /1/ adopted the compound nucleus statistical model
to treat the cascade decay process. Its calculational results
are not very good. Quite strong excitations for K = 6 (above
energy level 4.4 MeV) have not been observed experimentally.
A possible reason of this discrepancy is the statistical treatment of the first step.

This paper discusses the contributions of different direct
processes.

2. DIRECT INELASTIC SCATTERING

We consider the nucleus ${}^9Be$ consisting of two clusters: a core
${}^8Be$ and a neutron n_2; the last one is described by shell model
and its state determines the state of the nucleus ${}^9Be$. The incident neutron is described by plane wave with wave vector $\vec{k}$.
The initial state of ${}^9Be$ is I: (n,l,j,m), the state after scattering is k: (n´,l´,j´,m´). The scattered neutron is described
by the plane wave with $\vec{k´}$.

The differential cross-section for two-body scattering in
the c-m system is /2/

$$\sigma(\theta´) = \frac{k´}{k}\left(\frac{\mu}{2\pi\hbar^2}\right)^2 \frac{1}{2j+1} \sum_{m,m´} |T_{fi}|^2 ,$$

$$T_{fi} \equiv \langle f|V_D|i \rangle \tag{1}$$

$$= (4\pi)^{5/2} \sum_{PP´L} \sum_{N\mu´} (-1)^N \sqrt{2P+1}\; Y_P^{\mu´}(\hat{k´}) \langle P´\mu´|Y_L^{-N}|P0\rangle \cdot$$

$$\cdot \langle l´j´m´|Y_L^N|ljm\rangle R_{PP´L}.$$

The neutron-nucleus interaction exists only between the
neutrons n_1 and n_2. Paying attention to the peripheral character of inelastic direct reaction we introduce $\gamma\delta(r-R_0)$ into
the potential $V_D(r_1,r_2)$. As nucleus wave function we choose
the single particle wave function in the oscillator potential
/3/

$$R_{nl}(r) = (-1)^{\frac{n-1}{2}} \sqrt{\frac{2}{1+\frac{1}{2}} \left(\frac{\frac{n+l+1}{2}}{\frac{n-1}{2}} \right)} \; r^l \; e^{-\frac{\lambda r^2}{2}} \; {}_1F_1\left(\frac{1-n}{2}, 1+\frac{3}{2}, \lambda r^2\right).$$

So we can obtain the radial integral as follows

$$R_{PP'L} = \frac{-\gamma}{4\pi} R_{11}(R_o) R_{n'l'}(R_o) j_P(kR_o) j_{P'}(k'R_o).$$

We choose the potential parameter γ to fit the experimental excitation function for 2.43 MeV level.

The probability of the second step $\eta_{k\to\beta}$ for decay ${}^9Be^{*K}$ $\to {}^8Be^{*M} + n_2$ is taken from /1/. Hence the cross-section for this two-step reaction is

$$\sigma_1^{K\to M} = \sigma_1^K \sum_{\beta\in\{M\}} \eta_{K\to\beta} \; ,$$

where σ_1^K is the cross-section of the first step (expression (1)

TABLE 1. Calculated cross-sections (in mb) for two-step processes

E_o (MeV) K, M		3	4	6	8	10	12	14
K = 2	M = 1	1.1	1.0	0.6	0.4	0.3		
	M = 2	0	0	0	0	0	0	0
K = 3	M = 1	91	182	255	243	221	-	189
	M = 2	0	0	0	0	0	0	0
K = 4	M = 1		4.0	5.0	3.5	2.0	-	1.0
	M = 2	0	0	0	0	0	0	0
K = 5	M = 1		137	107	67	50	-	34
	M = 2	0	0	0	0	0	0	0
K = 6	M = 1		<1	<1				
	M = 2	~0	~0	~0	~0	~0	~0	~0
K = 7	M = 1			1.3	1.4	1.3	1.0	0.8
	M = 2	0	0	0.7	0.75	0.7	0.6	0.5
K = 8	M = 1				16	13.5	9	6
	M = 2	0	0	0	35	29.5	20	15
$\sigma_{n,2n}^{inel.} = \sum_{\substack{M \\ K\geq2}} \sigma_1^{K\to M}$		92	324	369	366	318	-	246

integrated over θ´). The nuclear spectrum parameters are taken from /4/.

The results are given in Table 1 and Figs. 1 and 2.

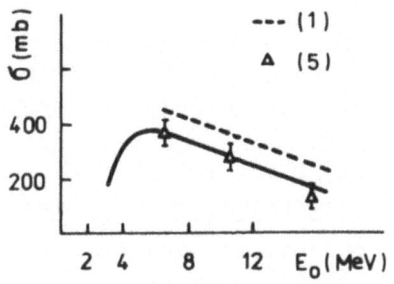

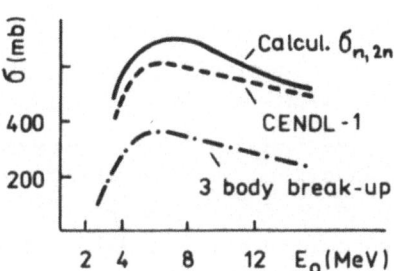

FIG. 1. Excitation of (K = 2, 3, 4, 5) levels.

FIG. 2. Cross-section $\sigma_{n,2n}$.

3. DIRECT BREAK-UP PROCESSES

By regarding ^{8}Be as spectator, the QFS model and the PWIA approximation are adopted in the calculation. The indexes 0, 1, 2, 3 represent an incident neutron, two outgoing neutrons and the recoil ^{8}Be, respectively. For the sake of easy comparison with the experiments, we have derived the following formula for the double differential cross-section

$$\frac{d\sigma}{dE_1 \, d\Omega_1} = \frac{6u_{23}(m_1+m_2)^2}{\pi h^3 m_2^2} \sqrt{\frac{E_1}{E_2}} \; \tilde{q} \; w_q \left(\frac{d\sigma}{d\Omega}\right)_{nn} ,$$

where

$$\tilde{q} = \sqrt{2u_{23}(E_o - Q - E_1 - \frac{u^2}{2(m_2+m_3)})} ,$$

$$u = \sqrt{2m_1(E_o + E_1 - 2\sqrt{E_o E_1} \cos \theta_1)} ,$$

$$w_q = \sum_{l=0}^{l_m} [(l+1)M_l^+ + lM_l^-] ,$$

$$M_l^{\pm} = \left| \int_{R_o}^{\infty} j_1 \left(\frac{m_3}{m_2+m_3} \frac{u}{\hbar} x\right) j_{l\pm1} \left(\frac{\tilde{q}}{\hbar} x\right) B \frac{\beta x+1}{\beta^2} e^{-\beta x} dx \right|^2 .$$

111

The wave function of relative motion of n and ^{8}Be in the nucleus ^{9}Be is chosen as follows

$$\Psi(\vec{R}) = \begin{cases} Aj_1(\alpha R)Y_{1m}(\hat{\vec{R}}) & R < R_o \\ B\dfrac{\beta R+1}{\beta^2 R^2} e^{-\beta R} Y_{1m}(\hat{\vec{R}}) & R \geq R_o \end{cases}.$$

For incident energy from 2 MeV to 20 MeV we have calculated the differential cross-section and obtained the energy spectrum and angular distribution of outgoing neutrons and the reaction cross-section. Some of the results are given in Fig. 2.

By using Sacks' semiempirical formula we also estimate the cross-section for the 4-body break-up process. The order of magnitude is ~ mb, hence its contribution can be neglected.

4. DISCUSSION AND CONCLUSION

Compared with the experiment /5/ our theoretical calculations give the rational results. The main contribution to the (n,2n) reaction through inelastic scattering is that from excitation of the 2.43 and 3.03 levels (K = 3, 5). The agreement with experiment is better than in /1/ and /6/.

In ref. /7/ the ejection mechanism and the compound nucleus cascade are considered to be the main mechanisms of this reaction. However, when compared with the latest evaluated data our results (direct inelastic scattering and decay with neutron emission plus 3-body break-up) seem better.

We prefer to conclude that the first step of the two-step process is not through compound but direct inelastic scattering. This mechanism gives correct energy spectrum and angular distribution of secondary neutrons. The one-step 3-body break-up also gives an important contribution to the continuous cross-section.

In the energy range 5—9 MeV the calculated $\sigma_{n,2n}$ is greater than evaluated about 20%, it may relate to the rough choice of the wave functions and needs to be improved.

REFERENCES

1. Zhang Ben Ai, Report on Cheng-Du Conference on Few-body Physics, 1980.
2. N. K. Glendenning, Phys. Rev. 114 (1959) 1297.
3. J. M. Eisenberg and W. Greiner,"Nuclear Theory, Vol. 1, Nuclear Models", North-Holland, Amsterdam, 1970, p. 188.
4. T. Lauritsen, Nucl. Phys. 78 (1966) 84.
5. D. M. Drake, Nucl. Sci. Eng. 63 (1977) 401.
6. S. T. Perkins, Report UCRL-50520.
7. R. Balian et al., Nucl. Phys. 17 (1960) 448.

CONSIDERATION OF THE (n,d) AND (p,d) SCATTERING AND BREAK-UP REACTIONS THROUGH SINGULAR EQUATIONS

Tian-Yuan Zhang

Institute of Applied Physics and Computational Mathematics, Beijing, China

ABSTRACT. The paper studies (n,d) and (p,d) processes through singular integral equations. We treat the two-body off-shell T-matrix by the limit operator of the screening Coulomb field. In this case we can still use the previous method /1/.

1. INTRODUCTION

In the reference /1/ we proposed a method to study the three-body scattering problem for the local potential. Now we hope to generalize the method to the (p,d) scattering and break-up reaction. However, when the long-range force is present, we shall consider the pure Coulomb operator as the limit of the screening Coulomb operator. Our three-body and the off-shell two-body problems are understood based on the limit of the screening operator. That is, first of all, we consider the problems in terms of the screening operator, and we make the results obtained go to their limits. Then we can completely solve the problem in which there are the long-range Coulomb and local nuclear forces.

2. THE THREE-BODY PROBLEM FOR LOCAL INTERACTIONS

We choose a proper complete set of basic vectors to expand Faddeev's equations of T-matrix form. Considering local two-body interactions, we obtain a set of two-variable integral equations. If the energy parameter S is negative and larger than the two-body binding energy S_0, or if S is positive and real $+i\varepsilon$, $\varepsilon \longrightarrow 0^+$, then the integral equation obtained is the Cauchy singular integral equation.

In order to solve the equation, we must make it become the standard form of the singular integral equation after the variable transformations. The index of the singular integral equation is very important, because the solution form of equation is well defined by the index. From the equation coefficient we can construct the two functions a and b, which are required to satisfy the following conditions

$$0 < a + \lambda_1 < 1,$$
$$-1 < b + \lambda_2 < 0,$$
(1)

where λ_1, λ_2 = 0, ± 1, ± 2, ... , then index of the integral equation equals

$$K = -(\lambda_1 + \lambda_2).$$
(2)

From this we get an algebra equation system:

$$W(\xi_{i,m})\phi(\xi_{i,m},\tau_j;\alpha) + \sum_{\alpha_2} \frac{B(\xi_{i,m},\tau_i)}{\pi} \sum_{l=1}^{N} w_l I_n^{(\alpha,\beta)} (\tau_1,$$

$$\xi_{i,m},\tau_j) \frac{\phi(\xi_{i,m}+\tau_j-\tau_1;\tau_1;\alpha_2)}{P_n^{(\alpha,\beta)}(\xi_{i,m}+\tau_j-\tau_1)} + \sum_{\alpha_2} \frac{B(\xi_{i,m},\tau_i)}{\pi} \cdot$$

$$\cdot \sum_{k=1}^{n} \sum_{l=1}^{N} w_l u_{k,n} \frac{\phi(\xi_{k,n},\tau_1;\alpha_2)}{(\xi_{k,n}+\tau_1-\xi_{i,m}-\tau_j)} + \frac{1}{\pi} \sum_{\alpha_2} \sum_{k=1}^{n} \sum_{l=1}^{N}$$

$$w_l u_{k,n} K_0(\xi_{i,m},\tau_j;\xi_{k,n},\tau_1;\alpha_2) \phi(\xi_{k,n},\tau_1;\alpha_2) =$$

$$= f(\xi_{i,m},\tau_j;\alpha) .$$
(3)

It is not difficult to solve the system /1/.

3. COULOMB MODIFIED PROBLEM FOR THE LOCAL NUCLEAR FORCE

The above-mentioned problem of the two-body short-range force can also be used to discuss the Coulomb problem. We remember that the Coulomb interaction is of long range. Therefore L-S equation cannot hold. For the sake of showing the problem we assume the transition operator T_c to satisfy L-S equation

$$T_c(z) = V_c G_0(z) T_c(z). \tag{4}$$

In the momentum representation the kernel $V_c G_0(z)$ of the equation (4) has the pole of multiplicity 2, if the momenta take the on-shell values. Therefore the equation (4) does not hold.

But, we know from the functional analysis that the Coulomb Hamiltonian

$$H_c = H_0 + V_c \tag{5}$$

is an unbound operator, and is self-adjoint. For any $I_m Z \neq 0$, if the norm of the resolvent operator $\| G_c^m(z) - G_c(z) \| \rightarrow 0$, then the operator H_c is the limit operator of H_c^m in the norm resolvent sense, i.e., $H_c^m \rightarrow H_c$. Thus our aim is to properly choose a sequence of the operators $\left\{ H_c^m \right\}_{m=0}^{\infty}$ which are solved without any difficulty. For this reason we introduce a screening factor into the Coulomb potential:

$$V_m = \frac{2k\gamma}{r} e^{-r/R_m}. \tag{6}$$

We can see that $V_m \rightarrow V_c$, if $m \rightarrow \infty$ and $R_m \rightarrow \infty$. Therefore we have a sequence such that

$$0 < R_0 < R_1 < \ldots < R_m \tag{7}$$

and we can immediately follow the Hamiltonian sequence $\left\{ H_c^m \right\}_{m=0}^{\infty}$, where

$$H_c^m = H_0 + V_m. \tag{8}$$

It is obvious that the equation (8) is a satisfactory operator which describes Yukawa-type interactions with the finite range. Therefore the problem solved is not difficult.

It can be seen from the above analysis that when the off-shell two-body T-matrix for the nuclear and Coulomb forces is considered, we first start from $V = V_m + V_s$ and then let $m \rightarrow \infty$ in the T-matrix. The limit transition is easy to complete, except the on-shell scattering matrix and the wave function. For the on-shell matrix and the wave function we get the limit by terms of renormalization /2/.

Thus the screening Coulomb potential satisfies the L-S equation. It is easy to prove that the partial T-matrix satisfies the following equation

$$T_1^m(k'',k;q) = - \int_0^\infty dr \; rj_1(k'r)U_1^m(r)u_1^m(k,q;r), \tag{9}$$

where $U_1^m(k,q;r)$ satisfies the integral equation

$$U_1^m(k,q;r) = rj_1(kr) - \int_0^\infty dr' \; G_{1,q}^m(r,r')U_1^m(r')u_1^m(k,q;r). \tag{10}$$

The kernel of the equation (10) is the H-class function. Therefore it is obviously bound and the equation (10) is the Fredholm integral equation solution of which is easy to find.

In order to determine Coulomb modified strong partial amplitudes, we must consider the problem from the two-potential scattering theory. Interaction of the system consists of two parts as follows

$$V = V_m + V_s , \tag{11}$$

correspondingly, T-matrix is such that

$$T = T_m + T^{(m)}. \tag{12}$$

First, we shall define Møller operator of the Coulomb screening field

$$\Omega_m(z) = 1 + G_0(z)T_m(z). \tag{13}$$

Making use of the completeness of momentum eigenstates and Coulomb wave functions we get the following equation

$$t_s^{(m)}(k',k;q) = \langle \vec{k}' | v_s^{(m)} | \vec{k} \rangle +$$

$$+ \frac{2}{\pi} \int_0^\infty p_2^2 \, dp_2 \, \mathcal{V}_m(k',p_2;q) t_s^{(m)}(p_2,k;q), \qquad (14)$$

where

$$\mathcal{V}_m(k,p_2;q) = \frac{2}{\pi} \int_0^\infty p_1^2 \, dp_1 \, \langle \, \vec{k}' \, | \, v_s^{(m)} \Omega_m(q) | \, \vec{p}_1 \, \rangle \, .$$

$$\cdot \, \frac{\langle \vec{p}_1 | \Omega_m(q) | \vec{p}_2 \rangle}{q^2 - p_1^2} \, . \qquad (15)$$

Thus we can obtain Coulomb modified strong amplitudes

$$T_s^{(m)}(z) = \Omega_m^+ t_s^{(m)}(z) \Omega_m \, . \qquad (16)$$

It is easy seen that $\langle \, \vec{p}_1 | \Omega_m(q) | \vec{p}_2 \rangle$ is the screening Coulomb wave function in the momentum representation. If the Coulomb bound state does not exist, the kernel (15) is bound, therefore the equation (14) is the Fredholm equation which is easy to solve. But if the Coulomb bound state is present, the equation is the singular equation which is solved by terms of the previous method.

After having considered the Coulomb screening modification, the three-body Faddeev equation still holds. Thus we can use the above method to solve the (p,d) scattering and the break-up reaction.

REFERENCES

1. Zhang Tian-Yuan, Proc. 4th Int. Conference on Nuclear Reaction Mechanisms, Varenna, Italy, June 10—15, 1985.
2. E. O. Alt, W. Sandhas and H. Ziegelmann, Phys. Rev. C17 (1978) 1981.
3. H. van Haeringen, J. Math. Phys. 24 (1983) 247.
4. M. Reed and B. Simon, "Methods of Modern Mathematical Physics", Vol. 1, Academic Press, New York - London, 1972.

ALPHA CLUSTERIZATION

O. Dumitrescu

Department of Fundamental Physics, Institute of Physics
and Nuclear Engineering, Central Institute of Physics,
Bucharest, Romania

ABSTRACT. The aplha clusterization and alpha decay are investigated within the framework of the Fermi liquid model and compared to other earlier or later models. Satisfactory agreement
with experimental data concerning the alpha-decay process is
obtained. Starting from the fact that the nuclear superfluidity
leads to a strong enhancement of the alpha clusterization amplitudes, the problem of superfluid condensate of alpha clusters (quadruplets) is investigated as the next step in describing fermion condensates after the well-known theory of superfluidity.

1. INTRODUCTION

The concept of clustering in nuclear systems is almost as old
as the nuclear physics itself /1–6/. In this respect the alpha
clusterization is of particular interest. Reflecting the satu-
ration of the nuclear forces and large binding energy of the
alpha particle, much of the early work focussed on it as an
essential cluster. The alpha cluster nuclear structure models
were surprisingly successful in reproducing the experimental
data concerning the spectra of a number of light 4n nuclei
/6, 7/, despite the difficulties in describing some states as,
e.g., the ground state of ^{16}O. This state behaves like a closed
shell model state rather than a state corresponding to a pyramidal alpha particle structure. The transition /7/ from a <u>shell
model phase</u> to an <u>alpha cluster molecular phase</u> seems to be
described by the strongness of the four-nucleon correlations
/8/, as, for instance, in the case of the first excited 0$^+$
state in ^{16}O or the fourth 0$^+$ state in ^{20}Ne.

Recently /9—11/, the Fermi liquid model for alpha clusterization and alpha decay has been proposed as a result of a comprehensive analysis /12/ of the current alpha-decay models.
This model introduces a new type of interaction for the irreducible reaction amplitude of the alpha-particle formation in
the four-particle channel, based on the prescriptions resulting from the Landau-Migdal theory /13/ of quantum liquids. Assuming that the nucleus is a Fermi liquid, one can expect that
the alpha decay is a strong collective phenomenon that takes
place in two steps: the clusterization and the barrier penetration. The clusterization process may be viewed as a _phase_
transition from a many-body Fermi liquid state to an alpha
cluster molecular state. In the Fermi liquid state the nucleons are more or less uniformly distributed over the whole volume of the nucleus while in the alpha cluster state the nucleons belong to the two Fermi liquid fragments (alpha and the
daughter nucleus) in relatively weak interaction.

Similar ideas can be found in the paper /14/ where such
four-nucleon interactions have been used to describe the Coulomb energy difference of mirror nuclei. The existence of alpha-cluster states has also been recently discovered in heavy
nuclei /15/ and successfully described in the framework of the
vibron model /16—18/ as a new kind of dipole states in the a-
tomic nuclei. Such alpha-clustering modes could co-exist with
the familiar quadrupole or octupole excitations of heavy nuclei. Whereas the quadrupole deformation of, e.g., ^{218}Ra ground
state give rise to the familiar $K^{\pi} = O^{+}$ rotational bands, the
alpha-cluster state with an alpha cluster external to an assumed ^{214}Rn-core is characterized by the radius vector linking
the cluster centroids and thus by a dipole degree of freedom.
This asymmetric configuration leads to the rotational band
that shows alternating even and odd parity states.

Because the nucleus is a finite system, the Fermi liquid
state $<=>$ alpha-cluster state transition, as a phase transition, is expected to be a gradual transition from systems
conserving the symmetry to systems that violate the symmetry.
As an example we may look at the structure of the O_1^{+}, O_2^{+}, O_3^{+}
and O_4^{+} states in ^{20}Ne. Along this sequence the four-nucleon

correlations increase their contribution till the O_4^+ state having the pyramidal alpha structure and high momentum of inertia.

Another type of four-nucleon correlations could be that recently proposed /19/ in the framework of the interacting boson model with the aim to describe the alpha cluster in nuclei as a bound state of two bosons, each boson corresponding to a pair of nucleons. However, the underlying fermionic structure of these bosons, ignored in this approach, seems to be of great importance as far as genuine four-fermion correlations are concerned. Such an investigation raises the question of whether a condesed state could directly be obtained by starting with purely fermionic Hamiltonian and incorporating four-fermion correlations just at the outset. Doing so, the ensuing condensed state might be viewed as corresponding to genuine four-fermion correlations rather than to a bound state of two already condensed pairs of fermions. Early attempts to account for alpha cluster in nuclei as arising from four-fermion correlations run /20—23/ into difficulties which on one hand are inherent of the complex nuclear structure, and on the other, come from much too general framework in terms of which the problem has been stated. A special attempt has been made to include the four-fermion correlations into the trial wave function of the condensed state /20/ and thereby simulate the formation of four-fermion condensate. A different point of view is assumed in our recent work /24/ by using the well-known /25/ BCS-like pairing wave function and accounting for four-fermion correlations by two-pair (proton and neutron) correlations. It has been shown that, within this simple pairing approximation, the four-fermion correlations lead to a condensed state of the Fermi gas model which consists of correlated fermion pairs. Within this model the alpha cluster is viewed as a correlated four-fermion object rather than a spatially localized cluster, as was the case in /20—23/. It results from our work /24/ that the nuclear superfluidity is not caused by the pairing interaction only, the contribution of the four-fermion correlations seems to be of about 30%.

2. FERMI LIQUID MODEL FOR ALPHA CLUSTERIZATION AND ALPHA DECAY

In the framework of the many-body theory an equation for the four-particle Green's functions can be deduced in the usual way /13/

$$
T_{4 \to \alpha} = \quad = \quad + \quad \tag{1}
$$

where

$$
T_{4 \to \alpha}^{(0)} = \tag{2}
$$

is the irreducible amplitude for the alpha-particle formation in the four-particle channel. For this amplitude we have used a contact interaction form /10, 11/ as in ref. /13/

$$
T_{4 \to \alpha}^{(0)} \cong \kappa \, \delta(\vec{f}_1) \delta(\vec{f}_2) \delta(\vec{f}_3) t \tag{3}
$$

in the position co-ordinates representation. $\vec{f}_i$ are the internal Jacobi co-ordinates of the alpha cluster. The $\delta(\vec{f}_i)$ - Dirac delta functions describe the packing process of the four nucleons in a small volume of the alpha particle volume - a process in which large momenta are transferred. The t-operator selects the terms containing two neutrons and two protons among the four-fermion orbitals. The constant κ is assumed to have a unique value for all such transitions. In the second quantization form we can write the operator (2) as follows

$$
T_{4 \to \alpha}^{(0)} = \sum_{s_4 s_\alpha \, j\mu} T_{s_4 s_\alpha}^{j} \, b_{s_\alpha j\mu}^{+} \, A_{s_4 j\mu} . \tag{4}
$$

Here

$$A_{s_4 j\mu} = ((a_\nu a_{\nu'})_{j_p} (a_\omega a_\omega)_{j_n})_{j\mu} ,$$

where $\nu, \nu'(\omega, \omega')$ stand for the single-particle proton (neutron) shell model quantum numbers (e.g. $\nu \equiv \{n_\nu l_\nu j_\nu\}$)

$$\langle \hat{R}, \xi_\alpha | b^+_{s\alpha j\mu} | 0 \rangle = (Y_{l_\alpha}(\hat{R}) \varphi_{s_\alpha}(\vec{\xi}_\alpha))_{j\mu} , \qquad (5)$$

$$\varphi_{s_\alpha \tau_\alpha}(\vec{\xi}_\alpha) = (\beta/\sqrt{\pi})^{9/2} e^{-\beta^2/2(\xi_1^2 + \xi_2^2 + \xi_3^2)} | \tfrac{1}{2} \tfrac{1}{2}(s_p) \tfrac{1}{2} \tfrac{1}{2}(s_n) s_\alpha \zeta_\alpha \rangle .$$

The quantities $T^j_{s_4 s_\alpha}$ are the reduced matric elements when eq. (3) is used.

If we use the initial state and the alpha-channel state wave functions of a given angular momentum, the alpha-decay width can be written as follows

$$\Gamma_\alpha(I_i^{\pi_i}, I_f^{\pi_f}) = 2\pi\kappa^2 \sum_j \int_0^\infty dR\, \tilde{u}_{j,\varepsilon}(R) g_j^{I_i^{\pi_i}, I_f^{\pi_f}}(R), \qquad (6)$$

where

$$g_j^{I_i^{\pi_i}, I_f^{\pi_f}}(R) = \sum_{s_4} RT^j_{s_4 s_\alpha} \langle \tilde{o} | [\Omega_{I_f^{\pi_f}} A_{s_4 j}, \Omega_{I_i^{\pi_i}}] | \tilde{o} \rangle \qquad (7)$$

while

$$\phi_{I_{i(f)}^{\pi_{i(f)}}} = \Omega^+_{I_{i(f)}^{\pi_{i(f)}}} | \tilde{o} \rangle .$$

The g-amplitudes from eq. (7) are the alpha-clusterization amplitudes and the $\tilde{u}$-functions from eq. (6)

$$\tilde{u}_{j,\varepsilon}(R) = u_{j,\varepsilon}(R) - \int_0^\infty dR'\, K_j(R,R') u_j(R') \qquad (8)$$

describe the relative motion of the alpha cluster around the core. For this motion we used an interaction potential obtained within the double-folding Yukawa (M3Y) procedure /26–29/. The K-kernel results from accurate normalization of the relative alpha core wave function /10, 11/ when the Pauli principle is correctly taken into account.

TABLE 1

Alpha transition	E_i(exp) (MeV)	$I_i^{\pi_i} T_i$	E_f(exp) (MeV)	$I_f^{\pi_f} T_f$	E_α(exp) (MeV)
^{16}O	9.85	2^+ 0	0	0^+0	2.69
	11.52	2^+ 0	0	0^+0	4.36
$\rightarrow ^{12}C$	13.02	2^+ 0	0	0^+0	5.86
	8.87	2^- 0	0	$0^+_.0$	1.71
	0	1^-	0.304	1^-	4.649
^{210}Bi	0	1^-	2.265	2^-	4.686
	0.268	9^-	0.304	1^-	4.908
$\rightarrow ^{206}Tl$	0.268	9^-	0.634	2^-	4.568
	0.268	9^-	1.221	2^-	4.224
	0.268	9^-	0.800	3^-	4.413
^{210}Po	0	0^+	0	0^+	5.305
$\rightarrow ^{206}Pb$	0	0^+		2^+	4.525
^{202}Rn $\rightarrow ^{198}Po$	0	0^+	0	0^+	6.336
^{206}Rn $\rightarrow ^{202}Po$	0	0^+	0	0^+	6.258
^{210}Rn $\rightarrow ^{206}Po$	0	0^+	0	0^+	6.041
^{208}Ra $\rightarrow ^{204}Rn$	0	0^+	0	0^+	7.131
^{210}Ra $\rightarrow ^{206}Rn$	0	0^+	0	0^+	7.018
^{212}Ra $\rightarrow ^{208}Rn$	0	0^+	0	0^+	6.869
^{214}Ra $\rightarrow ^{210}Rn$	0	0^+	0	0^+	7.136

*The n-values in the brackets for Γ represent the powers of

TABLE 1 (Continued)

Γ_{exp} (n)* (MeV)		Γ_{th} (n)* (MeV) (other refs.)		Γ_{th} (n)* (MeV) (Fermi liquid)	
0.625±0.1	(−3)	0.6	(−3)	0.62	(−3)
73±5	(−3)	7.4	(−2)	71.6	(−3)
120±8	(−3)	4.53	(−1)	36.3	(−3)
(1.03±0.28)	(−16)	0.54	(−16)	0.43	(−16)
7.90	(−34)	5	(−36)	7.18	(−34)
5.26	(−34)	1.03	(−35)	2.63	(−33)
1.62	(−36)	1.51	(−38)	5.22	(−36)
1.98	(−37)	4.77	(−40)	0.82	(−37)
4.13	(−40)			2.58	(−40)
1.23	(−38)	1.33	(−40)	3.07	(−38)
3.8	(−29)	2.83	(−31)	2.75	(−29)
4.56	(−34)	4	(−36)	3.32	(−34)
3.26	(−23)			1.05	(−22)
8.35	(−25)			5.22	(−25)
5.07	(−26)			3.25	(−26)
3.8	(−22)			6.78	(−22)
1.2	(−22)			1.2	(−22)
3.5	(−23)			2.24	(−23)
1.75	(−22)			0.59	(−22)

10, by which the entries in the table should be multiplied.

3. ALPHA-DECAY CALCULATIONS IN THE ^{16}O AND TRANSLEAD REGIONS

The self-consistency of our Fermi liquid model is upheld by
the fact that we do not use free parameters as in the R-matrix
theory. The only parameter in our model is the strength of the
alpha four-nucleon vertex which is taken to be equal to
1.43×10^6 MeV fm^9. All the other parameters are those entering the nuclear structure model wave functions. The adopted .
strategy was to use different nuclear structure models in different regions of nuclei. In the ^{16}O region we have used the
Zucker, Buck, McGrory large-scale shell model wave functions
/30/, for Po-Bi region the Leningrad RPA wave functions
/31–33/ and BCS wave functions for Rn-Ra region, where the
pairing-coupling constants G_Z and G_N were fitted to reproduce
the experimental pairing energies. The densities for double-
folding potentials and K-kernels were determined by the same
appropriate single particle potential used in the above-mentioned nuclear structure models. The results are given in Table 1 together with other values calculated in the framework
of Mang's R-matrix model (for ^{16}O) /34/ and Dubna Fermi gas
model /35/ (for Bi and Po isotopes). One can see that in order
to obtain good agreement with the experiment in ref. /34/ the
channel radius has been taken much too large, while the Dubna
Fermi gas model has not reproduce any experimental data.

4. ALPHA VIBRATIONS AND ALPHA SUPERFLUIDITY

The introduction of the effective four-nucleon interaction of
the type (3) makes us to consider whether this type of interaction does not have an analogous effect on the nuclear structure as the pairing interaction, i.e. whether there are alpha
vibrations and alpha superfluidity similar to pairing vibrations and pairing superfluidity, respectively.

In the nuclear systems assumed to be in the normal phase
/36/ the process of addition or removal of nucleon pair from
such a system constitutes a vibrational mode in the gauge
space in which the particle number plays a role similar to

that of the angular momentum in the geometrical space. When
the particle number does not commute with the (Mean Field type)
Hamiltonian, the _rotational_ symmetry in the gauge space is
broken and the nuclear system is assumed to be in a superfluid
phase. The process of addition or removal of a correlated pair
of nucleons from a superfluid nucleus constitutes a rotational
mode in the mentioned gauge space. Such pair rotational spectra, involving families of states in different nuclei, appear
as a prominent feature in the study of two-particle transfer
or alpha cluster (decay) processes. As is well-known the superfluidity of nuclear systems increases 10 to 3000 times /12, 37/
the probabilities of alpha-decay and alpha-transfer processes.
A little smaller increase is obtained by the two-nucleon transfer reaction cross-sections /38/. In spite of the fact that
the pairing superfluidity leads to a large increase of the alpha decay probabilities, there are cases where such an increase
is not large enough. Such cases involve isotopes with only one
magic (or about) number of protons (neutrons).

It seems that the nuclear superfluidity is not caused by the
pairing interaction only /19/.

In the following we evaluate the contribution of the four-
nucleon interaction term to the nuclear superfluidity.

By working with a BCS-like pairing wave function /24/:

$$|\tilde{o}> = \prod_{s=\nu,\omega} (u_s + v_s b_s^+)|\tilde{o}> ; \quad b_s = a_{s-}a_{s+}; \quad u_s^2 + v_s^2 = 1 \qquad (9)$$

and inserting an interaction Hamiltonian of the two-pair correlation type:

$$H_4 = -\frac{1}{4} G_4 \sum_{\nu\nu'} \sum_{\omega\omega'} b_\nu^+ b_\omega^+ b_{\omega'} b_{\nu'} \qquad (10)$$

in addition to the pairing Soloviev's /37/ Hamiltonian

$$H_o = \sum_{\nu\tau} E_\nu a_{\nu\tau}^+ a_{\nu\tau} + \sum_{\omega\sigma} E_\omega a_{\omega\sigma}^+ a_{\omega\sigma} - G_Z \sum_\nu b_\nu^+ b_\nu -$$

$$- G_N \sum_\omega b_\omega^+ b_\omega \qquad (11)$$

(here the notations are those adopted in ref. /37/), except for the s.p. proton (ν) and neutron (ω) orbital quantum numbers).

It is worth mentioning that the H_4-interaction Hamiltonian (10) may appear equivalent under the condition of the constant number of particles to the local one-boson-four-fermion interaction Hamiltonian which generates the vertex in (2), (3).

The minimum of $H_o + H_4 - \lambda_p Z - \lambda_n N$ has the following expression for an even-even nuclear system

$$W = <\tilde{o}| H_o + H_4 - \lambda_p Z - \lambda_n N |\tilde{o}> =$$

$$= \sum_\nu 2(E_\nu - \lambda_p) v_\nu^2 + \sum_\omega 2(E_\omega - \lambda_n) v_\omega^2 -$$

$$- G_Z \chi_p^2 - G_N \chi_n^2 - \tfrac{1}{4} G_4 \chi_p^2 \chi_n^2 \tag{12}$$

with

$$\chi_p = \sum_\nu u_\nu v_\nu , \qquad \chi_n = \sum_\omega u_\omega v_\omega . \tag{13}$$

The gap equations are obtained by applying the variational procedure /37/ $\delta W = 0$.

$$2 = \left\{ G_Z + \tfrac{1}{4} G_4 (\sum_\omega \frac{\Delta_n}{2\varepsilon_\omega})^2 \right\} \sum_\nu \frac{1}{\varepsilon_\nu} ,$$

$$Z = \sum_\nu (1 - \frac{E_\nu - \lambda_p}{\varepsilon_\nu}) ,$$

$$2 = \left\{ G_N + \tfrac{1}{4} G_4 (\sum_\nu \frac{\Delta_p}{2\varepsilon_\nu})^2 \right\} \sum_\omega \frac{1}{\varepsilon_\omega} ,$$

$$N = \sum_\omega (1 - \frac{E_\omega - \lambda_n}{\varepsilon_\omega}) , \tag{14}$$

where

$$\Delta_p = \chi_p (G_Z + \tfrac{1}{4} G_4 \chi_n^2) ,$$

$$\Delta_n = \chi_n (G_N + \tfrac{1}{4} G_4 \chi_p^2) ,$$

$$\mathcal{E}_\nu = \sqrt{(E_\nu - \lambda_p)^2 + \Delta_p^2} \, ,$$

$$\mathcal{E}_\omega = \sqrt{(E_\omega - \lambda_n)^2 + \Delta_n^2} \qquad (15)$$

and

$$\begin{pmatrix} u_s^2 \\ v_s^2 \end{pmatrix} = \frac{1}{2} \left(1 \pm \frac{E_s - \lambda}{\mathcal{E}_s} \right). \qquad (16)$$

For the odd-nuclear system we apply the procedure described in ref. /37/ by including blocking.

From the gap equations (14) we learn that when G_4 is switched off ($G_4 = 0$) we get the decoupled neutron and proton gap equations /37/.

The H_4 interaction Hamiltonian induces mutual superfluidity between neutron and proton pairs. For example, when we have no superfluidity for the proton system and have for neutron system in the absence of H_4-term, by including H_4-term in the total Hamiltonian we may obtain proton superfluidity due to the term $1/4\ G_4 x_n^2$ in the proton gap equations.

To find the value of the G_4-strength constant, besides the experimental odd-even mass difference (the eqs. 4.50, p. 164, ref./37/) we also take into account the following quantity

$$P_4 = \sum S - B(Z + 2, N + 2) + B(Z,N), \qquad (17)$$

where

$$\sum S = S_p(Z + 3, N + 2) + S_p(Z + 2, N + 2) +$$

$$+ S_n(Z + 2, N + 3) + S_n(Z + 2, N + 2) \qquad (18)$$

which seems to describe the internal energy of the alpha cluster in the superfluid nucleus.

When the densities are large enough, the sums can be replaced by integrals in the eqs. (12) — (14) and the experimental quantity (17) can be reproduced by the following theoretical quantity

$$\Delta b = b(G_2, G_4) - b(G_2, G_4=0) \qquad (19)$$

with

$$b(G_2, G_4) = \frac{2}{9E_c} (W - W_o) =$$

$$= \frac{2}{9E_c} \left\{ - 9E_c^2 \left(\sqrt{1 + \frac{\Delta^2}{E_c^2}} - 1 \right) + \frac{1}{4} G_4 \chi^4 \right\}. \qquad (20)$$

Here we have assumed the identity of neutron and proton systems, i.e.

$$9_p = 9_n = \tfrac{1}{2} 9, \quad E_p = E_n = E_c, \quad \Delta_p = \Delta_n = \Delta,$$

$$\chi_p = \chi_n = \chi, \quad G_Z = G_N = G_2,$$

where 9 is the density of single particle levels, E_c — the cut-off energy, Δ — the energy gap, $W_o = W (G_2=0, G_4=0)$ and $9E_c/4$ is the number of alpha clusters.

Inserting a new variable:

$$x = 4 \left[9G_2 + \tfrac{1}{4} 9G_4 \chi^2 \right]^{-1} \qquad (21)$$

the quantity (19) becomes

$$\Delta b = \frac{2E_c}{r} f(x),$$

$$f(x) = \tfrac{1}{2} \left(\tfrac{1}{x} - s \right)^2 - r \left(\frac{e^{-x/2}}{sh(x/2)} - \frac{e^{-1/2s}}{sh(1/2s)} \right), \qquad (22)$$

where

$$S = \tfrac{1}{4} 9G_2, \qquad r = \frac{1}{128} 9^3 G_4 E_c^2.$$

The G_2-parameter was taken to be /37/ $G_2 = 25/A$ MeV and G_4 is obtained by fitting the experimental P_4-quantity with our Δb (see eq. (22)), both of then measuring the excess binding energy of an alpha cluster due to the four-fermion interaction H_4.

Studying the same nuclei as in ref. /19/, in the region from ^{160}Gd–^{184}Os for the experimental average value of P_4 (eq.

TABLE 2

Nucleus	P_{4exp} (MeV)	$\triangle b$ (MeV)	$(\chi_{24}/\chi_2)^2$
$^{140}_{54}$Xe	−0.480	−0.760	1.14
$^{144}_{56}$Ba	0.380	−0.706	1.23
$^{148}_{58}$Ce	0.282	−0.656	1.33
$^{152}_{60}$Nd	−0.184	−0.608	1.43
$^{156}_{62}$Sm	−0.096	−0.563	1.54
$^{160}_{64}$Gd	−0.5856	−0.521	1.65
$^{164}_{66}$Dy	−0.43393	−0.481	1.76
$^{168}_{68}$Er	−0.354	−0.444	1.88
$^{172}_{70}$Yb	−0.4237	−0.409	2.00
$^{176}_{72}$Hf	−0.015	−0.376	2.13
$^{180}_{74}$W	−0.8236	−0.345	2.26
$^{184}_{76}$Os	−0.096	−0.317	2.38
$^{188}_{78}$Pt	−0.490	−0.290	2.51
$^{192}_{80}$Hg	−0.530	−0.265	2.65
$^{196}_{82}$Pb	−0.320	−0.242	2.78
$^{200}_{84}$Po	−0.370	−0.221	2.91
$^{204}_{86}$Rn	−0.350	−0.201	3.044

(17)) = -0.4 MeV, we have found /24/ the strength G_4 = = 1.4 x 10^{-3} MeV, which is consistent with our κ constant introduced in the preceding paragraph and refs. /10, 11/.

In analogy to the procedure given in ref. /37/ eq. (4.32) we find that the quantity (22) describing the experimental quantity (17) is given by the following expression

$$\Delta b \cong 4(\Delta_{24} - \Delta_2),$$

where Δ_{24} and Δ_2 are the energy gaps obtained with and without the H_4-interaction.

Working with G_2 = 25/A MeV and G_4 = 1.4 x 10^{-3} MeV we have calculated /24/ the Δb quantity (22) for a number of nuclei and compared it with the experimental P_4 quantity (17). The results are given in Table 2, where the enhancement correlation factor due to H_4-interaction $(\chi_{24}/\chi_2)^2$ entering the alpha-decay and alpha-transfer reactions is given as well.

Figure 1 shows that the quantity P_4 is quite well reproduced by our Δb, if we take into account the experimental errors.

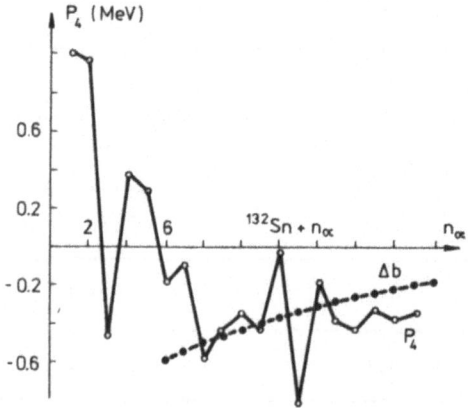

FIG. 1.

As mentioned in the beginning of this paragraph the pairing vibrational concept proposed by Bohr and Mottelson /36/ can be

enlarged by including the alpha-vibrational collective mode
associated with the possibility of collective fields that create or annihilate "alpha clusters". Such fields may be connected with the H_4-component in the nuclear interactions which
tend to bind two protons and two neutrons into a highly correlated state with angular zero momentum. The addition (substrac-
tion) of such a cluster to (from) a closed shell constitutes
an excitation than can be repeated and thus viewed as a quan-

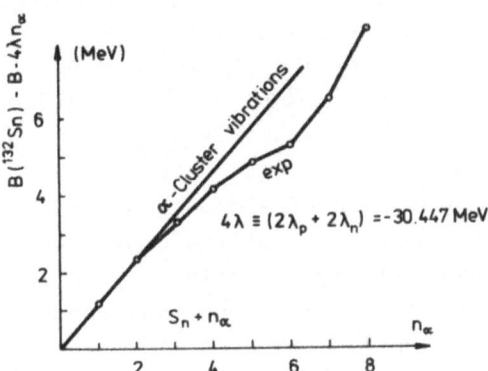

FIG. 2.

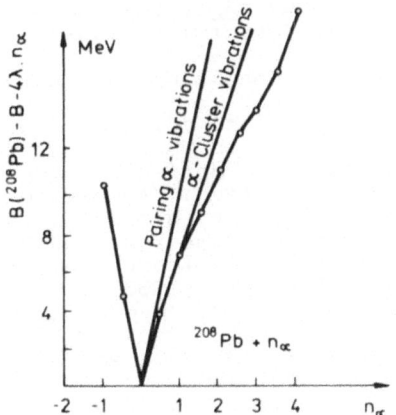

FIG. 3.

tum of a vibrational mode. Figures 2 and 3 show the alpha vibrational spectrum similar to that obtained by Bohr and Mottelson /36/ for pair vibrational spectrum.

5. CONCLUSIONS

Here we have reported on two types of implications of introducing four-nucleon interactions to study the structure of atomic nucleus; the clusterization process and the alpha superfluidity.

The Fermi liquid model for the alpha clusterization process
seems to remove the discrepancy between the experimental and
theoretical alpha-decay probabilities, which is one of the
long-standing problems in nuclear physics.

The second implication of accounting for four-nucleon interactions is the appearance of the alpha superfluidity besides the well-known pairing superfluidity in atomic nuclei,
which seems to be of a 30 per cent enhancement order when compared the correlation functions.

REFERENCES

1. W. Wefelmeir, Naturwissenschaften $\underline{25}$ (1937) 525.
2. J. A. Weeler, Phys. Rev. $\underline{32}$ (1937) 1083.
3. C. F. von Weiszacker, Naturwissenschaften $\underline{26}$ (1938) 209, 225.
4. K. Wildermuth and Th. Kanellopoulos, Nucl. Phys. $\underline{7}$ (1958) 150.
5. K. Widermuth and W. McClun, "Cluster Representations of Nuclei", Springer Tracts in Modern Physics, Vol. 41, Springer-Verlag, Berlin - Heidelberg - New York, 1966.
6. K. Wildermuth and Y. C. Tang, "A Unified Theory of the Nucleus", Vieweg Verlag, Braunschweig, 1977.
7. "Alpha-like Four-body Correlations and Molecular Aspects in Nuclei", Suppl. of Prog. Theor. Phys. $\underline{52}$ (1972).
8. H. Morinaga, Phys. Rev. $\underline{101}$ (1956) 254.
9. A. Bulgac and O. Dumitrescu, "Progrese în Fizică ICEFIZ", Bucharest, 1979, p. 1.
10. A. Bulgac, S. Holan, F. Cârstoiu and O. Dumitrescu, Nuovo Cimento $\underline{70A}$ (1982) 409.
11. F. Cârstoiu, O. Dumitrescu, G. Stratan and M. Braic, Nucl. Phys. $\underline{A441}$ (1985) 221.
12. O. Dumitrescu, Fiz. Elem. Chastits At. Yadra $\underline{10}$ (1979) 377; Sov. J. Part. Nucl. $\underline{10}$ (1979) 147.

13. A. B. Migdal, "Theory of Finite Fermi Systems", Moscow, 1965.
14. G. E. Brown, V. Horsfjord and K. F. Liu, Nucl. Phys. A205 (1973) 73.
15. J. F. Enis et al., Phys. Rev. Lett. 51 (1983) 646.
16. F. Iaohello and A. D. Jackson, Phys. Lett. 108B (1982) 151.
17. F. Iachello, Nucl. Phys. A396 (1983) 233c.
18. H. Deley and F. Iachello, Phys. Lett. 131B (1983) 282.
19. Y. K. Gambhir, P. Ring and P. Schuck, Phys. Rev. Lett. 51 (1983) 1235.
20. B. H. Flowers and M. Vujič, Nucl. Phys. 49 (1963) 586.
21. M. Baranger, Phys. Rev. 130 (1963) 1244.
22. J. Eichler and M. Yamamura, Nucl. Phys. A182 (1972) 33.
23. M. Kamimura, T. Marumari and K. Takada, Prog. Theor. Phys. Suppl. 52 (1972) 282.
24. M. Apostol, I. Bulboacă and O. Dumitrescu, in press.
25. J. R. Schreiffer, "Theory of Superconductivity", Benjamin, New York, 1964.
26. A. Bulgac, F. Cârstoiu and O. Dumitrescu, Rev. Roum. Phys. 27 (1982) 331.
27. G. R. Satchler and W. G. Love, Phys. Rep. 55 (1979) 123.
28. G. Bertsch, J. Borysowicz, H. McManus and W. G. Love, Nucl. Phys. A284 (1977) 399.
29. J. P. Elliott, A. D. Jackson, H. A. Mavromatis, E. A. Sanderson and B. Singh, Nucl. Phys. A121 (1968) 241.
30. A. P. Zucker, B. Buck and J. B. McGrory, BNL report No. 14085, 1969, Phys. Rev. Lett. 21 (1968) 39.
31. V. I. Isakov, Yu. I. Haritonov, S. A. Artamonov and L. A. Sliv, Preprint LNPI 276, Leningrad, 1976.
32. V. I. Isakov, S. A. Artamonov and L. A. Sliv, Izv. Akad. Nauk, Ser. Fiz. 41 (1977) 2074.
33. S. A. Artamonov and V. I. Isakov, Preprint LNPI 420, Leningrad, 1978.
34. B. Apagyi, G. Fai and J. Nemeth, Nucl. Phys. A272 (1976) 303, 317.
35. S. A. Artamonov, V. I. Isakov, S. G. Kadmenski, I. A. Lomachenkov and V. I. Furman, Preprint LNPI 620, Leningrad, 1980.
    S. G. Kadmenski and V. I. Furman, Fiz. Elem. Chastits At. Yadra 6 (1975) 409; Sov. J. Part. Nucl. 6 (1975) 189.
36. A. Bohr and B. Mottelson, "Nuclear Structure", Benjamin, New York, Vol. 1, 1970; Vol. 2, 1974.
37. V. G. Soloviev, "Teorya slozhnykh yader", Nauka, Moscow, 1971; "Theory of Complex Nuclei", Pergamon Press, New York, 1976.
38. D. R. Bès and R. A. Broglia, in "Elementary Modes of Excitation in Nuclei", ed. by A. Bohr and R. A. Broglia, North-Holland Publ. Co., Amsterdam - Oxford - New York, 1977, p. 55.

ISOVECTOR EFFECTIVE INTERACTIONS. AN EMPIRICAL APPROACH[*]

J. Rapaport

Department of Physics and Astronomy, Ohio University, Athens, Ohio, USA

ABSTRACT. In the last few years new data has been used to test more accurately models of the effective interactions which drive nucleon-nucleus scattering processes. At intermediate energies (E ≻ 100 MeV) where the impulse approximation is well justified, the effective interaction is identified with the free nucleon-nucleon t-matrix. At lower energies (E ≺ 100 MeV) nuclear medium effects become important and these effects need to be incorporated in the effective interaction via the G-matrix formalism. Charge exchange reactions uniquely select the isovector terms of the effective interaction and may be used to test the different models with data. The isovector terms may be empirically evaluated from a comparison of neutron and proton elastic scattering. If done in a model independent analysis, it provides a unique way to compare empirical and theoretical results.

1. INTRODUCTION

A main part of understanding nucleon-nucleus scattering is that of relating the reaction mechanism to the underlying nucleon-nucleon (N-N) interaction. Unfortunately the "effective N-N interaction" which results from a transformation of the "bare N-N interaction" is rather complicated to be used routinely in calculations of nucleon-nucleus scattering. It is invariably non-local, depends on energy, relative angular momentum, density, etc.

Several nucleon-nucleus effective interactions (V^{eff}) have

---

[*]Supported in part by the National Science Foundation.

been developed for the calculation and interpretation of scattering data. They are made "realistic" and "simple". Usually these V^{eff} are represented in each N-N channel (triplet-even, singlet-even, triplet-odd and singlet-odd) by a local operator of the form:

$$V^{eff}(r) = V_c(r) + V_{LS}(r) \vec{L} \cdot \vec{S} + V_T(r) S_{12},$$

where V_c denotes central interactions and $\vec{L}.\vec{S}$ and S_{12} are the usual spin-orbit and tensor operators. Each one of these terms has the usual spin and isospin dependences, and for instance the central term is written:

$$V_c^{eff}(r) = V_0 f_0(r) + V_\sigma f_\sigma(r)\sigma_1\sigma_2 + V_\tau f_\tau(r)\tau_1\tau_2 +$$
$$+ V_{\sigma\tau} f_{\sigma\tau}(r)\sigma_1\sigma_2 \tau_1\tau_2 .$$

This form for V^{eff} becomes non-local when exchange terms are included, but they are routinely calculated. As it is desirable, simple analytical forms for the radial dependences are assumed, the most common one being a superposition of Yukawa of different ranges.

At energies above 100 MeV a representation of V^{eff} as a t-matrix, and in particular the parameterization of Franey and Love /1/, has been quite successful in the description of a variety of scattering data.

At lower energies (E $\prec$ 65 MeV) the V^{eff} is chosen to represent as well as possible the G-matrix generated by some N-N potential which describes N-N scattering /2/. Values for G-matrix effective interaction strengths known as the M3Y interaction are given by Love /2/ and by Bertsch et al. /3/. Another one is the one generated with the Paris potential /4/. These interactions are usually modified with the addition of imaginary and spin-orbit terms. Density and energy dependences in these V^{eff} have been studied by Geramb et al. /5/ and a complex effective N-N interaction by Yamaguchi et al. /35/.

Until a few years ago, less sophisticated analyzes were

used to determine the properties of V^{eff}. A real interaction
with a single Yukawa (range = 1.0 fm) form factor was assumed
and several strength parameters represented the various components of the interaction. A typical study of such V^{eff} has
been given by Austin /6/. The energy dependence of the isovector term, V_τ, from that study is compared with the energy dependence of the isovector term, V_1, obtained in empirical analyzes of nucleon scattering data in ref. /7/.

An attempt to give a comprehensive picture of the present
state of our understanding of V^{eff} is beyond the scope of this
presentation. Instead I would like to present some ideas about
the low-energy isovector effective interaction (E $\prec$ 50 MeV).
In the last few months we have developed a model independent
analysis (MIA) of nucleon scattering. The result is that we
obtain, after a least χ^2 search, radial dependences for the
real and absorptive terms of the optical model potential (OMP).
A comparison of neutron and proton OMP results in an empirical
evaluation of the isovector term of the OMP. This isovector
potential may then be used to calculate charge exchange (p,n)
reactions. Isovector potentials generated with different effective interactions may then be compared directly with these
empirical results. Density, non-locality and exchange effects
included in the calculations may then be directly tested in
these comparisons.

2. ISOVECTOR EFFECTIVE INTERACTION

In the single-scattering distorted approximation and assuming
local, density independent interactions, the transition amplitude for nucleon scattering /8/ is given by

$$T_{fi}^{DW} = \int d\vec{r}_p \, \chi^{(-)}(\vec{k}',\vec{r}_p) \langle I'M',S_p\nu' \mid \sum_{i=1}^{A} V_{ip} \mid IM,S_p\nu\rangle \chi^+(\vec{k},\vec{r}_p),$$

where the $\chi^{(\pm)}$ are distorted waves and V_{ip} is the V^{eff} between
projectile and each target nucleon. The initial and final tar-
get and projectile spin projection are denoted by (MM') and
($\nu\nu'$); $\vec{k}(\vec{k}')$ is the initial (final) projectile-nucleus momen-

tum. In the above expression exchange effects are not explicitly included but are routinely calculated.

The transition potential $U_{fi}^{(\alpha)}(\vec{r}_p)$ may be defined by

$$U_{fi}^{(\alpha)}(\vec{r}_p) = \langle I'M' \mid \sum_{i=1}^{A} v_{ip}^{(\alpha)} \theta_\alpha(i) \mid IM \rangle \, ,$$

where

$$V_{ip} = \sum_\alpha v_{ip}^{(\alpha)} \theta_\alpha(i) \theta_\alpha(p) \, ,$$

$$\theta_\alpha(i) = \vec{\sigma}_i , \vec{\tau}_i , \vec{\sigma}_i \vec{\tau}_i \, , \text{ etc.}$$

and α denotes the type of excitation being considered. The transition potential may also be rewritten as:

$$U_{fi}^{(\alpha)}(\vec{r}_p) = \langle I'M' \mid \int d\vec{r} \, v^{(\alpha)}(\vec{r}_p - \vec{r}) \, \delta(\vec{r} - \vec{r}_i) \, \theta_\alpha(i) \mid IM \rangle$$

$$= \int d\vec{r} \, v^{(\alpha)}(\vec{r}_p - \vec{r}) \varrho_{I'I}^{(\alpha)}(\vec{r}) \, ,$$

where $\varrho_{I'I}^{(\alpha)}$ is the nuclear transition density defined by

$$\varrho_{I'I}^{(\alpha)}(r) = \langle I'M' \mid \sum \delta(\vec{r} - \vec{r}_i) \, \theta_\alpha(i) \mid IM \rangle \, .$$

Thus

$$T_{fi}^{DW} = \sum_\alpha \int d\vec{r}_p \, D_\alpha(\vec{k}, \vec{k}' ; \vec{r}_p) U_{fi}^{(\alpha)}(\vec{r}_p) \, , \qquad (1)$$

where $D_\alpha(\vec{k}, \vec{k}', \vec{r}_p)$ is a distortion function which with the structure of v^{eff} and the transition density (nuclear structure) are needed to calculate the transition amplitude.

The isovector effective interaction contains terms which depend on τ and $\vec{\sigma}\tau$, terms which may be appropriately filtered from all other terms in v^{eff} in a charge exchange (p,n) or (n,p) nuclear reaction.

We have used the (p,n) zero degree differential cross sections measured /9/ up to E_p = 200 MeV to provide an empirical energy dependence of the interaction strength $V_{\sigma\tau}/V_\tau$ at momentum transfer q ~ 0. The (p,n) reaction on even-A, T $\neq$ 0 targets leads to $0^+ \rightarrow 1^+$ and $0^+ \rightarrow 0^+$ transitions that are analogous to Gamow-Teller (GT) and Fermi (F) beta decay, respectively. In cases where the GT and F strengths are known (β-decay) the ratio

$$R(E_p)^2 = \frac{\sigma_{GT}(0^0)/B(GT)}{\sigma_F(0^0)/B(F)}$$

(where σ_α and $B(\alpha)$ are the measured cross section and strengths, respectively) is interpreted as:

$$R(E_p) \cong \left| \frac{J_{\sigma\tau}}{J_\tau} \right|$$

where J_α is the Fourier transform of the effective interaction. A value $R(E_p) = (E_p/55)$, where E_p is in MeV fits the data rather well between 50–200 MeV and some data points as low as 20 MeV /9/. This energy dependence can be understood in terms of meson exchanges. Results of calculations by Brown et al. (BSW) /10/, values obtained from the t-matrix interaction by Franey and Love /1/ and Picklesimer and Walker (PW) /11/ are compared with empirical results (shaded area) in Fig. 1. A

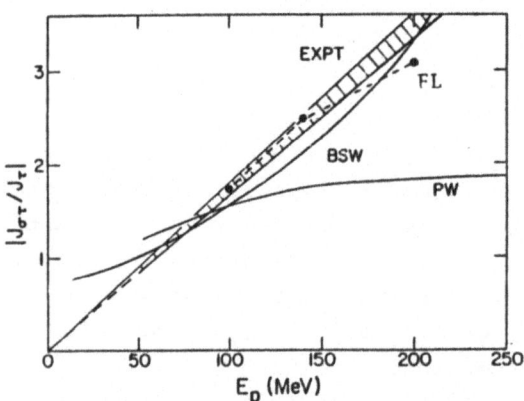

FIG. 1. Comparison of the empirical ratio $|J_{\sigma\tau}/J_\tau|$ (shaded area) to several predictions based on free N-N interaction studies.

similar energy dependence is obtained for this ratio by Orihara et al. /12/ for proton energies between 18 and 40 MeV.

Between 60 and 200 MeV empirical evidence seems to indicate that $J_{G\tau}(q \approx 0)$ is rather energy independent. This indicates, therefore, a rather sharp decrease of $J_\tau(q \approx 0)$ with increasing energy. Petrovich has calculated the energy dependence of the $G\tau$ and τ terms of V^{eff} between 25 and 150 MeV. Both the terms of the V^{eff} derived from the G-matrix or from the OPEP potential seem to be rather constant with energy. This seems to imply that in the energy range in which $V_{G\tau}$ is constant

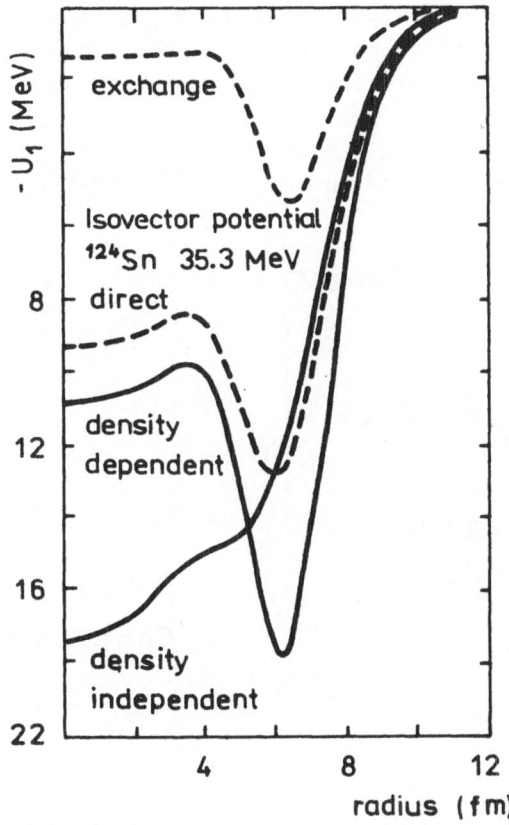

FIG. 2. Calculated real isovector potential, using a folding model /13/.

141

(50—200 MeV) the value of V_τ is inversely proportional to the incident energy.

Attempts to describe the (p,n) IAS transition microscopically at $E_p < 50$ MeV are given by Schery /13/. Distortion effects, $D(\vec{k},\vec{k}',\vec{r}_p)$ in eq. (1), are calculated using a folding model in terms of matter distribution folded with components of the N-N interaction. The isovector potential, $U_{fi}(\vec{r}_p)$ in eq. (1), is calculated within the same model and a local density approximation is used to estimate exchange effects. The angular distributions were improved by adding an absorptive isovector potential to the calculations. The real term of the isovector potential for the ^{124}Sn(p,n)^{124}Sb (IAS) is shown in Fig. 2. More recently, Dietrich and Petrovich /14/ have presented microscopic calculation results using several theoretical V^{eff},

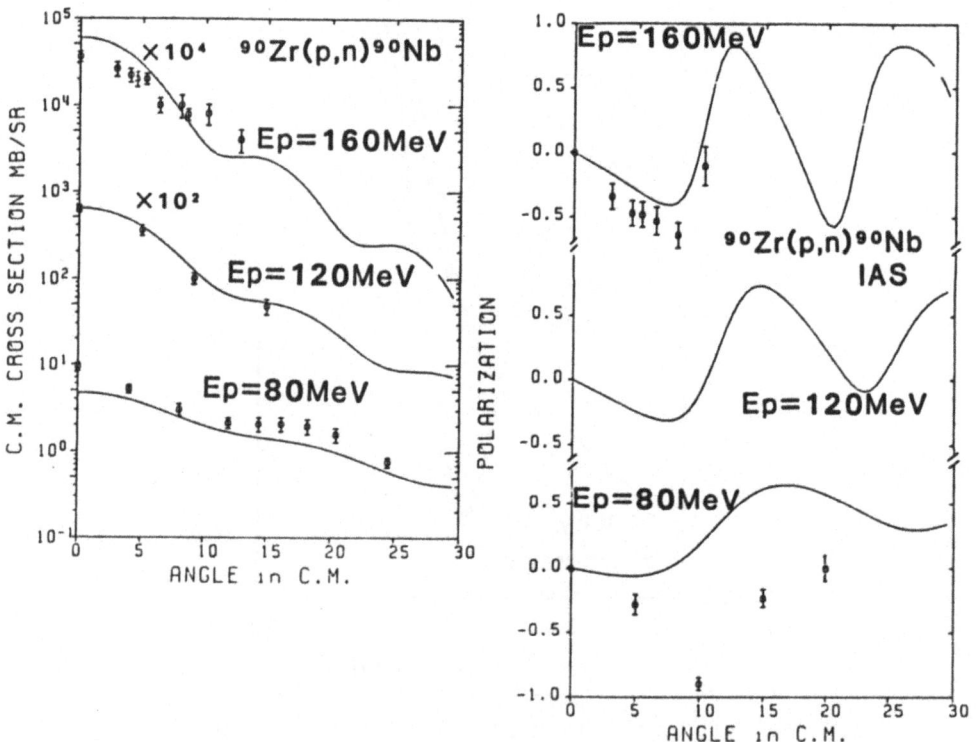

FIG. 3. Comparison between empirical and DWBA 70 calculated cross sections and Ay(θ) for the ^{90}Zr(p,n)^{90}Nb (IAS) transition.

together with the necessary local-density and exchange approximations. A systematic comparison is made with experimental
results for elastic, inelastic and isobaric analog charge-exchange reactions. Both the results of refs. /13, 14/ are very
encouraging but further studies need to be done.

At intermediate incident energies the impulse approximation
and the free N-N V^{eff} of Franey and Love /1/ seem to do quite
well. In Fig. 3 we present calculations at 80 MeV, 120 MeV and
160 MeV compared with data obtained at Indiana University. The
80 MeV data is from Lind et al. /15/, at 120 MeV from Bainum
et al. /16/ and the 160 MeV data has been reported by Clark et
al. /34/.

3. EMPIRICAL ISOVECTOR POTENTIAL

The simplest charge exchange reaction, particularly for even
A-nuclei, is the excitation of the isobaric analog state of
the ground state of the target. From a microscopic point of
view the isovector potential may be obtained from the effective interaction and the transition density, $U_{fi}(\vec{r}_p)$, in eq.
(1). In many cases the transition densities are well known,
thus an empirical evaluation of the isovector potential corresponds to a good test of our knowledge of the isovector terms
of V^{eff}.

An empirical isovector potential may be obtained if we assume a Lane /17/ consistent description of nucleon scattering.
Then we write the non-diagonal term

$$U_1 = \frac{1}{2\varepsilon} (U_p - U_n) ,$$

where the U's are nuclear optical potentials for protons and
neutrons and $\varepsilon = (N-Z)/A$. All these potentials have real (V)
and absorptive (W) components.

The empirical OMP is usually decomposed as:

$$-U(r,E) = U_N(r,E) - V_c(r) - U_{so}(r,E) ,$$

where $V_c(r)$ is the Coulomb potential, U_{so} the spin-orbit term
and U_N is expressed as suggested by Lane /17/

$$U_N(r,E) = U_0(r,E) + \frac{4}{A} U_1(r,E) \; \vec{t}.\vec{T} \; ,$$

where U_0 is the isoscalar and U_1 the isovector terms; $\vec{t}$ and $\vec{T}$ are the isospins of the incident nucleon and target, respectively. The isospin interaction splits the radial part of the potential into two diagonal terms which describe the proton and neutron elastic scattering while the non-diagonal or coupling term describes the (p,n) quasi-elastic scattering. The latter is given by

$$U_{pn}(r,E) = 2 \sqrt{\frac{E}{A}} \; U_1(r,E) \; .$$

The diagonal terms may be written

$$U_n(r,E) = U_0(E)f_0(r) - U_1(E)f_1(r) \; ,$$

$$U_p(r,E) = U_0(E)f_0(r) + U_1(E)f_1(r) + \Delta U_c f_c(r)$$

for neutrons and protons, respectively. In general U is a complex quantity and the term $\Delta U_c f_c(r)$ is the so-called Coulomb correction term first introduced by Lane [18]. Its real component is usually parameterized as $\Delta V_c f_c(r) = \beta (Z/A^{1/3})f_0(r)$. The OMP analyses of neutron and proton scattering on T = 0 nuclei give a unique way to obtain empirically the Coulomb correction term. Nucleon scattering on ^{40}Ca [19] has been used to obtain a value $\beta = 0.46 \pm 0.07$. A comparison of neutron and proton global OMP parameters, results in a value $V_1 f_1(r)$ = $(22.7 - 0.19 \text{ E})f_0(r)$ for the isovector term [19]. This energy dependence [7] (10–40 MeV region) is in good agreement with the empirical energy dependence of the isovector V^{eff} as determined by Austin [6].

A different approach may be taken in the analysis of elastic scattering data if sufficient data points are available. The Woods-Saxon geometries used empirically in OMP analyses are certainly not those geometries dictated by theoretical calculations; rather they are used for convenience. When comparisons are made between different models and empirical potentials, and because of different geometries it is not easy to arrive at definite conclusions. A more realistic comparison

144

may be provided if the data are analyzed in what is called a model-independent analysis (MIA), an idea similar to that used for electron scattering. This model has been used in the analysis of alpha scattering by Friedman and Batty /20/, in the analysis of neutron scattering near 11 MeV on [40]Ca by Tornow et al. /21/ and was suggested by Austin /22/ as a good alternative in the comparison of empirical and theoretical nucleon OMP. The method consist in adding to the conventional Woods-Saxon potential an extra potential given by a Fourier–Bessel series, coefficients of which are obtained in a least x^2 fit to the data.

We have applied this method to nucleon scattering data from [40]Ca (T = 0) and [208]Pb. In the [40]Ca case the comparison of neutron and proton optical potentials gives information about the "Coulomb correction potential". Data published /23, 24/ for p+[40]Ca in the 20–50 MeV range and recently neutron scattering

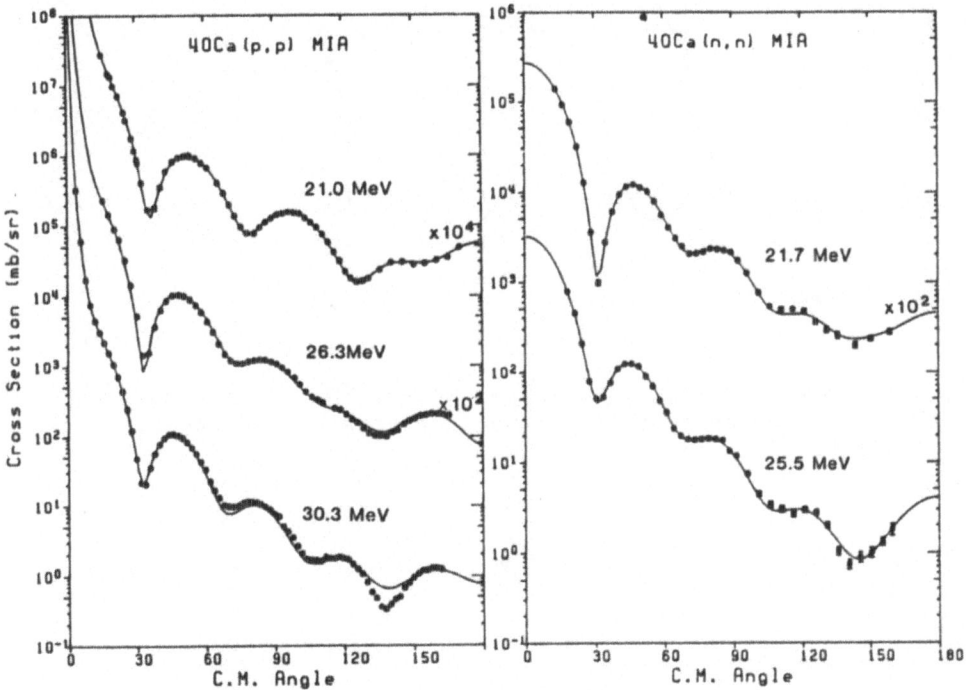

FIG. 4. Calculated fits to the nucleon elastic scattering data on [40]Ca using a MIA.

145

data taken at Ohio University at 19, 21.7 and 25.5 MeV /25/ have been used in this analysis. Some of the obtained fits are compared with data in Fig. 4. The resulting real and absorptive optical potentials are shown in Fig. 5. In all cases the solid line represents a best fit Woods-Saxon potential. The shaded areas represent the error bands determined from the co-variance matrix of the Fourier-Bessel coefficients.

Some preliminary results for "Coulomb correction terms" are

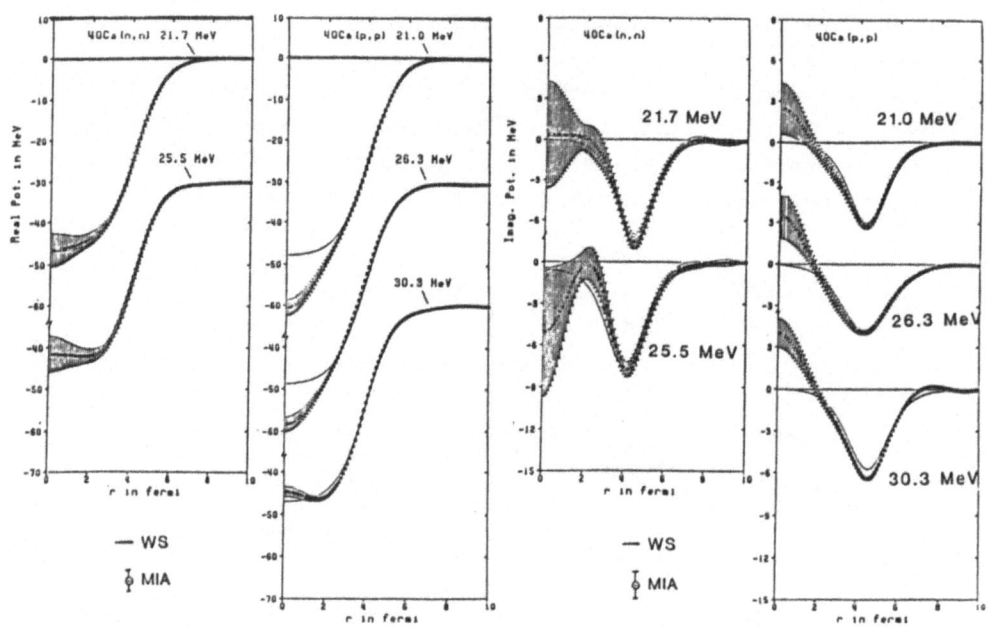

FIG. 5. Optical potentials obtained from the MIA (shaded) and Woods-Saxon geometry (solid line) analyses of nucleon elastic scattering on ^{40}Ca.

shown in Fig. 6. The real term with a volume integral of
$(J_{\Delta V_c}/A) = (15\pm10)$ MeV fm^3 is in agreement with previous empirical results /19/. The absorptive term has both a repulsive
(near the origin) and a small attractive region (near the surface).

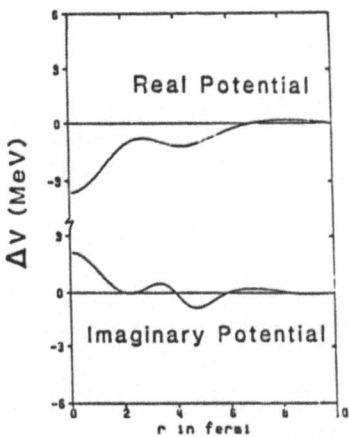

FIG. 6. "Coulomb correction" potentials obtained from a comparison of MIA neutron and proton elastic scattering data.

To obtain the isovector potential, the comparison of neutron
and proton OMP analyses may be done at the same lab incident
energy or at the same nucleon energy inside the nucleus. The
second method results in reducing the Coulomb correction term
and thus any potential differences in T = 0 nuclei may be mainly attributed to different neutron and proton densities inside
the nucleus. These effects for ^{40}Ca have been calculated by
Brown et al. /26/ and the present results agree qualitatively
with these calculations.

Several calculations of the absorptive term of the optical potential have been recently reported, in particular for
^{40}Ca. Vinh Mau /27/ presented local equivalent potentials for
n+^{40}Ca at 30 MeV. Similarly, Osterfeld and Madsen /28/ have
calculated the absorptive n+^{40}Ca potential. Those are compared
in Fig. 7 with the empirically determined MIA potential.

In the case of ^{208}Pb we have obtained empirical optical

147

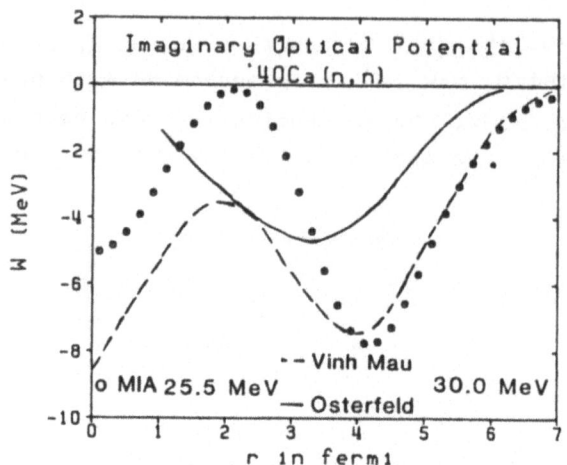

FIG. 7. Absorptive optical potential for n+^{40}Ca. The empirical MIA results are compared with microscopic calculations.

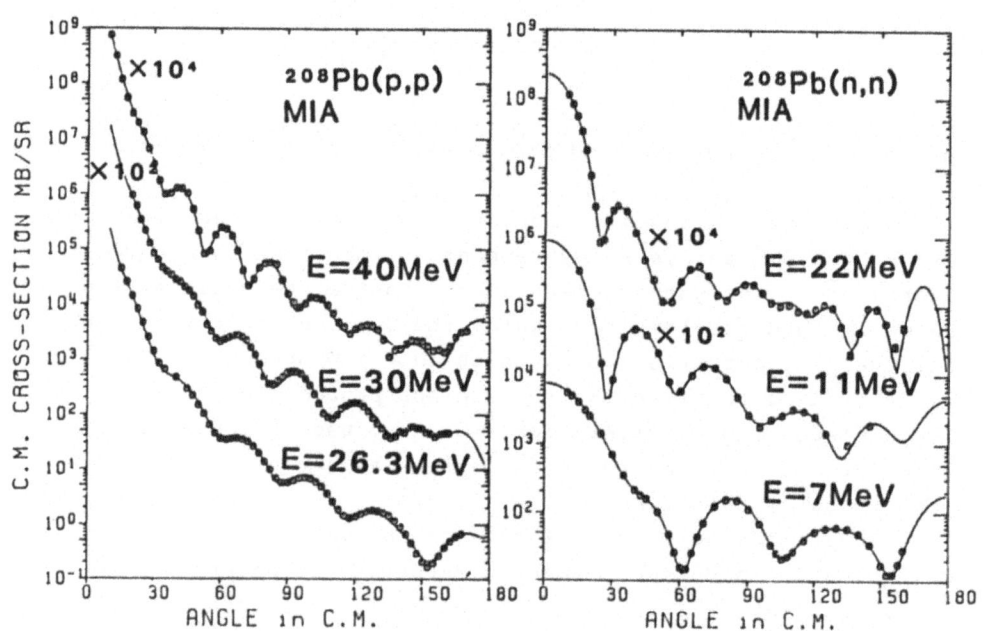

FIG. 8. Calculated fits to the nucleon elastic scattering data on ^{208}Pb using a MIA.

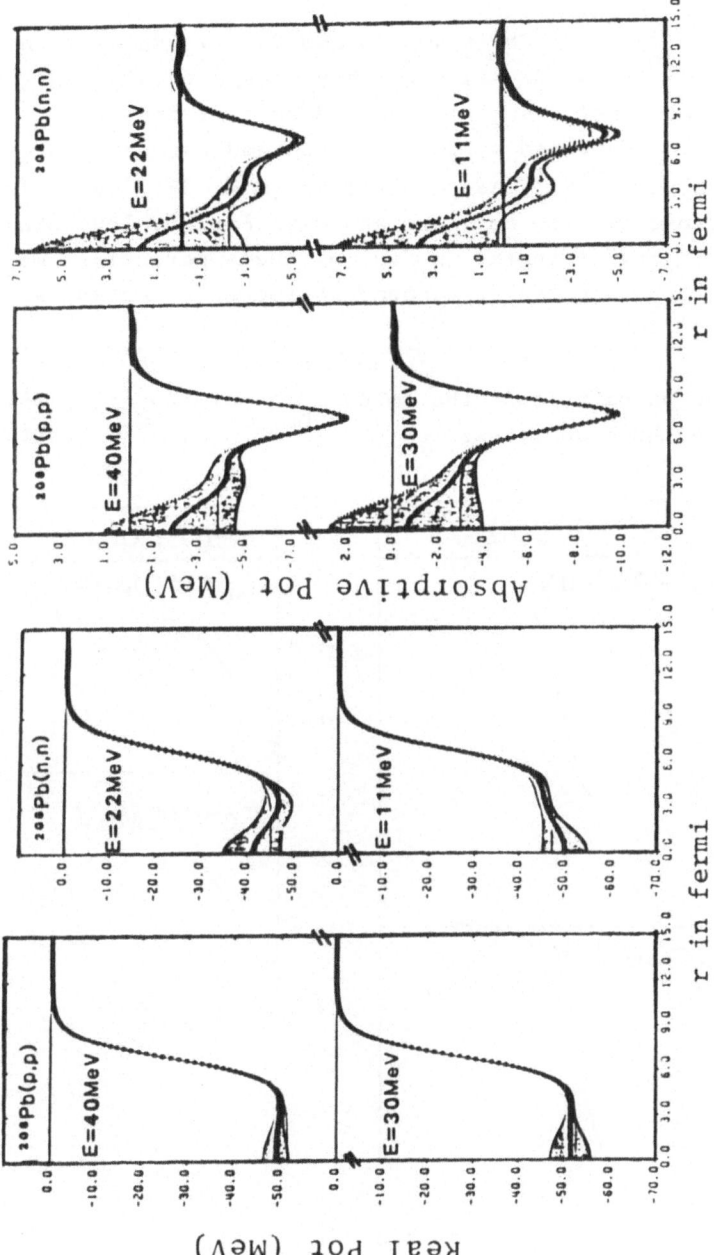

FIG. 9. Optical potentials obtained from the MIA (shaded) and Woods-Saxon geometry (solid line) analyses of nucleon elastic scattering on 208Pb.

potentials for p+^{208}Pb at 40, 30 and 26 MeV (data of ref. /29/),
and for n+^{208}Pb at 22, 11 and 7 MeV, respectively (data of ref.
/30/). These energies were selected so that they would closely match incident and outgoing proton and neutron energies for
the ^{208}Pb(p,n)^{208}Bi (IAS) transition (Q = -18.8 MeV). The fits
to the cross section data are shown in Fig. 8. The obtained
potentials are indicated in Fig. 9. These potentials may then
be used to obtain the corresponding isovector potentials (eq.
(2)) (Fig. 10).

The consistency of this method is realized by using these
isovector potentials in the calculation of the (p,n) IAS transition. We have done this for the ^{208}Pb(p,n)^{208}Bi reaction at

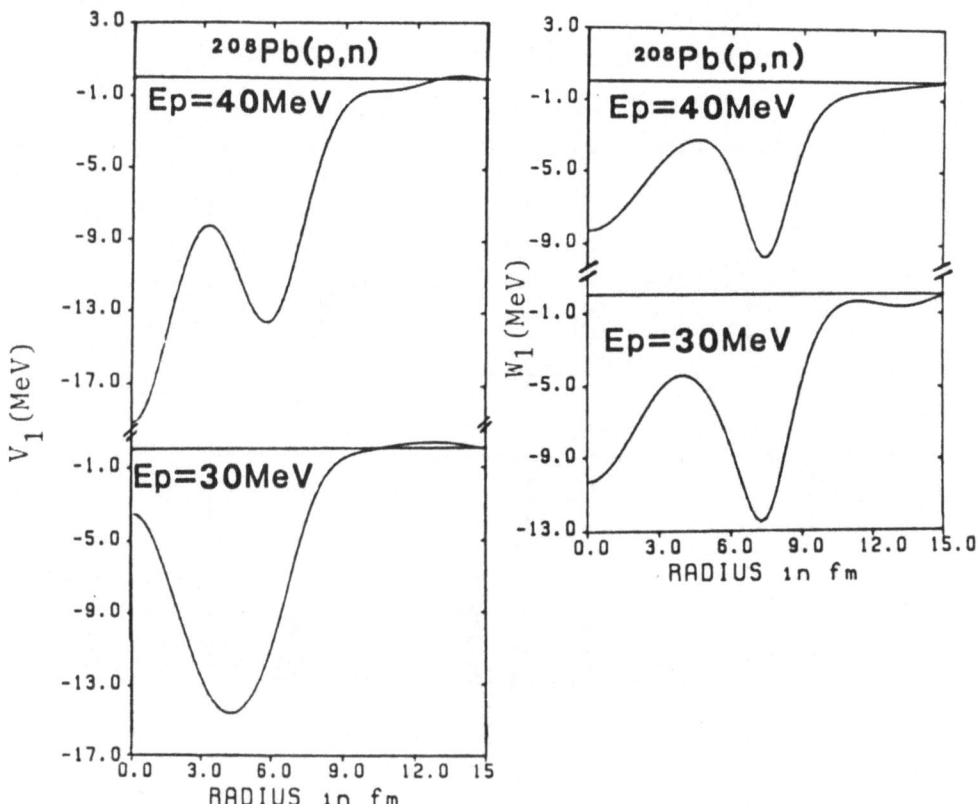

FIG. 10. Isovector potentials, real (V$_1$) and absorptive (W$_1$)
obtained from a comparison of MIA neutron and proton elastic
scattering data.

E_p = 45 and 35 MeV. The calculations are compared with the
data of ref. /31/ in Fig. 11. The agreement is very good.

These empirical isovector potentials may be compared with
those predicted from effective interactions. We have used the
Franey-Love interaction at 50 MeV /1/ to calculate the differential cross section for the $^{208}Pb(p,n)^{208}Bi$ reaction at 45 MeV.
The DWIA calculations were made by assuming that the neutron
excess may be described by the several single particle configurations above Z = 82, with an oscillator length of 2.459 fm.
The shape of the calculated cross section is in very good
agreement with the data but the magnitude is too large by a
factor of 5. Contributions to the cross section from the direct
term and exchange term in the real V^{eff} are in opposite phase

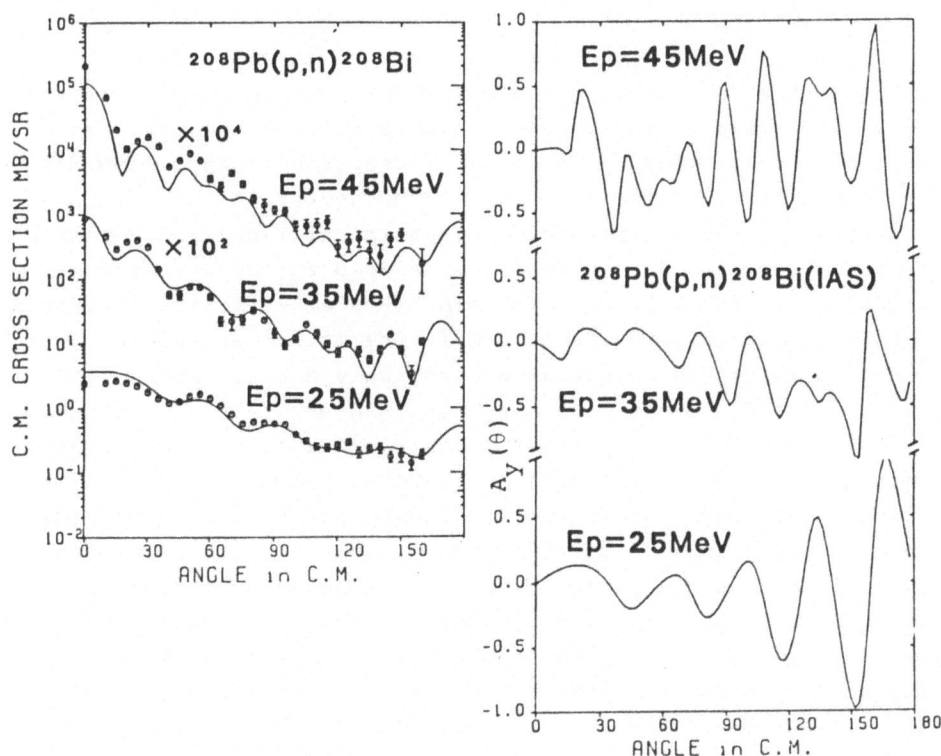

FIG. 11. Comparison between data and predictions for the
$^{208}Pb(p,n)^{208}Bi$ (IAS), using the isovector potentials shown
in Fig. 10. Predicted $A_y(\theta)$ are shown.

151

resulting in a very small net contribution. This is not the case in the absorptive V^{eff} with the result that the total calculated cross section is mainly due to the absorptive V^{eff}.

It is interesting to note that global parameterization of the proton OMP indicate that the absorptive term above 50 MeV is mainly of a volume form /32/. The empirical absorptive term of the isovector potential is mainly surface /19/ and thus it is assumed to be zero above 50 MeV. However, as noted by Lind et al. /15/, the (p,n) data for IAS transitions require both real and absorptive terms. The comparison presented here clearly indicates the need of a $W_1(r)$ term. Elastic scattering data and $A_y(\theta)$ values above 50 MeV should be able to help in determining this term of the isovector potential.

4. CONCLUSIONS

It seems that at energies above 100 MeV the isovector effective interactions used in DWIA calculations reproduce quite well IAS transitions. However, at energies between 30—100 MeV, the situation is different. In this presentation, I have concentrated on the determination of empirical isovector potentials at E = 40 and E = 30 MeV that may be used to compare with calculations of these potentials using several effective interactions. The density independent M3Y interaction or the one derived from the Paris potential provide only a real interaction. If supplemented by an imaginary coupling such as that obtained from empirical global analyses, their predictions are in very good agreement with the data /33/. Other calculations using a density dependent interaction including absorptive terms /14/ seem to come close to the data but further studies are needed. This study in a MIA approach provides shapes and strengths of local optical potentials which may be compared directly with predictions. Future high energy neutron scattering data and analyzing power measurements for the ($\vec{p}$,n) reaction to IAS transitions should be very valuable for our understanding of isovector effective interactions in the 30—100 MeV region.

REFERENCES

1. M. A. Franey and W. G. Love, Phys. Rev. C31 (1985) 488.
2. W. G. Love, in "The (p,n) Reaction and the Nucleon-Nucleon Force", ed. by C. Goodman, S. Austin, S. Bloom, J. Rapaport and G. Satchler, Plenum Press, New York, 1980, p. 23.
3. G. Bertsch, J. Borysowicz, H. McManus and W. G. Love, Nucl. Phys. A284 (1977) 399.
4. M. Lacombe et al., Phys. Rev. C21 (1980) 861.
   N. Anantaraman et al., Nucl. Phys. A398 (1983) 269.
5. H. V. von Geramb, F. A. Brieva and J. R. Rook, in "Microscopic Optical Potentials", Lect. Notes in Physics, Vol. 89, ed. by H. V. von Geramb, Springer-Verlag, Berlin, 1979, p. 104.
   H. V. von Geramb, in "Neutron-Nucleus Collisions, AIP Conf. Proc., Vol. 124, New York, 1984, p. 14.
6. S. M. Austin, see ref. /2/, p. 203.
7. J. Rapaport, see ref. /2/, p. 233.
8. W. G. Love, A. Klein and M. A. Franey, in "Telluride Conference on Antinucleon and Nucleon-Nucleus Interactions", 1985, to be published.
   J. Varr and F. Petrovich, in "Neutron-Nucleus Collisions", AIP Conf. Proc., Vol. 124, New York, 1984, p. 230.
9. T. N. Taddeucci, J. Rapaport, D. Bainum, C. Goodman, C. Foster, C. Gaarde, D. Horen and E. Sugarbaker, Phys. Rev. C25 (1982) 1094.
10. G. E. Brown, J. Speth and J. Wambach, Phys. Rev. Lett. 46 (1981) 1057.
11. A. Picklesimer and G. E. Walker, Phys. Rev. C17 (1978) 237.
12. Orihara et al., Annual Report CYRIC 1980, unpublished, 44 p.
13. S. D. Schery, see ref. /2/, p. 409.
14. F. S. Dietrich and F. Petrovich, in "Neutron-Nucleus Collisions", AIP Conf. Proc., Vol. 124, New York, 1984, p. 90.
15. D. Lind et al., in Proc. Conf. "Light Ion Reaction Mechanism", Osaka University, 1983, p. 206.
16. D. Bainum et al., Phys. Rev. Lett. 44 (1980) 1751.
17. A. M. Lane, Phys. Rev. Lett. 8 (1962) 197; Nucl. Phys. 35 (1962) 676.
18. A. M. Lane, Rev. Mod. Phys. 29 (1957) 191.
19. J. Rapaport, Phys. Rep. 87 (1982) 25.
20. E. Friedman and C. I. Batty, Phys. Rev. C17 (1978) 34.
21. W. Tornow et al., Nucl. Phys. A385 (1982) 373.
22. S. M. Austin, see ref. /14/, p. 527.
23. R. H. McCamis et al., Manitoba-OSU preprint, 1984.
24. B. Ridley and J. Turner, Nucl. Phys. 58 (1964) 497.
25. R. Alarcon et al., to be published.
26. B. A. Brown, S. E. Massen and P. E. Hodgson, J. Phys. G5 (1979) 1655.
27. N. Vinh Mau, in "Microscopic Optical Potentials", Lect. Notes in Physics, Vol. 89, ed. by H. V. von Geramb, Springer-Verlag, Berlin, 1979, p. 40.
28. F. Osterfeld and V. Madsen, see ref. /14/, p. 26.
29. W. T. H. Van Oers et al., Phys. Rev. C10 (1974) 307.
30. R. W. Finlay et al., Phys. Rev. C30 (1984) 796.

31. R. R. Doering, D. M. Patterson and A. Galonsky, Phys. Rev. C12 (1975) 378.
32. P. Schwandt et al., Phys. Rev. C26 (1982) 55.
33. W. G. Love, Phys. Rev. C15 (1977) 1261.
34. B. C. Clark et al., Phys. Rev. C30 (1984) 314.
35. N. Yamaguchi, S. Nagata and T. Mutsuda, Prog. Theor. Phys. 17 (1983) 459.

DESCRIPTION OF NEUTRON AND RADIATIVE STRENGTH FUNCTIONS IN THE QUASIPARTICLE-PHONON NUCLEAR MODEL

V.G. Soloviev

Laboratory of Theoretical Physics, Joint Institute for
Nuclear Research, Dubna, USSR

ABSTRACT. The basic assumptions of the quasiparticle-phonon nuclear model (QPNM) are expounded as well as its advantages in
describing nuclear states at low, intermediate and high excitation energies. It is shown that the spectroscopic factors of
reactions of the type (d,p) and the neutron strength functions
are determined by the fragmentation of one- and two-quasiparticle states. The QPNM calculations of neutron s-, p- and d-wave
strength functions in spherical and deformed nuclei are presented. A correct description of the relevant data is obtained.
The energy dependence of s- and d-wave strength functions for
the tin isotopes is explained. The results of calculations of
the spin splitting of s-wave strength functions are given. It
was found to be small, that is in agreement with experimental
data. The specific features of the calculations of radiative
strength functions for $E\lambda$ and $M\lambda$ transitions from neutron resonances to low-lying states of spherical and deformed nuclei are
discussed.

1. INTRODUCTION

The neutron and radiative strength functions define some properties of states lying near the neutron binding energy B_n. The
generally accepted description is statistical. An ever increasing amount of experimental data on non-statistical effects in
neutron resonances is reported every year /1,2/. Most of these
data refer to magic nuclei /3/. However, non-statistical effects
are clearly seen in some spherical and deformed nuclei /5/.

As is known, low-lying non-rotational states and high-lying
collective states of the giant resonance-type are well described within the microscopic models based on the introduction of

155

an average field, pairing and the effective interactions between quasiparticles. The states lying below and above the neutron resonances are successfully described microscopically
without the statistical treatment. The inclusion of neutron
resonances into the general scheme of microscopic calculations
of few-quasiparticle components of the excited-state wave functions has been performed within the quasiparticle-phonon nuclear model (QPNM) /5, 8/. The microscopic non-statistical description of some characteristics of the neutron resonances within
the QPNM has been recognized in ref. /9/ in the general study
of the damping of nuclear excitations.

In this report we dwell upon the basic assumptions of the
QPNM, review the calculations of s-, p- and d-wave neutron
strength functions and point out the specific features of the
calculations of radiative strength functions.

2. SPECIFIC FEATURES OF THE QUASIPARTICLE-PHONON NUCLEAR MODEL

The Hamiltonian of the QPNM includes the terms describing the
average nuclear field as the Saxon-Woods potential, pairing interactions, multipole and spin-multipole isoscalar and isovector including charge-exchange interactions. The general characteristics of the model Hamiltonian is given in refs. /6, 7/ for
deformed nuclei and in ref. /8/ for spherical nuclei.

Transforming the model Hamiltonian by the canonical Bogolubov transformation, one passes from nucleon operators to quasi-
particle operators α^+_{jm} and α_{jm}. The pairs of operators $\alpha^+_{jm}\alpha^+_{j_2m_2}$
and $\alpha_{j_2m_2}\alpha_{jm}$ are expressed through phonon operators, and the
quasiparticle operators remain only in the form $\alpha^+_{jm}\alpha_{j'm'}$. Such
an inclusion of the phonon operators overcomes difficulties
with double counting of some diagrams, which take place in the
nuclear field theory /10/. Then the RPA equations are solved
to determine the energies and wave functions of one-phonon
states. At this stage all the model parameters are fixed. Using
the experimental data to fix the constants of pairing, multi-
pole-multipole and spin-multipole - spin-multipole isoscalar
and isovector interactions.

The first specific feature of the QPNM is the use of one-

<u>phonon states as a basis</u> since the RPA provides a unique des-
cription of collective, weakly collectivized and two-quasiparticle states. By using the RPA the secular equations of the
model Hamiltonian are transformed to the form

$$H_M = \sum_{jm} \varepsilon_j \alpha^+_{jm} \alpha_{jm} + H_{\mathcal{V}} + H_{\mathcal{V}q} \, , \qquad (1)$$

which contains free quasiparticles and phonons and the quasiparticle-phonon interaction $H_{\mathcal{V}q}$. Formula (1) includes also the
np phonon operators describing charge-exchange giant resonances
and T excited states.

   <u>The second specific feature of the model is</u>: the quasipar-
ti-cle-phonon interaction is responsible for the fragmentation
of quasiparticle and collective·motion and thus for the compli-
cation of the nuclear state structure with increasing excitation energy.

   The excited state wave functions are represented as a series
in a number of phonon operators; in odd-A nuclei each term is
multiplied by a quasiparticle operator. The approximation consists in the cut-off of this series, that is the <u>third specific</u>
<u>feature of the model</u>. The cut-off of the series is the approxi-
mation similar to the cut-off of the chain of equations in the
Hartree-Fock-Bogolubov approximation. At present our expansion
is limited to two phonons. To elucidate the influence of many-
phonon terms of the wave functions on the calculated effects
is as difficult as to evaluate the role of neglected in the
Hartree-Fock-Bogolubov approximation chains of equations of the
many-body problem. It is stated in both the cases that approximate equations describe correctly the properties of nuclear excitations, and the terms neglected are partially taken into account by using constants fixed from the experimental data. In
the calculations the Pauli principle is taken into account
using exact commutation relations between the phonon and quasiparticle operators.

   <u>The fourth specific feature of the model is the use of the</u>
<u>strength function method</u>. By using a version of the strength
function method developed in ref. /6/, one can directly calculate the reduced transition probabilities, spectroscopic fac-

tors, transition densities, cross sections and other nuclear characteristics without solving the relevant secular equations. The use of the strength function method reduces the computer time by 10^3 times, and it becomes possible to calculate the fragmentation of one-quasiparticle, quasiparticle ⊗ phonon and one-phonon states for many nuclei. Calculations of characteristics of highly excited states are performed for spherical nuclei with closed and open shells and for deformed nuclei.

The general scheme of calculations within the QPNM is the following. The wave functions of the excited states of odd-A and doubly even spherical nuclei are written as

$$\Psi_\nu(JM) = C_{J\nu}\left\{ \alpha^+_{JM} + \sum_{\lambda i j} D^{\lambda i}_j(J\nu)\left[\alpha^+_{jm}Q^+_{\lambda\mu i}\right]_{JM} + \right.$$

$$+ \sum_{\lambda_1 i_1 \lambda_2 i_2}\sum_{jI} F^{\lambda_1 i_1 \lambda_2 i_2}_{jI}(J\nu) \cdot$$

$$\left. \cdot \left[\alpha^+_{jm}\left[Q^+_{\lambda_1\mu_1 i_1}Q^+_{\lambda_2\mu_2 i_2}\right]_{IM}\right]_{JM'}\right\}\Psi_0 , \qquad (2)$$

$$\Psi_\nu(JM) = \left\{ \sum_i R_i(J\nu)Q^+_{\lambda\mu i} + \sum_{\lambda_1 i_1 \lambda_2 i_2} P^{\lambda_1 i_1}_{\lambda_2 i_2}(J\nu) \cdot \right.$$

$$\left. \cdot \left[Q^+_{\lambda_1\mu_1 i_1}Q^+_{\lambda_2\mu_2 i_2}\right]_{JM}\right\}\Psi_0 , \qquad (3)$$

where Ψ_0 is the ground state wave function of a doubly even nucleus (phonon vacuum); $Q^+_{\lambda\mu i}$ is the phonon creation operator. Then we find an average value of (1) over (2) or (3). Using the variational principle and taking into account the normalization of the wave function (2) or (3) we get the secular equation for finding the energies η_ν of excited states, and write it down as

$$F(\eta_\nu) = 0. \qquad (4)$$

We also get the system of equations for the coefficients of the wave functions (2) or (3).

3. SPECTROSCOPIC AND NEUTRON STRENGTH FUNCTIONS

The spectroscopic factors of the one-neutron transfer reactions of the type (d,p) and the neutron strength functions for doubly even targets-nuclei are defined by the fragmentation of one-quasiparticle states. We present the spectroscopic strength functions of excited particle states in odd-A spherical nuclei. According to ref. /11/ the strength function for the fragmentation of the j subshell is

$$c_j^2(\eta) = \sum_\nu c_{j\nu}^2 \frac{1}{2\pi} \frac{\Delta}{(\eta - \eta_\nu)^2 + \Delta^2/4} =$$

$$= \frac{1}{\pi} \operatorname{Jm} F^{-1}(\eta + i\Delta/2) , \tag{5}$$

where $C_{j\nu}$ enters into (2), Δ is the averaging parameter. The spectroscopic function of the excited state j in the (d,p) reaction is

$$c^2 S = (2j + 1)U_j^2 \int_{\Delta E} c_j^2(\eta)d\eta , \tag{6}$$

where U_j is the Bogolubov transformation coefficient.

The fragmentation of one-quasiparticle states in odd-A spherical nuclei has been calculated within the QPNM in refs. /11-14/ and other papers. A qualitatively correct description of the relevant experimental data was obtained. The calculated fragmentation in some cases has been confirmed by the experimental data obtained later. The fragmentation of particle neutron states has been calculated in ref. /15/. The fragmentation of neutron subshells $2f_{5/2}$ in ^{121}Sn and $2h_{11/2}$ in ^{209}Pb is shown in Fig. 1. It is seen from this figure that at the neutron binding energy B_n in ^{121}Sn there is a maximum in the $2f_{5/2}$ strength distribution and in ^{209}Pb at $\eta = B_n$ there is a tail of $2h_{11/2}$ strength distribution. The results of calculations /15/ provide a qualitatively correct description of the available experimental data /16/ on the fragmentation of high-lying neutron one-quasiparticle states.

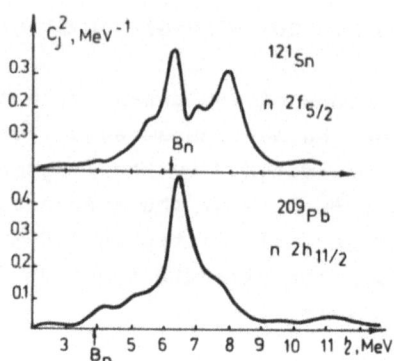

FIG. 1.

Now we give the neutron strength functions. If a neutron
with orbital momentum 1 is captured by a target with spin J_0
the neutron strength function is determined as

$$S_1 = \sum_{Jj} g(J,1)S_1^{Jj} , \qquad (7)$$

where

$$g(J,1) = \frac{1}{21 + 1} \frac{2J + 1}{2(2J_0 + 1)} ,$$

S_1^{Jj} is the partial strength function with the spin of the compound-nucleus equal to $\vec{J} = \vec{J}_0 + \vec{1} + \vec{1/2} = \vec{J}_0 + \vec{j}$ in the channel j. For the doubly even target $J_0 = 0$ and $J = j$. Then S_1^{Jj}
is expressed through $C_j^2(\eta)$ as follows

$$S_1^{Jj} \equiv S_1^j = \frac{1}{\Delta E} \sum_{\nu \Delta E} \Gamma_{n\nu}^{ol}(j) = \frac{\Gamma_{s.p.}^{ol}}{\Delta E} U_j^2 \int_{\Delta E} C_j^2(\eta)d\eta, \qquad (8)$$

where $\sum_{\nu \Delta E} \Gamma_{n\nu}^{ol}$ is the sum of reduced neutron widths in the energy interval ΔE, and $\Gamma_{s.p.}^{ol}$ is the single-particle neutron width
for the Saxon-Woods potential. The comparison of formulae (6)
and (8) shows that the spectroscopic functions of the (d,p) re-

160

action and the neutron strength functions are determined by the fragmentation of one-quasiparticle states. Therefore, in the QPNM they are calculated in a similar way.

If a neutron is captured by an odd-N target with the wave function $\alpha^+_{J_0 M_0} \Psi_0$, the partial strength function with spin of the compound-nucleus has the form

$$S^{Jj}_1 = \frac{\Gamma^{ol}_{s.p.}}{\Delta E} \sum_{\nu \Delta E} \left| R_i(J\nu) \phi_i \right|^2 , \qquad (9)$$

$$\phi_i = \sum_n U_{nlj} \Psi^{Ji}_{nlj,n_0 l_0 j_0} , \qquad (9')$$

where quantum numbers of the single-particle states are denoted by nlj and for the ground state of the target-nucleus by $n_0 l_0 j_0$; $\Psi^{Ji}_{nlj,n_0 l_0 j_0}$ are the phonon amplitudes /17/ which are found from the RPA solutions. It is seen from (9) and (9') that S^{Jj}_1 depends, first, on the value of one-phonon terms R in the wave functions (3), second, on the amplitudes $\Psi^{Ji}_{nlj,n_0 l_0 j_0}$ entering into the phonon Q^+_{JMi}, and third, on the coefficient U_{nlj}. The U_{nlj} is close to unity for high-lying particle states and small for hole states. Following ref. /18/ we get

$$S^{Jj}_1 = \frac{\Gamma^{ol}_{s.p.}}{\Delta E} \gamma^2_{Jj}(\eta) = \frac{\Gamma^{ol}_{s.p.}}{\Delta E} \frac{1}{\pi} .$$

$$\cdot \, \mathrm{Im} \left\{ \frac{\sum_{ii'} M_{ii'}(\eta + i\Delta/2) \phi_i(\eta + i\Delta/2) \phi_{i'}(\eta + i\Delta/2)}{F(\eta + i\Delta/2)} \right\} , \qquad (10)$$

where $M_{ii'}$ are the minors of eq. (4). The rank of the determinant (4) is equal to the number of one-phonon terms in the wave function (3). It should be noted that the imaginary part in formula (10) coincides with that in formula (2.20) of ref. /19/ describing the fragmentation of two-quasiparticle states in spherical nuclei.

4. NEUTRON STRENGTH FUNCTIONS IN SPHERICAL NUCLEI

The dependence of neutron strength functions on mass number A is defined by the corresponding single-particle states. The neutron strength function is maximal when the subshell is in the B_n region and minimal when the subshell energy is below the Fermi energy. The QPNM calculations of s- and p-wave neutron strength functions of spherical nuclei have been made /14, 18, 20—23/ for nuclei where the strength functions decrease from maximal value to minimal one, i.e. where the corresponding neutron subshells are bound. The results of calculations, the experimental data and the neutron binding energies B_n for compound nuclei are given in Table 1. The experimental data are taken mainly from ref. /23/, and from recent papers /24—26/. It is seen from the table that a correct description of the experimental data is obtained for most of the cases. In contrast with the optical model, there are no difficulties in describing s-wave function in the region of its minimum at $A \approx 120$. It should be noted that there are no free parameters when calculating the neutron strength functions within the QPNM. Some freedom is due to the averaging parameter Δ, in most calculations $\Delta = 0.4$—0.6 MeV. Now we proceed to discuss the spin-dependence of neutron strength functions. If s-wave neutron is captured by an odd-N target with spin J_0, the compound states with spins $J = J_0 \pm 1/2$ are excited. The corresponding strength functions are denoted by $S^{\pm} = S^{J_0 \pm 1/2, 1/2}$. They may differ from each other due to different fragmentation of two-quasiparticle states with $J_0 + 1/2$ and $J_0 - 1/2$. The calculations /18/ of the functions $\gamma_{J1/2}^2(\eta)$ with $J = 2$ and 3 for the compound nucleus ^{96}Mo are shown in Fig. 2. It is seen that the fragmentation of two-quasiparticle states $\left\{ 2d_{5/2}, 3s_{1/2} \right\}$ with $J = 2$ and $J = 3$ is different in the low-energy region and close at energies above 5 MeV. The latter fact caused the proximity of values of S^+ and S^- for ^{96}Mo. The calculations /18/ of the spin dependence of s-wave strength functions are given in Table 2. For most nuclei $S^+ \approx S^-$, that is in agreement with the analysis of experimental data in ref. /28/.

In ref. /26/ the p-wave neutron strength functions have been

TABLE 1. The s- and p-wave neutron strength functions

Target A element	B_n (MeV)	$S_0 \cdot 10^4$ Exp.	$S_0 \cdot 10^4$ Calc.	$S_1 \cdot 10^4$ Exp.	$S_1 \cdot 10^4$ Calc.
53 Cr	9.72	4.5±1.1	4.5		
54	6.25	2.8±1.0	2.3	0.67±0.11[a]	0.08
54	9.29	8.7±2.4	8.8	0.58±0.11	0.18
56	7.64	2.6±0.6	3.9	0.45±0.05	0.20
58 Ni	9.00	2.8±0.6	2.0	0.5±0.1	0.08
60	7.82	2.7±0.6	3.1	0.3±0.1	0.20
61	10.6	3.2±0.8	2.5	–	0.10
62	6.84	2.8±0.7	2.5	0.3±0.1	0.03
73	10.2	1.5±0.4	1.6	–	–
74	6.5	1.5±0.7	3.0	–	–
91 Zr	8.63	0.36±0.08	0.6	–	–
92	6.76	0.50±0.10	2.0	–	–
95	9.15	0.35±0.07	0.5	–	–
96	6.82	0.43±0.14	0.8	–	–
116 Sn	6.94	0.18±0.04[b]	0.19	3.76[b]	3.8
118	6.48	0.16±0.05[b]	0.15	3.23[b]	4.2
120	6.18	0.06±0.04[b]	0.11	2.2[b]	5.8
122	5.94	0.17±0.05[b]	0.15	3.1[b]	5.6
124	5.73	0.19±0.03[b]	–	4.1[b]	4.6
124 Te	6.57	0.63±0.20	0.20	2.0	1.6
126	6.35	0.28±0.10	0.12	1.6	1.4
204 Pb	6.73	0.93[c]	1.0	0.23±0.04	0.2
206	6.74	1.06±0.26	0.8	0.32±0.04	0.2
207	7.37	0.76±0.31	1.1	–	–
208	3.94	0.85[c]	1.3	–	–

a - Ref. /25/, b - Ref. /26/, c - Ref. /27/.

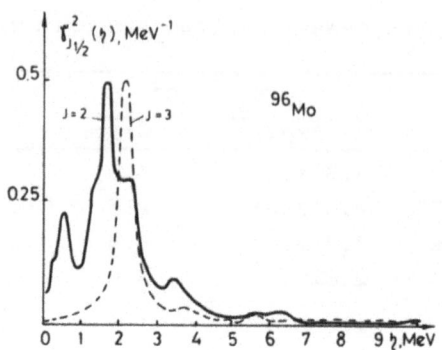

FIG. 2.

TABLE 2. Spin dependence of s-wave strength functions

Target	$S_o^+ \cdot 10^4$	$S_o^- \cdot 10^4$	$2\dfrac{S_o^+ - S_o^-}{S_o^+ + S_o^-}$
^{53}Cr	4.64	4.37	0.06
^{61}Ni	2.46	2.40	0.02
^{87}Sr	0.7	1.1	-0.44
^{91}Zr	0.65	0.58	0.11
^{95}Mo	0.55	0.42	0.27
^{97}Mo	0.86	0.75	0.24
^{177}Hf	2.0	1.9	0.06
^{179}Hf	1.2	1.1	0.10
^{183}W	1.3	1.2	0.08

experimentally determined for spins $\mathtt{J} = 1/2$ $S_1^{1/2}$ and $\mathtt{J} = 3/2$ $S_1^{3/2}$ in the region $A \approx 50{-}130$ and the spin-orbital splitting of the 2p maximum was observed. The experimental data /26/ and the results of calculations performed by V.V.Voronov for $S_1^{1/2}$ and $S_1^{3/2}$ for the tin isotopes are given in Table 3. It is seen from the table that $S_1^{3/2}$ is considerably less than $S_1^{1/2}$, be-

TABLE 3. Strength functions $S_1^{1/2} \cdot 10^4$ and $S_1^{3/2} \cdot 10^4$

Target	$S_1^{1/2} \cdot 10^4$		$S_1^{3/2} \cdot 10^4$		$S_1^{1/2}/S_1^{3/2}$	
	Exp.	Calc.	Exp.	Calc.	Exp.	Calc.
^{116}Sn	7.0±1.3	5.3	2.14±0.15	3.1	3.27	1.7
^{118}Sn	5.7±1.7	5.3	1.96±0.22	3.7	2.91	1.4
^{120}Sn	2.4±1.5	6.8	2.02±0.18	5.3	1.19	1.3
^{122}Sn	4.9±1.5	7.8	2.07±0.21	4.5	2.37	1.7
^{124}Sn	10.5±1.0	7.9	1.35±0.18	3.0	7.75	2.6

cause the $3p_{3/2}$ subshell is far from B_n than $3p_{1/2}$. The calculations also indicate the existence of the spin-orbital splitting.

Usually the neutron strength functions are assumed to be independent of energy. It follows from the fragmentation of one- and two-quasiparticle states that the neutron strength functions may depend on the excitation if there are substructures in the strength distribution of the relevant subshells near B_n. Indeed, according to the calculations /14/ in ^{207}Pb there are substructures in the strength distribution of the subshells $4s_{1/2}$, $3d_{5/2}$ and $3d_{3/2}$ at the energy 0.4—0.6 MeV above B_n, i.e.

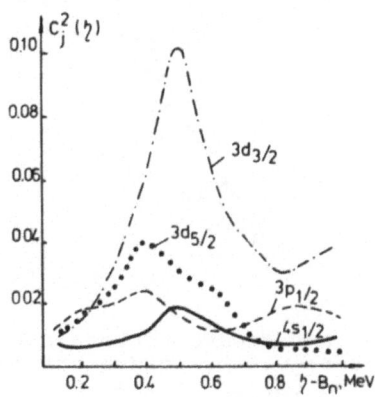

FIG. 3.

at the excitation energies 7.1–7.3 MeV. These substructures comprise a smaller part of the strength of the subshells $4s_{1/2}$ $3d_{5/2}$, $3d_{3/2}$ and $3p_{1/2}$; They are shown in Fig. 3. The experimental /29/ and calculated /14/ sums of reduced neutron widths $\sum \Gamma_n^{ol}$ for the s- and d-wave strength functions in ^{206}Pb+n are presented in Fig. 4. The strength function S_l^j is determined

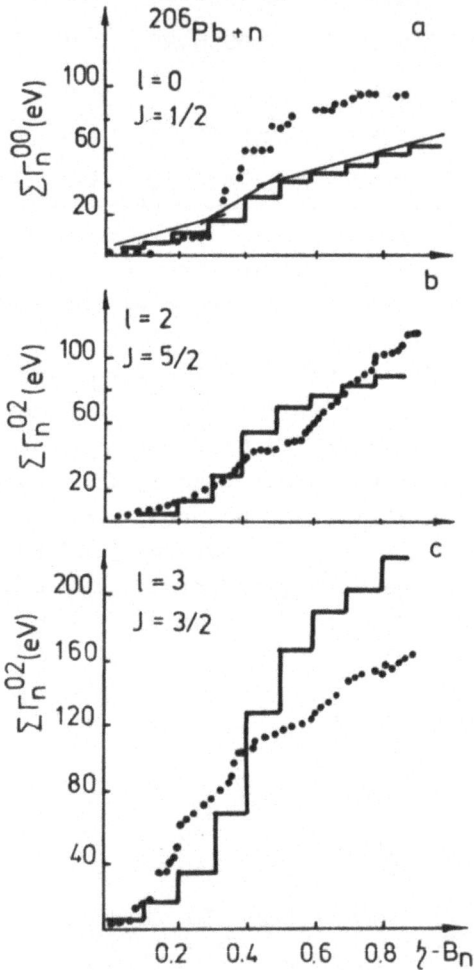

FIG. 4. Sum of the s- and d-wave reduced neutron widths $\sum \Gamma_n^{ol}$ plotted vs. neutron energy, the points are experiment /29/, the solid lines are the calculation within the QPNM.

from the slope of $\sum \Gamma_n^{ol}$ versus the neutron energy. The s-wave neutron strength function in ^{207}Pb changes sharply near 0.3 and

0.5 MeV. The calculations predict a smaller change in the s-wave strength functions at these energies. The change in slope of $\sum_n \Gamma_n^{02}$ vs. $\eta - B_n$ occurs in d-wave strength functions at 0.4 MeV. The QPNM provides a correct description of the energy dependence of the s- and d-wave neutron strength functions in ^{207}Pb. According to the calculations /22/ shown in Fig. 5, the sum widths $\sum_n \Gamma_n^{00}$ for the s-wave functions change

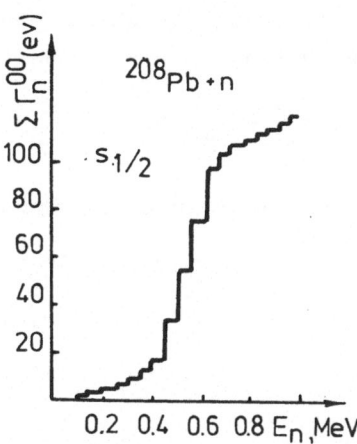

FIG. 5. Calculation within the QPNM of the sum of s-wave reduced neutron widths $\sum_n \Gamma_n^{00}$ vs. neutron energy.

in slope versus the neutron energy at 0.5 and 0.7 MeV in the ^{208}Pb+n. This change is caused by the substructure at the excitation energy 4.5 MeV in ^{209}Pb in the $4s_{1/2}$ strength distribution. According to the calculations /22/ this substructure in the ^{204}Pb+n is smeared. For the s-wave strength function this result is confirmed experimentally /27/. The change in slope of $\sum_n \Gamma_n^{01}$ vs. E_n is observed /3/ for the s- and d-wave functions in the ^{207}Pb+n. According to the calculations /14/ the change in d-wave function is caused by a substructure in the strength distribution of two-quasiparticle states $3p_{1/2}$, $3d_{3/2}$ in ^{208}Pb.

5. NEUTRON STRENGTH FUNCTIONS IN DEFORMED NUCLEI

The neutron strength functions in deformed nuclei have been

167

calculated in refs. /30, 31/. The wave function of an odd-N deformed nucleus is used in the form

$$\Psi_\nu(K) = \frac{1}{\sqrt{2}} \sum_{\mathsf{G}} \left\{ \sum_{q_0} C_{q_0\nu} \alpha^+_{q_0\mathsf{G}} + \sum_{\lambda\mu i q} D_q^{\lambda\mu i} \alpha^+_{q\mathsf{G}} Q^+_{\lambda\mu i} \right\} \Psi_0 .$$ (11)

The single particle wave functions of the deformed Saxon-Woods
potential are defined by the quantum numbers $q\mathsf{G}$, $\mathsf{G} = \pm 1$ and are
expressed as an expansion over the shell functions φ_{nlj} of
spherical-symmetric potential

$$\varphi_q^K = \sum_{nlj} a^{qK}_{nlj} \varphi_{nlj}, \quad a^{qK}_{1j} = \sum_n a^{qK}_{nlj} .$$

In case of a doubly even target nucleus the neutron strength
function has the form

$$S_1 = \Gamma^1_{s.p.} \sum_{Kj} S^K_{1j} ,$$ (12)

where

$$S^K_{1j} = \frac{1}{\Delta E} \sum_{\nu\Delta E} \left| \sum_{Kj} \sum_{q_0} a^{q_0K}_{1j} U_{q_0} C_{q_0\nu} \right|^2$$ (13)

is the spectroscopic function of the (d,p) reaction. The explicit form of the strength function S^K_{1j} is presented in ref. /30/.

The calculated /30, 31/ and experimental /24/ s-, p- and d-
wave neutron strength functions are listed in Tables 4 and 5.
The results of calculations are in agreement with experimental
data. The spin dependence of s-wave functions has been calculated in ref. /31/; the results of calculations are shown in
Table 2. It is seen from this table that $S^+ \approx S^-$, that is in
agreement with experimental data obtained in ref. /32/ for
other deformed nuclei.

TABLE 4. The s- and p-wave neutron strength functions

Target A .element	B_n (MeV)	$S_0 \cdot 10^4$ Exp.	Calc.	$S_1 \cdot 10^4$ Exp.	Calc.
150 Nd	5.33	3.0 ± 0.4	1.5	0.8 ± 0.2	1.8
154 Sm	5.81	1.8 ± 0.4	1.0	0.8 ± 0.2	1.0
156 Gd	6.36	1.7 ± 0.3	1.1	0.55±0.10	1.2
158	5.94	1.5 ± 0.2	1.1	-	1.2
160	5.64	1.6 ± 0.3	1.0	0.5 ± 0.1	1.1
162 Dy	6.27	1.8 ± 0.3	1.8	1.1 ± 0.4	1.0
164	5.72	1.70±0.25	1.8	1.3 ± 0.3	0.9
162 Er	6.91	2.1 ± 0.5	3.0	-	0.7
164	6.65	1.4 ± 0.3	3.0	-	0.7
166	6.44	1.6 ± 0.2	3.0	0.94±0.16	0.7
168	6.00	1.5 ± 0.2	3.0	0.95±0.20	0.7
170	5.68	1.5 ± 0.2	3.0	0.94±0.20	0.9
170 Yb	6.61	2.4 ± 0.3	3.0	-	0.7
174	5.82	1.6 ± 0.2	2.5	-	0.6
176	5.57	2.3 ± 0.3	2.4	-	0.7
178 Hf	6.10	2.2 ± 0.7	2.0	0.51±0.03	0.7
180	5.70	1.9 ± 0.6	1.6	0.44±0.05	0.6
182 W	6.19	2.2 ± 0.3	2.6	0.72±0.06	0.6
184	5.75	2.5 ± 0.2	2.7	0.58±0.07	0.6
186	5.47	2.1 ± 0.3	2.3	0.37±0.05	0.6
188 Os	6.29	2.2 ± 0.3	2.6	0.25±0.02	0.4
230 Th	5.12	1.5 ± 0.4	1.1	-	0.7
232	4.79	0.84±0.07	0.8	1.48±0.07	1.0
232 U	5.75	0.91±0.20	0.9	-	1.0
234	5.30	0.86±0.11	1.0	-	1.2
236	5.13	1.1 ± 0.1	1.2	2.3 ± 0.6	1.6
238	4.81	1.2 ± 0.1	1.3	1.7 ± 0.3	2.0
238 Pu	5.65	1.3 ± 0.3	1.0	-	1.6
240	5.24	0.93±0.08	1.0	2.8 ± 0.8	1.6
242	5.03	0.9 ± 0.1	1.1	-	1.4
242 Cm	5.70	0.9 ± 0.3	1.0	-	1.0

TABLE 5. The d-wave neutron strength functions

| Target | $S_2 \cdot 10^4$ | |
	Exp.	Calc.
^{156}Gd	2.6±0.4	1.2
^{158}Gd	1.9±0.3	1.0
^{160}Gd	1.3±0.3	1.2
^{166}Er	4.7±0.7	5.7
^{168}Er	3.0±0.5	5.0
^{170}Er	2.4±0.3	4.0
^{178}Hf	1.7±0.1	3.0
^{180}Hf	1.8±0.1	2.0
^{182}W	1.8±0.1	2.0
^{184}W	1.4±0.1	1.3
^{186}W	0.9±0.1	1.2

6. RADIATIVE STRENGTH FUNCTIONS

The partial radiative strength functions for $E\lambda$ and $M\lambda$ transitions from neutron resonances to the ground states of doubly even nuclei are determined by the fragmentation of one-phonon configurations. Their description within the QPNM is performed in refs. /33, 34/ with the wave function in the form (2) for neutron resonances. A good description of the relevant experimental data was obtained. The influence of the giant dipole resonance on E1 transitions from neutron resonances has been studied in refs. /14, 33/. It was shown that the effect of the giant dipole resonance on the probabilities of E1 transitions is small in single- and double-closed-shell nuclei and it is considerable in open-shell spherical nuclei.

To describe the partial gamma transitions from neutron resonances to the ground and low-lying one-quasiparticle states in odd-mass nuclei one should know the fragmentation of one-quasiparticle and quasiparticle ⊗ phonon configurations. Within the QPNM the E1 and M1 transitions have been calculated in the Fe and Ni isotopes /23/. The role of valence transitions was analysed and agreement with the relevant experimental data was obtained. At present efforts are made to describe within the QPNM the partial radiative strength functions for gamma transitions from neutron resonances, deep hole and other states.

The radiative strength functions of E1 and M1 transitions from ground states of doubly even deformed nuclei to states lying near B_n have been calculated in ref. /35/ within the QPNM. The k_{E1}- and k_{M1}-values are calculated for some deformed nuclei of the rare-earth and actinide regions. The calculated values of k_{E1} are by 1.5—2 times larger and the values of k_{M1} are somewhat less than the average values obtained in /36/ from the analysis of the available experimental data.

In ref. /37/ the wave function of a highly excited state is represented as a series in the number of quasiparticles and phonons, and it is predicted in what cases one can observe the correlations between the partial radiative and reduced neutron widths as well as between the partial radiative widths at gamma transitions to different low-lying states. The corresponding experiments have been performed in refs. /4, 38/. The analysis of experimental data performed in refs. /4, 38/ has shown that for ^{154}Gd, ^{167}Er, ^{173}Yb, ^{176}Lu and ^{185}Re target nuclei there is a statistically significant correlation between the partial radiative widths and reduced neutron widths of neutron resonances. In case of neutron radiative capture in ^{173}Yb a strong correlation between various partial radiative widths has been observed. These experimental data are very important. They provide additional indications to the efficiency of representing the excited state wave functions as an expansion over the number of phonons, which underlies the QPNM.

7. CONCLUSION

The microscopic calculations of nuclear characteristics at low, intermediate and high excitation energies do not pretend to a "literal" agreement with experimental data. Such an agreement is achieved only in the phenomenological calculations by fitting the corresponding parameters. Within the QPNM the constants are fixed on the basis of experimental data in calculating the phonon basis. In calculating the neutron and radiative strength functions within the QPNM there are no free parameters.

The QPNM provides a correct description of the s-, p- and d-wave neutron strength functions in spherical and deformed nuclei and of the energy dependence of s- and d-wave neutron strength functions in the lead isotopes.

REFERENCES

1. A. M. Lane, in Neutron Capture Gamma-ray Spectroscopy, Reactor Centrum Nederland, Petten, 1975, p. 31.
   S. F. Mughabghab, ibid., p. 81.
2. A. M. Lane, Proc. Int. Conf. on Interactions of Neutrons with Nuclei, Univ. Lowell, Lowell, Mass., 1976, p. 525.
3. D. J. Horen et al., Phys. Rev. C18 (1978) 722.
4. F. Bečvář et al., Yad. Fiz. 33 (1981) 3.
   F. Bečvář, in Neutron Induced Reactions, edited by P. Obložinský, Institute of Physics, Bratislava, 1982, p. 171.
   F. Bečvář, in Capture Gamma-ray Spectroscopy and Related Topics 1984, AIP Conf. Proc. No. 125, edited by S. Raman, 1985, p. 345.
5. V. G. Soloviev, Izv. Akad. Nauk SSSR, Ser. Fiz. 35 (1971) 666.
6. V. G. Soloviev, Particles and Nuclei 9 (1978) 810; Nukleonika 23 (1978) 1149.
7. L. A. Malov and V. G. Soloviev, Particles and Nuclei 11 (1980) 810.
8. A. I. Vdovin and V. G. Soloviev, Particles and Nuclei 14 (1983) 237.
   V. V. Voronov and V. G. Soloviev, Particles and Nuclei 14 (1983) 1381.
   A. I. Vdovin et al., Particles and Nuclei 16 (1985) 245.
9. G. F. Bertsch et al., Rev. Mod. Phys. 55 (1983) 287.
10. D. Bes et al., Nucl. Phys. A260 (1976) 77.
11. V. G. Soloviev et al., Nucl. Phys. A342 (1980) 261.
12. Chan Zuy Khuong et al., J. Phys. G: Nucl. Phys. 7 (1981) 151.
13. Nguen Dihn Thao et al., Yad. Fiz. 37 (1983) 43.
14. V. G. Soloviev et al., Nucl. Phys. A399 (1983) 141.
15. A. I. Vdovin and Ch. Stoyanov, this volume.
16. S. Gales et al., Phys. Lett. 144B (1984) 323.
17. V. G. Soloviev, "Theory of Complex Nuclei", Nauka, Moscow, 1971 (translation: Pergamon Press, New York - Oxford, 1976).
18. V. V. Voronov et al., Yad. Fiz. 31 (1980) 327.
19. V. G. Soloviev et al., Nucl. Phys. A370 (1981) 13.
20. D. Dambasuren et al., J. Phys. G: Nucl. Phys. 2 (1976) 25.
21. V. V. Voronov, 4th School on Neutron Physics, JINR Publ., Dubna, 1982, p. 105.
22. V. V. Voronov and Ch. Stoyanov, Preprint JINR P4-85-3, Dubna, 1985.
23. V. G. Soloviev and Ch. Stoyanov, Nucl. Phys. A382 (1982) 206.
24. Neutron Cross Section, Vol. 1, Parts A and B, 4th ed., edited by S. Mughabghab et al., Academic Press, New York,

1981, 1984.
25. H. M. Agrawab et al., Phys. Rev. C30 (1984) 1880.
26. N. G. Nikolenko et al., in Nuclear Data for Science and
    Technology, Antwerpen, 1982, p. 781.
27. D. J. Horen et al., Phys. Rev. $\underline{C29}$ (1983) 2126.
28. L. Lason et al., Acta Phys. Pol. $\underline{B8}$ (1977) 1009.
29. D. J. Horen et al., Phys. Rev. $\underline{C20}$ (1979) 478; $\underline{C24}$ (1981)
    1961.
30. L. A. Malov and V. G. Soloviev, Nucl. Phys. $\underline{2}$ (1976) 87.
31. L. A. Malov and D. G. Yakovlev, Preprint JINR P4-85-270,
    Dubna, 1985.
32. V. P. Alfimenkov et al., Nucl. Phys. $\underline{A376}$ (1982) 229.
33. V. G. Soloviev et al., Nucl. Phys. $\underline{A304}$ (1978) 503.
34. V. G. Soloviev et al., Phys. Lett. $\underline{79B}$ (1978) 187.
35. L. A. Malov et al., Z. Phys. $\underline{A320}$ (1985) 521.
36. C. McCullagh et al., Phys. Rev. $\underline{C23}$ (1981) 1394.
37. V. G. Soloviev, Particles and Nuclei $\underline{3}$ (1972) 770.
38. F. Bečvář et al., Yad. Fiz. $\underline{34}$ (1982) 1158; $\underline{37}$ (1983) 1357.

A METHOD TO DETERMINE THE PAIRING ENERGY FOR EXCITED NUCLEI

G. Rohr

CEC-JRC, Central Bureau for Nuclear Measurements, Geel, Belgium

ABSTRACT. The method is based on the Bethe level density expression, where the parameters are used with constraints. It is applied to the level density of compound resonances observed at neutron separation energy. The change of the pairing energy compared to the ground state value determined by mass differences is discussed for different mass regions.

1. INTRODUCTION

The most important consequence of a residual interaction between nucleons is the existence of the pairing force. It is manifested in a special high separation energy of paired nucleons of the same kind, and in all even-even nuclei having a zero angular momentum in the ground state. The pairing force is also responsible for the energy gap in the excitation spectrum of nuclei, reflecting the need to break a pair of nucleons before the nucleus can be excited at low energy. The correlation of pairs in the single particle level scheme results in a lower ground state energy, so that all states in even nuclei are shifted by the pairing energy. Taking this so-called even-odd effect of the level density into account, the level density for different types of nuclei can be expressed:

$$\rho_{odd-odd}(E) = \rho_{odd-A}(E+\Delta) = \rho_{even-even}(E+2\Delta), \qquad (1)$$

where E is the excitation energy in an odd-odd nucleus. The Δ-values (for nuclei in the ground state) are determined empirically and the best available information of the gap parameter is obtained from a study of even-odd mass differences /1/

174

There is no empirical evidence, except for very high spin nuclear states (backbending), that the pairing changes with energy. However there are calculations in the framework of the
independent quasi-particle model (based on the BCS-treatment)
of a reduction in the pairing energy due to the blocking effect
/2/ and a breakdown in the pairing correlation (Δ = 0) at a
certain critical nuclear temperature /3/. Also the inclusion
of other residual interactions such as neutron-proton interaction (which counteracts the pairing correlation /4/) and alpha
clustering /5/ may influence the gap parameters. In this contribution a method will be described in order to determine the
pairing energy at neutron separation energy /6/. In doing this
the resonance data newly evaluated at Geel have been used /7/.

2. DESCRIPTION OF THE METHOD

According to the relationship (1) discussed in the introduction,
the level density can be used to determine the pairing energy.
For this purpose the level density of odd-odd nuclei will play
a central role. These nuclei have Δ = 0 in the ground state and
it is to be expected that this value will not change with energy. Therefore the level density for the odd-odd nuclei far
from closed shells can be used to define a base line for the
level density of all nuclei with no residual interaction. Deviations from the base line for even nuclei will be interpreted
as a change of the pairing energy compared to the ground state
value.

For the level density systematics at neutron separation energy the Bethe expression /10/ has been used for the following
reasons:

1. The observed level density of nuclear states beyond neutron separation energy for non-closed shell nuclei can be fitted very well with the Bethe level density expression /8/.

2. As shown by Grimes /9/ the use of a Gaussian distribution
with a large basis of eigenvalues, and the energy dependence of
the Bethe formula are almost indistinguishable over a wide range
of energy ($\Delta E \approx 20$ MeV) if the parameters are chosen properly.

The level density, at excitation energy $U = E_B + E_{max}/2$, and spin of the compound state J, is given by /10/

$$\rho(U,J) = (2J + 1)\exp\left\{2[a(U-\Delta)]^{1/2} - J(J+1)/2\sigma^2\right\}/(24 \times 2^{1/2}a^{1/4}\,^3(U-\Delta)^{5/4}), \qquad (2)$$

where E_B is the neutron separation energy and E_{max} the highest energy of the resonances used to determine the level density. The parameters in expression (2) are taken with the following constraints: spin cut-off factor: $\sigma = (0.1045 \quad atA^{2/3})^{1/2}$, nuclear temperature: $t = [(U-\Delta)/a]^{1/2}$.

The spin cut-off factor has been determined with $I_{eff} = 0.7 \times I_{rig}$ and takes into account the fact that the moment of inertia of the nucleus in the superfluid state is about 30% smaller than for a rigid body of the same radius. The pairing energy Δ is taken from ref. /1/.

According to the Fermi gas model the level density parameter a is proportional to the level density of single particle states at the Fermi surface energy $g = 6a/\pi^2$.

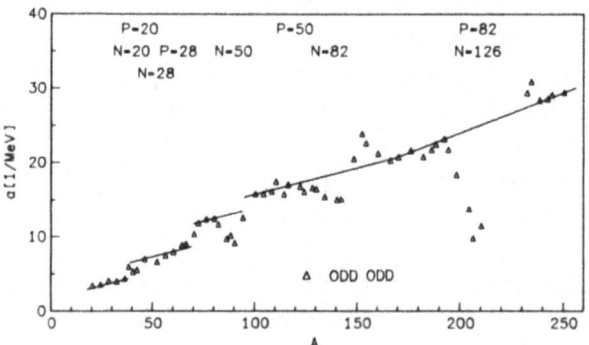

FIG. 1.

In Fig. 1 the level density parameter for odd-odd nuclei together with the step-like base line are given. Two different structures for a as a function of the mass number A can be distinguished:

1. The usual closed shell effects at specific numbers of neutrons and protons reflected in minima in the level density at mass numbers A = 40, 90, 140, 208, and pronounced maxima at

A = 155 (N = 90) and 230 (Z = 90) indicating an increase in the
level density of single particle states at the Fermi surface
energy.

2. Distinct steps in the level density parameter at A = 38,
69, and 93 which indicate that only a small number of p-h's are
participating in the excitation. It has been shown /6/ that the
calculated level densities for the doorway states 2p1h agree
with the level density observed below A = 39. With increasing
A and at constant excitation energy more p-h excitations become
energetically possible and the level density increases and
causing the jumps from 2p1h- to 3p2h-states at A = 38, from
3p2h- to 4p3h-states at A = 63 and from 4p3h- to 5p4h-states at
A = 94.

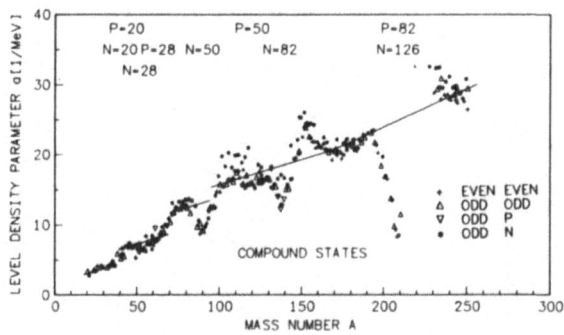

FIG. 2.

In Fig. 2 the level density parameter for all considered nuclei are presented. The deviations from the base line are discussed for a few examples in the following section.

3. APPLICATIONS OF THE METHOD

1. In Fig. 3 a part of the level density systematic at
A ≈ 75 is presented. All even nuclei have values which are
above (or on) the base line indicating an increase in the excitation energy due to a reduction of the pairing energy. The
average reduction of Δ is in agreement with the predicted
blocking effect and is for the first time reproduced experimentally.

177

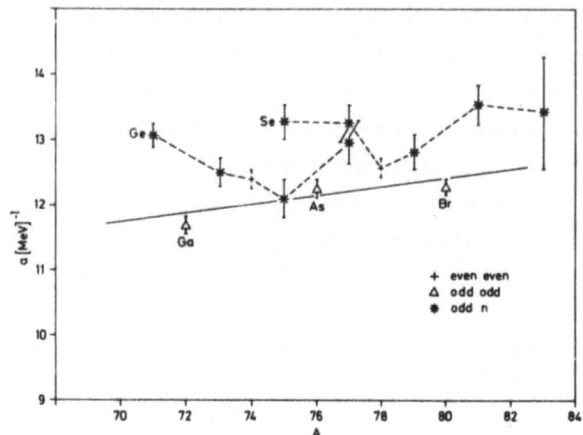

FIG. 3.

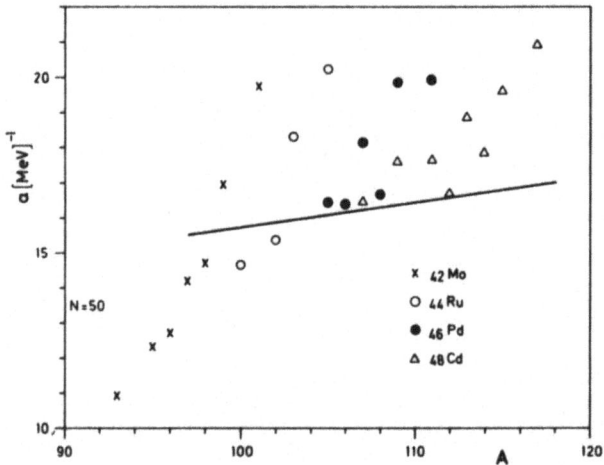

FIG. 4.

2. In Fig. 4 a very strong fluctuation of a-values around
A ≈ 105 is seen. The level density parameter for even-odd nuclei increases with the mass number for the isotope series Mo,
Ru, Pd, and Cd. The values with the largest deviation from the
base line at the end of the isotope series correspond to a
breakdown of the pairing correlation, i.e. assuming $\Delta = 0$, the
level density parameter approaches the base line. The shell pro-

perties of these nuclei and the increase in deformation for
some of the nuclei have led to an interpretation that this
breakdown may be caused by neutron-proton interaction which
counteracts the pairing force /6/.

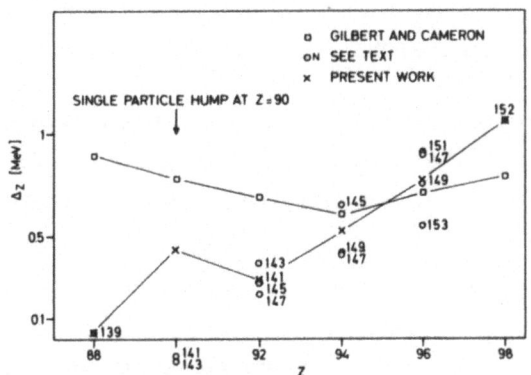

FIG. 5.

3. In Figs. 5 and 6 the pairing energies in the proton (Δ_Z)
and neutron (Δ_N) configurations determined by means of the
base line are presented as a function of the even proton and
even neutron number, respectively, for masses A ≻ 230. The number close to the experimental points is the neutron or proton
number of the nucleus for which the level density has been used.
The neutron pairing energy is determined from even-even nuclei
with Δ_Z taken from even-odd nuclei (Fig. 5). The pairing energy at high excitation is marked with a cross and is an average

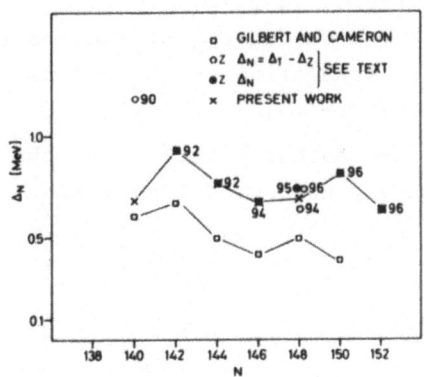

FIG. 6.

value of the experimental points. For Z = 90 the average value has been increased by 0.52 MeV due to single particle effects previously mentioned. Notable are the Δ_N values which are by about 50% larger than the ground state values. Is it an indication for a correlation of four nucleons in the nucleus?

4. REMARKS AND CONCLUSIONS

The proposed method to determine Δ at high excitation energy seems to work because:

1. Δ changes systematically with even neutron and with even proton number,

2. Δ is generally reduced, in agreement with theory (blocking effect),

3. largest deviation from the base line corresponds to a breakdown of the pairing correlation (Δ = 0); there are no $\Delta < 0$ values,

4. there is no influence on the level density due to the collective properties of states at neutron separation energy because:

a) no rapid change of ϱ from nucleus to nucleus exists as predicted by theory /11, 12/,

b) the base line, a measure of the absolute value of ϱ, can be explained by properties of intrinsic states and shell effects only /6/.

Without 4. the method would fail.

The application of the method is not limited to level densities observed at neutron separation energy but can be used for all excitation energies $E >> \Delta$.

The success of the method implies that the pairing correlation can be adequately accounted for by simply shifting the excitation energy:

$$\varrho_{odd}(E) = \varrho_{even}(E+\Delta).$$

Therefore Δ determined at neutron separation energy has to be used to describe the level density of unbound states correctly.

The results simplify the properties of the level density at high excitation energy and yield empirical values for pairing correlations which can be used to study level densities theoretically including residual interactions.

REFERENCES

1. A. Gilbert and A. G. W. Cameron, Can. J. Phys. 43 (1965) 1446.
2. S. Wahlborn, Nucl. Phys. 37 (1962) 554.
3. G. Kluge, Nucl. Phys. 51 (1964) 41.
4. P. Federman and S. Pittel, Phys. Rev. C20 (1979) 820.
5. Y. K. Gamblin, P. Ring and P. Schuck, Phys. Rev. Lett. 51 (1983) 1235.
6. G. Rohr, Z. Phys. A318 (1984) 299.
7. G. Rohr, L. Maisano and R. Shelley, Proc. of the Specialists' Meeting on Neutron Cross Section for Fission Product Nuclei, Bologna, 1979, p. 197. (Recently evaluated level densities are available on request.)
8. J. R. Huizenga and L. G. Moretto, Ann. Rev. Nucl. Sci. 22 (1972) 427.
9. S. M. Grimes, Proc. Conf. on Moment Methods in Many Fermion Systems Plenum Press, New York, 1980, p. 17.
10. H. A. Bethe, Phys. Rev. 50 (1936) 332; Rev. Mod. Phys. 9 (1937) 69.
11. S. Bjornholm, A. Bohr and B. R. Mottelson, in Physics and Chemistry of Fission, Rochester, 1973, p. 367.
12. L. A. Malov, V. G. Soloviev and V. V. Voronov, in Neutron Capture Gamma-ray Spectroscopy, Petten, 1974, p. 175.

ISOSPIN MIXING DURING THE DECAY OF THE GIANT DIPOLE RESONANCE

D. Ryckbosch, E. Van Camp, R. Van de Vyver, P. Berkvens,
E. Kerkhove, P. Van Otten and H. Ferdinande

Laboratory of Nuclear Physics, Kings University of Ghent,
Belgium

ABSTRACT. Isospin mixing is studied in the decay of the ^{28}Si
Giant Dipole Resonance. Mean isospin mixing coefficients are
deduced for the 1p-1h doorway states and for compound states.
A simple model for the equilibration of the $T_>$ GDR is presented,
which accounts fairly well for the observed isospin mixing.

1. INTRODUCTION

The Giant Dipole Resonance is one of the best known examples
of a doorway state. Indeed, when a nucleus absorbs a 15—25 MeV
photon, a coherent 1p-1h state is excited, which decays within
a short time ($\Gamma_{GDR} \simeq 4$ MeV) by different mechanisms: either it
may couple to more complicated (2p-2h) states with a probability
given by the spreading width $\Gamma^\downarrow$, or it may decay by emission of
a nucleon to the continuum (with a probability given by the
escape width $\Gamma^\uparrow$). The 2p-2h state may in turn decay to the continuum, couple to 3p-3h states (Γ_+) or couple back to the 1p-1h
doorway states (Γ_-; $\Gamma_+ + \Gamma_- = \Gamma^\downarrow$). Eventually the complexity of
the states reached in the intranuclear cascade is so great that
we may think of the nucleus as being thermally equilibrated,
and consequently it will decay by evaporation of low-energetic
nucleons. Moreover, the GDR may also act as a primary doorway
with respect to isospin. For self-conjugate nuclei, the isovector dipole operator can only excite T = 1 states starting
from the (T = 0) ground state. The Coulomb interaction between
nucleons will gradually mix $T_<$ components into the original $T_>$
state, thus opening decay channels which were otherwise isospin
forbidden. We present here a study of the decay of the ^{28}Si

GDR, in which we try to establish the amount of isospin mixing as the nucleus evolves from the 1p-1h doorway state towards a compound nucleus in thermal equilibrium.

2. DISCUSSION

Using monochromatic photons originating from the annihilation-in-flight of accelerated positrons, produced at the Ghent University Linac, we measured the decay of the ^{28}Si GDR into several proton and alpha channels. We were able to derive the absolute cross-sections for the (γ,p_0), (γ,p_3), (γ,α_0) reactions and for other proton decay channels leaving ^{27}Al in various excited states as well /1/. Decay to simple proton-hole states

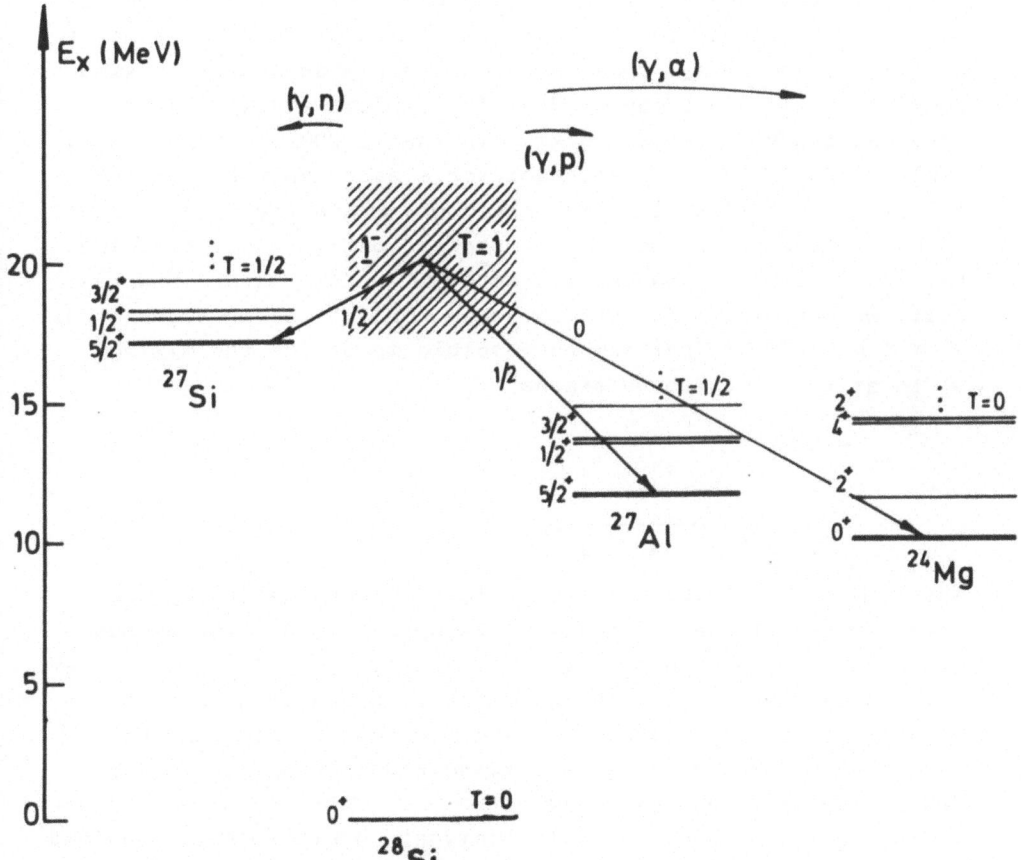

FIG. 1. Decay scheme of the ^{28}Si GDR, with squared isospin vector coupling coefficients.

in ^{27}Al originates predominantly from a direct decay of the primary doorway state, while decay to other states, and alpha decay occurs essentially after spreading of the 1p-1h strength over more complicated states. Moreover, emission of alpha particles to the low-lying (T = 0) states in ^{24}Mg can happen only after coupling of the $T_>$ doorway states to $T_<$ configurations (Fig. 1). Whereas a large proton pick-up spectroscopic factor was found for the ^{27}Al ground state, no proton-hole strength was identified for the third excited state /2/. Consequently, the (γ, p_0) cross-section is largely due to a direct decay process, while the (γ, p_3) cross-section (as well as (γ, α_0)) stems from a statistical decay of the compound nucleus.

We may then use the formalism developed in ref. /3/ for isospin mixing between compound states to deduce, on the basis of our (γ, p_3) and (γ, α_0) cross-sections, the mean isospin mixing coefficient μ^2. This may be done in complete analogy to our previous study of isospin mixing in the $T_>$ GDR of medium-heavy nuclei /4/. (Note, however, that for a self-conjugate nucleus the (γ, α) channel is isospin forbidden, while the (γ, n) channel is not.) As a result we find a mean isospin mixing coefficient between $T_>$ and $T_<$ compound states of μ^2 = 15%. Using this result, we can also calculate the mean Coulomb matrix element /4/ $H_c \simeq 1$ keV. It is furthermore possible to deduce the mixing width of the $T_>$ compound states

$$\Gamma_{mix} = 2\pi H_c^2 \, \rho_<$$

$$= \frac{\mu^2}{1 - \mu^2(1 + N_>/N_<)} \, \Gamma_>^\uparrow, \tag{1}$$

where $\rho_<$ is the level density of $T_<$ compound states, N_m the total number of open channels of isospin m and $\Gamma_>^\uparrow$ the escape width for $T_>$ compound states. Accordingly, $(\Gamma_{mix})_{comp} \simeq 7.5$ keV. This is consistent with the systematics given in a recent compilation of mixing widths /5/, where a roughly constant (i.e. independent of mass number A) mixing width of 30 keV ($\pm$ one order of magnitude) was found.

On the other hand, for self-conjugate nuclei, it is possible to deduce the amount of $T_<$ isospin impurity in the 1p-1h door-

way state, on the basis of the so-called Barker—Mann formalism.
In the case of a doorway state with pure isospin the direct
(γ,p) and their respective analogue (γ,n) cross-sections are
expected to be equal (apart from penetrability effects). However, when some $T_<$ configurations are mixed into the doorway
state, the two isospin components may interfere and do so in
a different way in the (γ,p) and (γ,n) channels. Specifically /6/

$$\frac{\sigma(\gamma,p_o)}{\sigma(\gamma,n_o)} = \frac{P_p}{P_n} \left| \frac{\alpha_o + \alpha_1}{\alpha_o - \alpha_1} \right|^2 , \tag{2}$$

where $\alpha_o = \sqrt{\mu^2}$, $\alpha_1 = \sqrt{(1-\mu^2)}$, and $P_p(P_n)$ are the proton (neutron)
penetrabilities. Using the (γ,n$_o$) data of Wu et al. /7/ and our
(γ,p$_o$) data, we thus find: $(\mu^2)_{GDR} \simeq 0.1\%$. It is clear that the
relatively weak isospin mixing Coulomb interaction has not had
time enough to mix an appreciable amount of $T_<$ configurations
into the short-lived ($\Gamma \simeq 4.5$ MeV) Giant Dipole state, whereas
it did in the long-lived ($\Gamma \simeq 40$ keV) compound states. We can
obtain an estimate for the mixing width of the $T_>$ doorway state
by using expression (1), taking into account, however, that we
should replace the escape width $\Gamma_>^\uparrow$ by the total non-mixing
width, being $\Gamma^\uparrow + \Gamma^\downarrow$ (the latter contribution obviously vanishes
for compound states). Experimentally, a width of 4.5 MeV is
found for the GDR, thus $(\Gamma_{mix})_{GDR} \simeq 4.5$ keV, remarkably close
to the value deduced for the mixing width of compound states.
We are thus led to consider a constant isospin mixing width,
of the order of 5 keV in ^{28}Si, regardless of the complexity
of the state involved.

   This may be used in a simple model for the equilibration of
the Giant Resonance in a self-conjugate nucleus (Fig. 2). The
nucleus is initially excited into a 1p-1h $T_>$ doorway state,
and this may decay in several ways, the probability for each
decay mode given by the appropriate width. The new feature in
this model, a direct generalization of the exciton model picture of precompound reactions, is the isospin mixing width,
which gives the probability for decay of a $T_>$ state into a $T_<$
one, or vice versa. This applies only between states of equal
complexity, in keeping with the assumption usually made that

an analog state couples only to its antianalogs /3/. Values for the various widths are estimated from exciton model transition rates /8/, while for the mixing width we assumed a constant

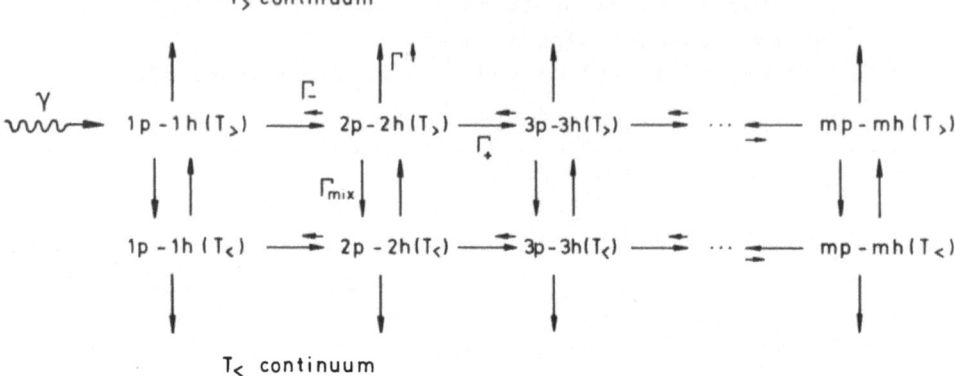

FIG. 2. Decay of the 1p-1h $T_>$ doorway state. The decay widths for a 2p-2h state are indicated.

(independent of exciton number) value of 5 keV. The decay of the initial 1p-1h $T_>$ state was followed in a Monte Carlo simulation of the decay process. (Such a procedure is perfectly adequate for our purpose, where we are only interested in the evolution of the isospin mixing coefficient, and not in the actual emission spectra of the reaction products.) As a result, we calculate an isospin mixing coefficient $\mu^2 - (\mu^2)_{GDR} = 0.4\%$ (and similar values for the mixing of 2p-2h and 3p-3h states), and $(\mu^2)_{comp} = 30\%$. These results are in fair agreement with the experimentally observed values, which indicates that the nucleus indeed loses its memory of the isospin in which it was prepared, in the way described by our simple model.

REFERENCES

1. Annual Report;Nuclear Physics Laboratory, Ghent, 1983.
2. P. M. Endt,At. Data Nucl. Data Tables, 19 (1972) 23.
3. H. L. Harney, H. A. Weidenmüller and A. Richter, Phys. Rev. C16 (1977) 1774.
4. E. Van Camp, D. Ryckbosch, R. Van de Vyver, E. Kerkhove, P. Van Otten and P. Berkvens, Phys. Rev. C30 (1984) 1176.

5. E. Kuhlmann and K. Glasner, Preprint Bochum, 1984.
6. F. C. Barker and A. K. Mann, Philos. Mag. $\underline{2}$ (1957) 5.
7. C.-P. Wu, F. W. K. Firk and T. W. Phillips, Nucl. Phys. $\underline{A147}$ (1970) 19.
8. D. Agassi, H. A. Weidenmüller and G. Mantzouranis, Phys. Rep. $\underline{22}$ (1975) 145.

STRENGTH FUNCTIONS OF HIGH-LYING SINGLE-NEUTRON STATES

A. I. Vdovin and *Ch. Stoyanov

Laboratory of Theoretical Physics, JINR, Dubna, USSR
*Institute for Nuclear Research and Nuclear Energy,
Bulgarian Academy of Sciences, Sofia, Bulgaria

ABSTRACT. Strength functions of single-neutron states with excitation energy from 7 to 12 MeV in nuclei ^{91}Zr, $^{117,121}Sn$, ^{145}Sm and ^{209}Pb are calculated. The quasiparticle-phonon nuclear model and strength function method are used. The results of calculations are compared with available experimental data for tin isotopes.

1. INTRODUCTION

Since the initial observations of deep-hole states in neutron
pick-up reactions in the early 70s /1/ the one-nucleon transfer reaction has been widely used to study the high-lying single-particle (or single-hole) strengths in medium and heavy nuclei. The growing amount of experimental data has accumulated
very rapidly and new exciting features of nuclear excitation
spectra have been found. In the proton pick-up reactions the
deep-hole proton states have been observed /2/ and in the proton stripping reactions the high-lying single-proton states
have been found /3/. Now, the high-energy resolution gives us
a possibility to study the fine structure of excited resonance-
like structures, and using a polarized incident beam experi-
mentalist can make unambiguous spin assignments to it.

So, the one-nucleon transfer reactions at high energies
made it possible to investigate the new kind of nuclear excitations which one may call "giant single-particle (or single-
hole) resonances". These giant resonances are the response of
nucleus to the external field acting on its single-particle

degrees. The comparison of the experimental data with the predictions of different theoretical models helps us to check our
knowledge on the average nuclear field and on the mechanisms
responsible for the spreading of the single-particle strength
in nuclei. On the other hand, to extract the single-particle
(or -hole) strength from an experimental cross-section it is
useful to have some theoretical predictions about the fragmentation of high-lying nuclear subshells because their strength
distributions are overlapped.

One of the latest experimental achievements is the observation of high-lying single-neutron states in tin isotopes with
mass numbers A = 117, 119 and 121 /4/. Analogous experiments
for the target-nucleus ^{90}Zr, ^{144}Sm and ^{208}Pb are now in progress /5/. So, we shall try to calculate the strength functions
of these high-lying single-neutron subshells in the corresponding N-odd nuclei. For the tin isotopes we compare our results
with the available experimental data.

2. MODEL

We use the quasiparticle-phonon nuclear model and the strength
function method in our calculations. The quasiparticle-phonon
nuclear model (QPM) has been expounded in detail in a series
of reviews /6—9/. The residual NN-interaction in the QPM consists of the traditional monopole pairing interactions in the
particle-particle channel and separable multipole and spin-multipole interaction in the particle-hole channel. The particle-
hole interaction is isotopically invariant. It is well known
now that damping widths of giant resonances and resonance-like
structures in nuclei are due to the coupling of simple nuclear
modes (in our case single-particle modes) with the complex configurations. The coupling with low-lying surface vibrations
plays the key role in the spreading of the strengths of simple
excitations /8—11/. But our previous studies /9, 10/ have shown
the importance of the coupling with collective phonon excitations at intermediate energies (e.g., M2-resonance or LEOR) to
describe the fine structure of the strength distributions. In
the framework of the QPM we take into account the interaction

of single-particle states with all types of phonons: low-lying vibrations, collective phonons of intermediate energies and noncollective phonons (i.e. almost pure two-quasiparticle states). We should like to stress that the quasiparticle-phonon interaction H_{qph} is calculated microscopically in the QPM. The interaction H_{qph} depends on the characteristics of phonons which are calculated in the RPA and on single-particle matrix elements of residual particle-hole interaction. In terms of the creation and annihilation operators of quasiparticles and phonons (α^+_{jm}, α_{jm} and $Q^+_{\lambda\mu i}$, $Q_{\lambda\mu i}$) the operator H_{qph} has the form:

$$H_{qph} = -\frac{1}{2\sqrt{2}} \sum_{\lambda\mu i} \left\{ (Q^+_{\lambda\mu i}(-)^{\lambda-\mu} + Q_{\lambda-\mu i}) \cdot \right.$$

$$\left. \cdot \sum_{jj'\tau} \frac{f^{(\lambda)}_{jj'} v^{(-)}_{jj'}}{\sqrt{\mathcal{J}^{\lambda i}_{\tau}}} B(jj'\lambda - \mu) + h.c. \right\} , \qquad (1)$$

$$B(jj'\lambda-\mu) = \sum_{mm'} (-)^{j'+m'} \langle jmj' m'| \lambda -\mu \rangle \alpha^+_{jm}\alpha_{j'-m'} ,$$

where $f^{(\lambda)}_{jj}$ is reduced single-particle matrix element of the operator $f(r)Y_{\lambda\mu}$;* $v^{(-)}_{jj} = u_j u_{j'} - v_j v_j$, ($u_j$, v_j are the coefficients of Bogolubov's transformation). The factor $\mathcal{J}^{\lambda i}_{\tau}$ is the normalizing factor in the one-phonon wave function. There are two of them for each phonon: one for the neutron part of the wave function ($\tau = n$) and the other for the proton one ($\tau = p$). The factors $\mathcal{J}^{\lambda i}_{\tau}$ depend on the collectivity of the phonon λi. The values of $\mathcal{J}^{\lambda i}_{\tau}$ are small for the collective phonons and large for the noncollective ones. Note, the same isotopic index τ as in $\mathcal{J}^{\lambda i}_{\tau}$ is included in the single-particle quantum number $j \equiv nlj\tau$. One can see from (1) that H_{qph} couples the states which differ from each other by one phonon, i.e.

---

*In this work $f(r) = dU/dr$ and U denotes a central part of the Saxon-Woods potential.

190

$$\alpha^+_{JM} \leftrightarrow [\alpha^+_{jm} Q^+_{\lambda\mu i}]_{JM} ; \quad [\alpha^+_{jm} Q^+_{\lambda\mu i}]_{JM} \leftrightarrow$$

$$\leftrightarrow [\alpha^+_{j'm'} [Q^+_{\lambda_1\mu_1 i_1} Q^+_{\lambda_2\mu_2 i_2}]_{I\mu}]_{JM} .$$

When calculating the matrix element of H_{qph} we don't take into account the anharmonic corrections for the phonon excitations. We believe these corrections affect slightly the fragmentation of high-lying single-particle strength. Moreover, we use a special approximate procedure to preserve the violation of the Pauli principle in the complex components of our model wave function. We have studied all these effects correctly in refs. /9, 12, 13/. But to solve numerically the full set of equations of the QPM /13/ for high-lying states in odd-mass nuclei is a heavy computational problem. At present we use model wave function which consists of one-quasiparticle, "quasiparticle+one phonon" and "quasiparticle+two phonon" components and solve a simpler set of equations than that in ref. /13/ (see ref. /12/). We take into account the coupling of an odd-quasiparticle with all the phonons with momenta and parities $\lambda^\pi = 1^\pm + 7^\pm$ and excitation energies less than 20 MeV.

We don't calculate the energy and wave function structure of each of the numerous high-lying excited states. Instead, we calculate the single-particle strength function $C_j^2(E_x)$ which describes the dependence of the smearing on the energy interval Δ and the square of one-quasiparticle amplitude C_j^2 on the excitation energy E_x. The strength function method has been used for the first time by Bohr and Mottelson /14/ for a schematic model. Later, the method has been developed for more complex problems /15/ (see also refs. /8, 9/), and now it is widely used in nuclear structure calculations /11/. The strength function method makes the calculation much simpler and the results more obvious. In the present work, in order to construct a single neutron strength function we use the Lorentz function with the width parameter $\Delta = 0.5$ MeV as a weight function.

We have discussed the model parameters in detail in refs. /7, 8, 10/. Here we display (Fig. 1) only the parts of single-neutron schemes of the nuclei in question. The single-particle

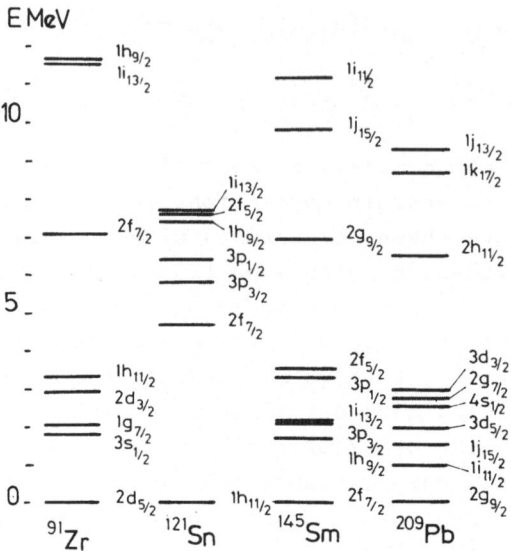

FIG. 1. Parts of single-neutron schemes in nuclei ^{91}Zr, ^{121}Sn, ^{145}Sm and ^{209}Pb. The Fermi levels are assumed to have zero energy.

energies and wave functions for all the nuclei except ^{209}Pb are calculated in the spherically symmetric Saxon—Woods potential with standard parameters from ref. /16/. These parameters have been used in our previous calculations. The energies of the levels of the shells which are closest to the Fermi surface in ^{209}Pb are taken from ref. /17/. In that paper they have been fitted in the QPM calculation to the experimental data on the energies and one-nucleon spectroscopic factors of low-lying states of the nuclei $^{207, 209}$Pb, ^{207}Tl and ^{209}Bi. The energies of single-particle (-hole) state well above (below) the Fermi surface are calculated in the Saxon—Woods well, too.

3. RESULTS AND DISCUSSION

We should like to start the discussion of our results with the tin isotopes. The clearly seen resonance-like structures at excitation energies of 4—10 MeV have been observed in the study of the ^{116,118,120}Sn(α, ^{3}He) reactions at 183 Mev incident energy /4/. The measured excitation energies and angular distributions support the assumption that these bumps arise from neutron striping to the $1h_{9/2}$ and $1i_{13/2}$ orbitals located well above the valence subshell. With decreasing mass number of the isotope the centroid energy and the spreading of the strength distribution increase. In ^{121}Sn, the resonance-like structure consists of a sharp peak at E_x = 4.89 MeV with a width less than 1 MeV (a range $\langle A \rangle$ in Table 1) and additional broader components ($\langle B \rangle$ and $\langle C \rangle$ in Table 1). In ^{119,117}Sn, this narrow peak smears and almost disappears. The quantitative data on the single-neutron strength distributions are only for ^{121}Sn.

TABLE 1. Experimental data and theoretical results for the strength distributions of $1i_{13/2}$ and $1h_{9/2}$ subshells in ^{121}Sn

Experiment /4/			Theory		
E_m, ΔE_x (MeV)	J^π	C^2S	E_m, ΔE_x (MeV)	J^π	C^2S
$\langle A \rangle$ E_m = 4.89±0.05	$9/2^-$	0.67	E_m = 5.1	$9/2^-$	0.12
	-	-	ΔE_x = 4.6-5.6	$13/2^+$	0.21
$\langle B \rangle \Delta E_x$ = 5.4–7.0	$9/2^-$	0.25	ΔE_x = 5.4–7.0	$9/2^-$	0.4
	$13/2^+$	0.16		$13/2^+$	0.20
$\langle C \rangle \Delta E_x$ = 7.0–10.0	-	-	ΔE_x = 7.0–10.0	$9/2^-$	0.34
	$13/2^+$	0.27		$13/2^+$	0.39

E_m - the energy of the peak; ΔE_x - the energy interval for which the value C^2S is found.

In the one-nucleon transfer reaction at intermediate incident energies the transitions with large transfer momenta are enhanced. Therefore, at first the strengths of the subshells with high orbital momentum are extracted from experimental cross-sections. In tin isotopes those neutron subshells are $1h_{9/2}$ and $1i_{13/2}$. These quasibound neutron states are located in the single-neutron spectrum very closely to each other. As

one can see in Fig. 2 and Table 2 in our calculation the main
strengths of the both subshells are spread over the same energy
interval $4 \leq E_x \leq 9$ MeV. The fragmentation of the subshells in
^{117}Sn is stronger than in ^{121}Sn to some extent. This difference is most prominent for the subshell $1h_{9/2}$. The strength
function maxima are located at $E_x \sim 5\text{-}6$ MeV, therefore the
relatively high subshell strengths would be observed here. So,
it seems to us that there is a qualitative agreement between
the experimental data and our results.

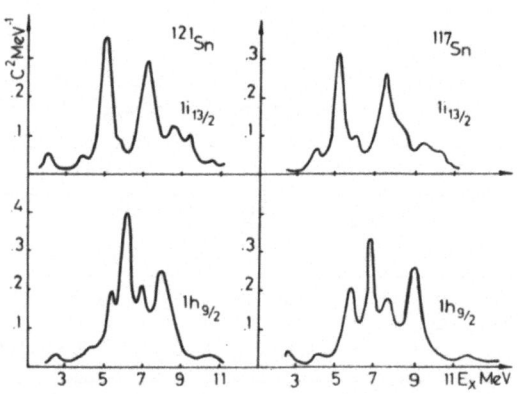

FIG. 2. Strength functions of the high-lying neutron states
$1i_{13/2}$ and $1h_{9/2}$ in ^{117,121}Sn.

But the quantitative agreement is somewhat worse (see Table 1). The most striking, though not so important, disagreement is that with the energy interval $\langle A \rangle$. According to the
experimental interpretation, in this region almost 2/3 of the
total $1h_{9/2}$ strength is concentrated. The maximum of the theoretical $1h_{9/2}$ strength function is placed at $E_x = 6.2$ MeV,
i.e. 1.3 MeV higher than the experimental peak. But there is
the main peak of the $1i_{13/2}$ strength function ($E_x = 5.1$ MeV)
very close to the experimental one. Of course, the accuracy of
the theoretical model is not high enough to make a definite
conclusion on the fine structure of the strength distributions
of high-lying states. However, some part of the $1i_{13/2}$ strength

194

TABLE 2. The theoretical characteristics of the high-lying single-neutron strength distributions in spherical nuclei

Nucleus	nlj	ΔE_x (MeV)	$\bar{E}$ (MeV)	σ (MeV)	c^2S
^{91}Zr	$2f_{7/2}$	2.5—13.5	8.1	1.76	0.93
		7.2—9.7	8.45	0.75	0.55
	$1i_{13/2}$	5.5—18.5	12.1	2.19	0.85
		9.4—13.9	11.5	1.22	0.54
	$1h_{9/2}$	7.5—16.5	12.6	1.60	0.83
		13.5—16.5	14.9	0.95	0.24
^{117}Sn	$1i_{13/2}$	2.0—11.0	6.9	1.9	0.79
		4.3—6.8	5.4	0.53	0.30
		6.8—8.9	7.8	0.53	0.32
	$1h_{9/2}$	2.0—13.0	7.5	2.0	0.91
^{121}Sn	$1i_{13/2}$	1.0—11.0	6.7	2.0	0.89
		3.0—6.0	5.0	0.57	0.31
		6.0—8.2	7.3	0.48	0.33
	$1h_{9/2}$	1.0—11.0	6.8	1.6	0.92
^{145}Sm	$1i_{11/2}$	5.0—16.0	12.3	1.94	0.90
		10.0—12.6	11.5	0.46	0.45
	$1j_{15/2}$	5.0—16.0	10.9	2.46	0.87
		7.0—10.0	8.7	0.72	0.31
		10.0—12.0	11.1	0.46	0.23
^{209}Pb	$1k_{17/2}$	2.5—15.5	8.9	2.32	0.94
	$1j_{13/2}$	2.5—15.5	9.6	1.8	0.94

$\bar{E}$ is the energy centroid, σ is the second moment, c^2S is the part of the total state strength which is exhausted in the interval ΔE_x.

must be located in the region $\langle A \rangle$ of the spectrum. We should like to point out that at scattering angles $\Theta \sim 10^{0}$ the experimental angular distribution is closer to the theoretical one calculated for transfers $l = 6$, although at $\Theta \prec 10^{0}$ the transfers $l = 5$ are preferable.

A more important disagreement concerns the wide energy intervals ΔE_x. From data of ref. /4/ (Table 1) it is seen that almost the total $1h_{9/2}$ strength (92%) is exhausted in the interval $\Delta E_x \sim 2$ MeV ($4.9 \le E_x \le 7$ MeV). At the same time only 43% of the $1i_{13/2}$ strength has been observed at wider interval $5.4 \le E_x \le 10$ MeV. From these data one can make a conclusion that the $1i_{13/2}$ subshell has the excitation energy much higher than that the $1h_{9/2}$ subshell and that it is fragmented stronger. This conclusion contradicts our results. Of course, the positions of both the $1h_{9/2}$ and $1i_{13/2}$ single-neutron states have been calculated in the phenomenological potential well and one can try to change potential parameters to improve the agreement with new experimental data. But the close positions of these subshells in a single-neutron scheme is a rule valid for all the nuclei in question (see Fig. 1). Moreover, strong concentration of $1h_{9/2}$ strength in the excitation spectrum seems improbable to us.

In view of these results we should like to stress once more that to extract the subshell strengths from experimental cross-sections correctly it is necessary to take into account the overlapping of their distributions. In some cases the contribution of subshells with relatively low orbital momentum numbers becomes important. For example, the agreement of theoretical and experimental strength distributions of the $1h_{9/2}$ and $1i_{13/2}$ single-proton states had been improved noticeably since the contribution of the $2f_{7/2}$ subshell to the one-proton stripping cross-section was taken into account /3, 18/. It means that in the present case experimenters would take into account the $2f_{5/2}$ and $2f_{7/2}$ strengths. The calculations of the strength functions of these subshells are in progress.

The calculated strength functions of the high-lying single-neutron states in ^{91}Zr, ^{145}Sm and ^{209}Pb are displayed in Figs. 3—5. The integral characteristics of the single-neutron strength

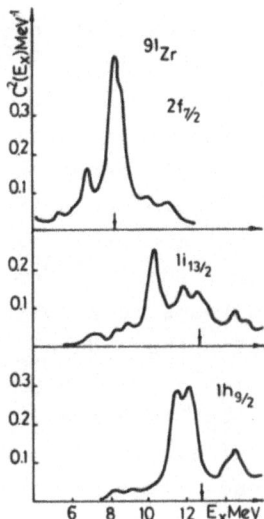

FIG. 3. Strength functions of the high-lying neutron subshells in ^{91}Zr. Arrows indicate the energies of the one-quasiparticle states.

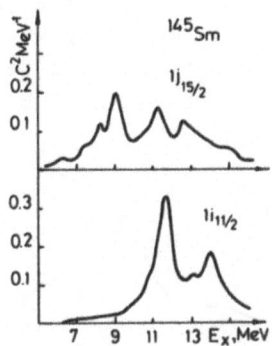

FIG. 4. Strength functions of the high-lying neutron subshells in ^{145}Sm.

distributions are given in Table 2. Now only preliminary experimental data on these states are available /5/ and we hope that our calculations will help to analyse the one-neutron stripping reaction cross-section. The results for ^{91}Zr have been published in ref. /19/.

In all three nuclei, ^{91}Zr, ^{145}Sm and ^{209}Pb, an odd neutron is coupled to an even-even core with magic neutron number (N = 50, 82 and 126, respectively). The valence shells beginning to be occupied and the excitation energies of quasibound neutron states are higher than in $^{117-121}$Sn. Therefore, the escape widths of the quasibound states with low orbital momentum are large and these states are not shown in Figs. 3–5. The displayed subshells have small escape widths. The patterns of the quasibound part of the single-neutron scheme in ^{91}Zr, ^{145}Sm and ^{209}Pb are similar. Two subshells with high l-numbers (in ^{91}Zr l = 6, in ^{145}Sm l = 7 and in ^{209}Pb l = 8) are located close to each other and there is a subshell with much less l-number several MeV below them. Note, that there are the narrow quasibound neutron states with higher orbital momentum in single-neutron spectra, but they are located at much higher excitation energy. We believe that due to very strong spreading the corresponding resonance-like structures will be hardly observable.

Our calculations show that all subshells studied are strongly fragmented. The second moments of the strength distributions are in the interval $1.6 \leq \sigma \leq 2.5$ MeV. The strength distributions of the high-l subshells are overlapped and this effect should be taken into account in analysing experimental data. The effect of the low-l subshells will be weaker than in tin isotopes because they are located well below the high-l quasi-

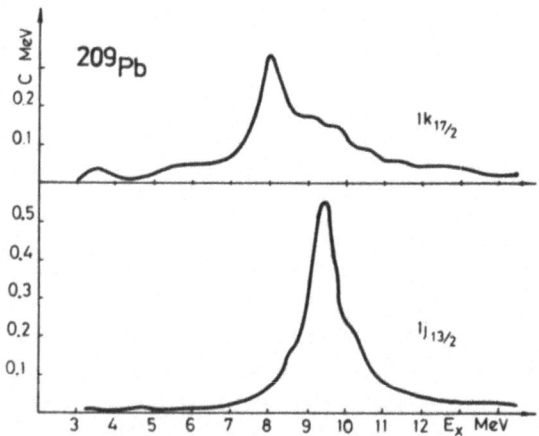

FIG. 5. Strength functions of the high-lying neutron subshells $1j_{13/2}$ and $1k_{17/2}$ in ^{209}Pb.

bound subshells. We assume that it is possible to observe the resonance-like structures which are due to the one-neutron stripping on the subshells $2f_{7/2}$, $2g_{9/2}$ and $2h_{11/2}$ in ^{91}Zr, ^{145}Sm and ^{209}Pb, respectively.

We are grateful to Prof. V. G. Soloviev for his interest in this work, Prof. H. Langevin-Joliot for fruitful discussions and supply with the experimental data on tin isotopes prior to publication and to Dr. V. Yu. Ponomarev for his assistance in the calculations.

REFERENCES

1. M. Sakai and K. Kubo, Nucl. Phys. A185 (1972) 217.
   S. Y. Van der Werf et al., Phys. Rev. Lett. 33 (1974) 712.
   E. Gerlic et al., Phys. Lett. 57B (1975) 338.
2. P. Doll et al., Phys. Lett. 82B (1979) 357.
   A. Stuirbrink et al., Z. Phys. A297 (1980) 307.
3. S. Gales et al., Phys. Rev. Lett. 48 (1982) 1593; Phys. Rev. C31 (1985) 94.
4. S. Gales et al., Phys. Lett. 144B (1984) 323.
5. S. Gales, Preprint IPNO-DRE/83-29 and private communication.
6. V. G. Soloviev, Particles and Nuclei 9 (1978) 810; Nukleonika 23 (1978) 1149.
7. A. I. Vdovin and V. G. Soloviev, Particles and Nuclei 14 (1983) 237.
8. V. V. Voronov and V. G. Soloviev, Particles and Nuclei 14 (1983) 1380.
9. A. I. Vdovin et al., Particles and Nuclei 16 (1985) 245.
10. V. G. Soloviev et al., Nucl. Phys. A342 (1980) 261.
11. G. F. Bertsch et al., Rev. Mod. Phys. 55 (1983) 287.
12. Chan Zuy Khuong et al., J. Phys. G7 (1981) 151.
13. Dao Tien Khoa et al., Report JINR E4-84-501, Dubna, 1984.
14. A. Bohr and B. Mottelson, "Nuclear Structure". Vol. 1. Benjamin, New York, 1969.
15. L. A. Malov and V. G. Soloviev, Nucl. Phys. A270 (1976) 87.
    L. A. Malov, Report JINR, P4-81-228, Dubna, 1981.
16. K. Takeuchi and P. A. Moldauer, Phys. Lett. 28B (1969) 384.
    V. A. Chepurnov, Yad. Fiz. 6 (1967) 955.
17. V. V. Voronov and Dao Tien Khoa, Izv. Akad. Nauk SSSR, Ser. Fiz. 48 (1984) 2008.
18. Ch. Stoyanov and A. I. Vdovin, Phys. Lett. 130B (1983) 134.
19. A. I. Vdovin and Ch. Stoyanov, Yad. Fiz. 41 (1985) 1134.

NEUTRON RESONANCE AVERAGING WITH FILTERED BEAMS[*]

R. E. Chrien

Brookhaven National Laboratory, Upton,
New York, USA

ABSTRACT. Neutron resonance averaging using filtered beams
from a reactor source has proven to be an effective nuclear
structure tool within certain limitations. These limitations
are imposed by the nature of the averaging process, which
produces fluctuations in radiative intensities. The fluctuations have been studied quantitatively. Resonance averaging
also gives us information about initial or capture state parameters, in particular the photon strength function. Suitable
modifications of the filtered beams are suggested for the
enhancement of non-resonant processes.

INTRODUCTION

Just above an excitation energy corresponding to particle binding in the nuclear system, there is a region, or energy window,
in which radiative deexcitation is a significant decay mode. In
this window, the metastable states or resonances are well separated and can be excited by slow neutrons with exceptionally
high cross sections. This slow neutron "window" has provided an
important source of information about nuclear structure on one
hand, and about highly-excited nuclear states on the other.
Much of this information has been provided by neutron capture
gamma-ray spectroscopy.

While the observation of gamma transitions from individual
resonances provides the maximum information, it appears that
for most questions arising in nuclear theory, a suitable averaging over resonances is sufficient. For example, in the

Research has been performed under contract DE-ACO2-
76CHO0016 with the United States Department of Energy.

assessment of photon strength functions, the averaging must
in any case be taken. Thus when Bollinger suggested /1/ in
1966, a direct experimental method for averaging, it soon became evident that for most applications of the (n,γ) method
to nuclear models, the averaging technique is preferred, since
it at once (a) avoids the necessity for tedious accumulation
of many discrete complete spectra, (b) gives directly the most
important quantity desired, i.e., the mean primary gamma-ray
strength, and (c) can produce an average over a very large number of resonances. As we shall see in subsequent sections, part
of the price to be paid for avoidance of the discrete resonance
measurements is a loss of the knowledge of the <u>absolute</u> values
of the partial radiative widths, $\Gamma_{\gamma\lambda f}$. Another important disadvantage is the loss of the ability to discriminate among different capturing states; states of varying spin and parity are
included.

Averaging Techniques

The basis of the direct experimental resonance-averaging technique is the use of a neutron spectrum which has a suitably defined and limited range in energy. Definition of this energy
range can often be accomplished by combining the properties of
the source with the properties of a transmission filter of some
kind, to produce a situation in which the capture reaction
occurs over a limited energy range.

   There are three major averaging techniques that have been
applied to (n,γ) reaction studies. These involve filters of
boron, scandium, and iron (or ^{56}Fe) applied to the spectrum of
reactor neutrons. It is interesting to observe that in these
averaging methods, the fission reactor remains the neutron
source of choice.

Boron Filter Technique

The boron filter technique was developed by Bollinger and his
colleagues /2/ at the Argonne CP-5 reactor in the period 1966
–1970. The arrangement features an <u>internal</u> target facility

surrounded by a ^{10}B absorber placed in a thru-tube passing the edge of the CP-5 core (thermal flux = 3×10^{13} n/cm^2/s at $R_{Cd} \simeq 7$).

The combination of the 1/E epithermal reactor spectrum and the 1/v ^{10}B absorption cross section results in the spectrum of Fig. 1, which shows a broad incident neutron spectrum peaking under 100 eV, but extending to 10 keV and beyond. The <u>captured</u>

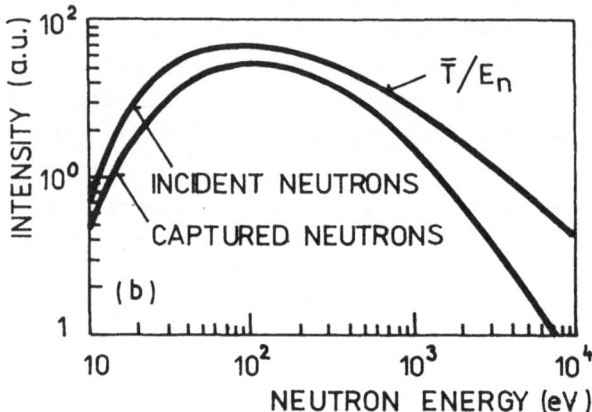

FIG. 1. Spectra of incident and captured neutrons in a holmium sample of the Argonne CP-5 facility /4/.

neutron spectra for a typical sample, however, differs from the <u>incident</u> spectrum because of the absorption properties of the sample. The combination of the E^{-1} reactor spectrum, the $E^{-1/2}$ absorber cross section, and the $E^{-1/2}$ sample cross section produce a crudely defined averaging interval <u>which is not large compared to the typical resolution of a gamma-ray detector</u>. Because of this fact, the detector conveniently averages over a number of contributing resonances in this interval, and a suitably averaged transition strength is obtained.

The spectroscopic utility of resonance-averaging has been well documented /2—5/. Subject to the usual statistical model assumptions that the gamma-ray strength function is independent of the initial and final state spins, then the population patterns for the final states are simply determined by how many ways that particular final state can be fed by a transition of a particular multipolarity (Fig. 2). Furthermore, there is a sizeable intensity difference to be expected depending on the

radiation 'multipolarity, and thus a parity dependence of the
final state population is evident.

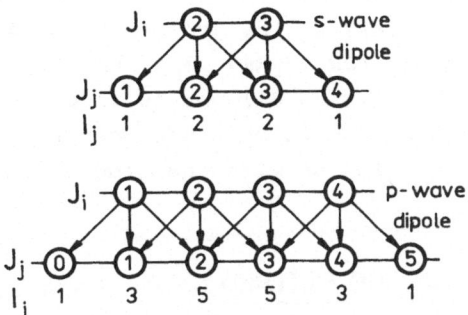

FIG. 2. Population paths for various final states reached from
neutron capture on a spin 5/2 target /3/.

Transmission Filters

While the internal target arrangement described above was effective for a limited number of targets, a number of drawbacks
are readily apparent.

1. Targets internal to a reactor are subjected to constraints
on size, and physical/chemical form, on purity, on temperature
stability, and other factors related to the hostile environment
in which they are placed. Activation of sample impurities can
also be a significant problem.

2. The use of the boron filter in a geometry in which it
surrounds the target makes it difficult to define the captured
neutron spectrum precisely. Closely coupled to this difficulty
is the fact that the boron-reactor spectrum results in a peak
capturing rate at low energy (see Fig. 1), typically near
100 eV. Thus, low energy resonances may contribute heavily and
distort the averaging. A very thorough attempt has been made
to evaluate fluctuations from capturing states located at low
energies (< 500 eV) in ref. /3/. However, the difficulties can
be gauged by noting that for the heavier elements, holmium and
gadolinium, the observed fluctuations exceed the calculated
fluctuations by a factor of 2.

3. It is difficult to alter the neutron spectrum.

These disadvantages are largely overcome using an external target combined with an appropriate transmission filter /3/. A number of such filters have been suggested /6/, but only two, scandium and iron, have been used for (n,γ) studies. Figure 3 shows the transmission peaks at 2 and 24 keV resulting from the scandium and ^{56}Fe filters at the Brookhaven National Laboratory High Flux Beam Reactor (BNL HFBR) /3/. This figure shows the superior spectrum definition alluded to in point 2 above. With regard to point 3, the BNL filters are mounted in a rotatable collimator assembly which permits changing a filter in a matter of a few seconds.

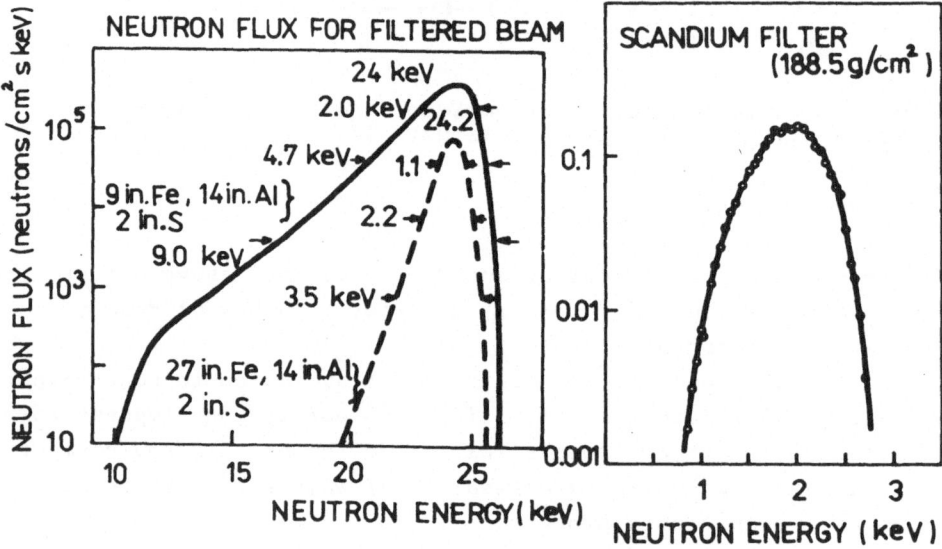

FIG. 3. Transmission curves for the BNL 2 keV and 24.3 keV scandium and ^{56}Fe filters.

It might be expected that a problem could arise from the transmission of neutrons at cross-section minima located at higher energies. The spectrum from the BNL HFBR ^{56}Fe filter has been measured by Greenwood /7/ using hydrogen filled recoil detectors (Fig. 4). Clearly distinct neutron groups are present at higher energies, but they present no problem because the average capture cross section falls, as does the reactor spectrum. The weighting of these higher groups goes roughly as E^{-2}, and they contribute only a few percent to the observed (n,γ) spectrum.

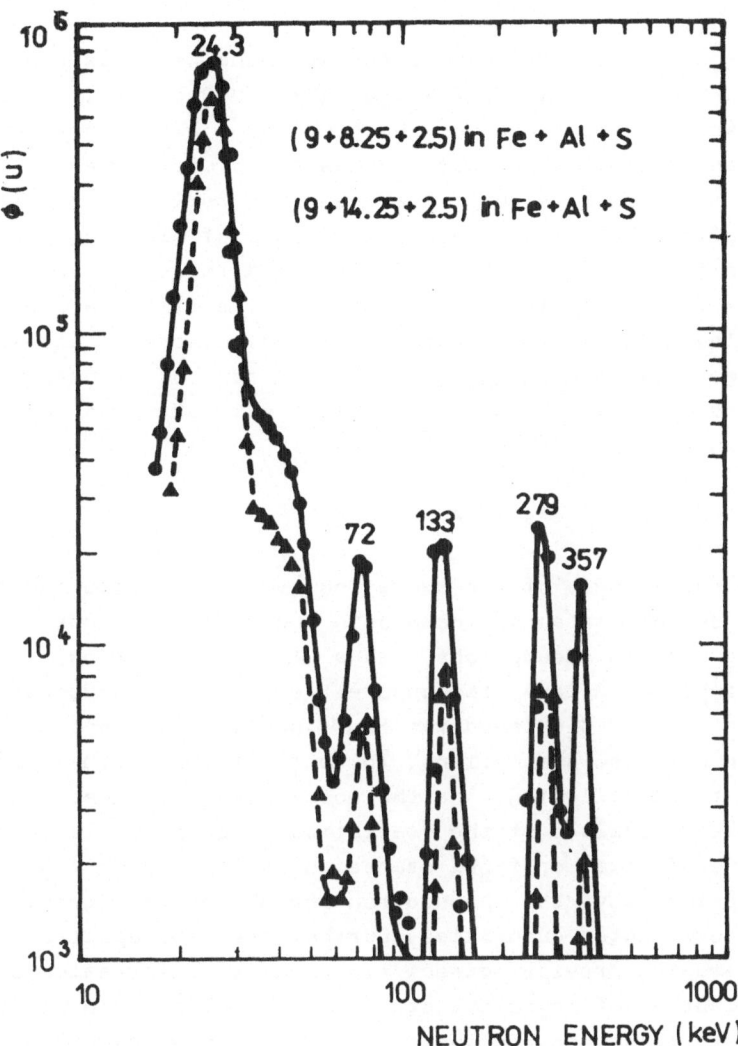

FIG. 4. Neutron group structure from the 24.3 keV filter.

Discrete Resonance Averaging

Using the neutron time-of-flight technique, it is of course possible to measure the partial radiative widths for the re-solved resonance region, and average them after the fact. This procedure gives the maximum possible information. The capturing states can be separated by spin and parity (in principle, at least). They can also be categorized according to size (small or large widths). Much discrete resonance work has been done at Brookhaven /8, 9/, Oak Ridge /10/ and Dubna /11/, with particular emphasis placed on the studies of width correlations and non-statistical effects. Perhaps the most relevant point to be made about the discrete resonance region is that these time-of-flight measurements are capable of yielding absolute values for partial radiative widths and therefore absolute values for photon strength functions. The filter averaging technique cannot yield absolute values. Discrete resonance average is limited usually to several tens of resonances. The most complete set is due to Sahal et al. /12/, who report 49 resonances for ^{173}Yb$(n,\gamma)^{174}$Yb.

Quantitative Aspects of Averaging

Since the success of the averaging method in producing a complete set of states of known spin and parity depends on combining certain assumptions with a purely statistical decay of the compound nucleus, the apparent experimental success which the method has displayed for heavy nuclei represents a vindication of these assumptions. In spite of this global vindication, there will always be the possibility of exceptions, for example, a failure of the statistical assumption for a certain class of initial or final states, or a failure of the various assumptions covering the neutron and photon strength functions, or other related quantities. Furthermore, the spectra obtained by averaging provides direct experimental information on the distribution of radiative strength in nuclei — the photon strength functions. In the remainder of this review, I will concentrate on the information that averaging gives us about the properties of the initial states — the resonances, rather than the nuclear structure aspects. The latter have been reviewed extensively by many others, much more qualified than I to treat this aspect.

The discussion starts with an expression of the averaged capture cross section to a specific final state

$$\langle \delta n, \gamma \rangle = 2\pi^2 \lambda^2 \sum_{J,l} g_J D_J^{-1} \langle \Gamma_n \Gamma_{\gamma if} / \Gamma \rangle .$$

The following assumptions are made to simplify the above expression:

a) Consider only $l = 0$, 1 and ignore higher partial waves.

b) Consider dipole transitions only, E1 and M1.

c) The neutron and photon strength functions are spin independent, and invariant over a limited range of excitation energy.

d) The reduced neutron and photon widths are uncorrelated Porter—Thomas variables.

e) The total radiation widths are independent of spin, parity, and excitation energy.

f) The level density of initial states is given by $\rho_J = D_J^{-1}$ $= \rho_o(2J+1) \exp[-J(J+1)/2\delta^2]$ where 0 indicates spin 0 states, and δ, the spin cutoff factor is taken to be 3.5.

The evaluation of the quantity $\langle(\Gamma_n \Gamma_{\gamma if}/\Gamma)\rangle$ over the resonances is conveniently accomplished by Monte Carlo methods, such as the code RACA at BNL /4/. Some model has to be assumed for the photon strength functions S(E1) and S(M1). For the former, the Brink—Axel /13/ hypothesis suggests a power law expression of the form E_r^n, where n is approximately 5. For the latter, a similar form can be assumed, with an n varying from 3 to 5. There is some evidence, as we shall see, for a Lorentzian form for both electric and dipole strength functions.

For filtered beam experiments, the energy-averaging intervals are nearly rectangular in form (Fig. 3). For scandium $\Delta E \approx 850$ eV, centered at 2000 eV, and for ^{56}Fe, $\Delta E \approx 1900$ eV centered at 24,300 eV. At these higher neutron energies, we can use the averaged cross section expressions directly to infer gamma-ray intensities, since sample self-absorption effects, for neutrons of gamma rays, are negligible.

Several simple observations concerning the above expressions may be in order. In the low energy limit, the correlation between Γ and Γ_n may be ignored, since $\Gamma \gg \Gamma_n$. There Γ is spin independent, and the pattern of Fig. 2 is accurate; this is the situation with respect to boron filtering. At higher energies, the Γ becomes quite spin dependent through the dependence on Γ_n.

Furthermore, the effects of p-wave capture cannot be ignored.
Both effects serve to lessen the spin and parity discrimination
of this method as the neutron energy increases. These effects
are illustrated in Fig. 5. It is clear that the separate
effects of the neutron strength functions S_0 and S_1 and the
photon strength functions $S(E1)$ and $S(M1)$ must be considered.

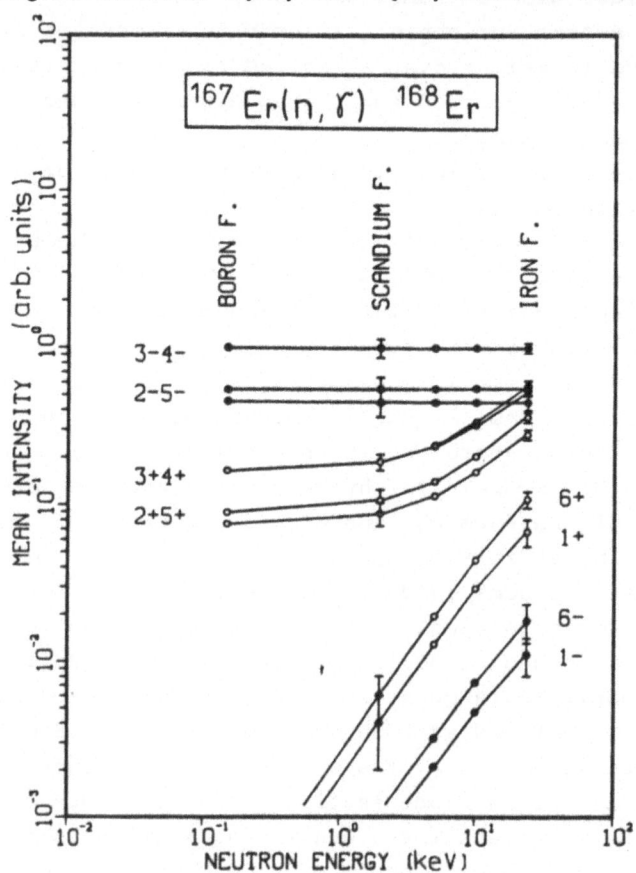

FIG. 5. Intensities for population of final states as a function
of neutron energy.

The use of <u>both</u> iron and scandium filtering, with their different s- and p-wave admixtures, can result in reliable parity
information. Such information is displayed in Fig. 6, for the
classical case of $^{167}Er(n,\gamma)^{168}Er$ /14/. A firm parity identification is provided over an excitation range of 2 MeV or more.

The failure to discriminate at high excitation energy is almost certainly attributable to the failure to resolve closely-spaced transitions.

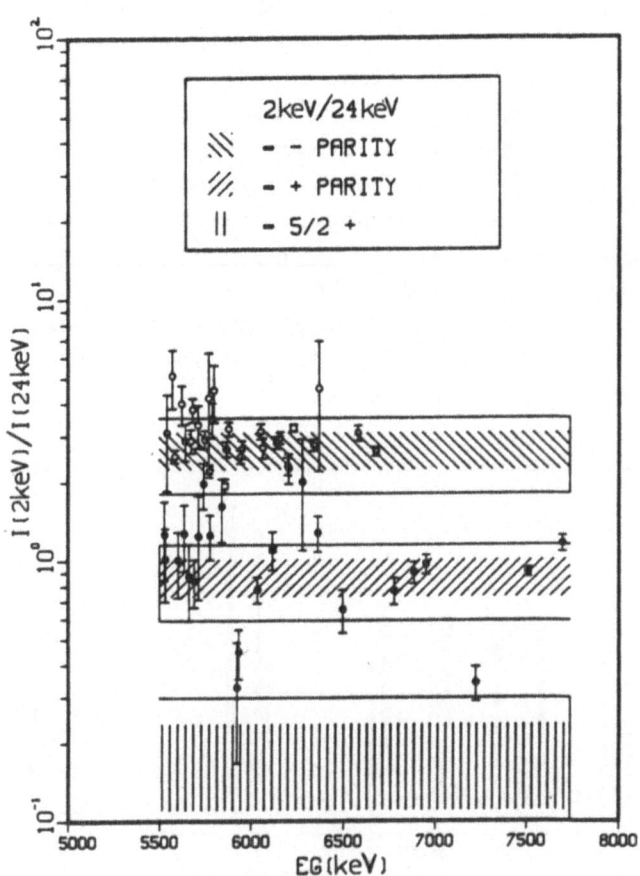

FIG. 6. Ratio of 2 and 24 keV populations.

Occasionally, the ability of resonance-averaging in identifying all states within a spin-parity range (completeness) has been questioned. In particular, Hamilton et al. /15/ cite a number of states in ^{146}Nd (levels at 1518, 1769, 1906, 2046, and 2168 keV) with suitable spin and parity, which were not reported in an earlier ^{10}B filtered experiment at Argonne by Bushnell et al. /16/. The absence of those states in an averaged spectrum would clearly violate the premise of resonance-averaging and render its usefulness questionable.

The ^{146}Nd spectrum was checked at the 2 keV scandium beam available at the Brookhaven HFBR. Figure 7 shows the appropriate energy region of the spectrum, and the arrows indicate transitions not reported by Bushnell. Aside from the weak 5799 keV transition, all the states cited by Hamilton appear clearly in the spectrum. These observations emphasize the fact that while high-quality averaged spectra are required to support the "completeness" claim, there is no evidence here that the method of resonance averaging is inadequate in identifying a complete level set. In fact we go further: there is evidence that resonance averaging is as reliable <u>as any other method of spin-parity determination</u>, including that of nuclear polarization, provided only that it be properly applied. In this case "proper application" means simply a careful analysis of the data coupled with a proper fluctuation analysis.

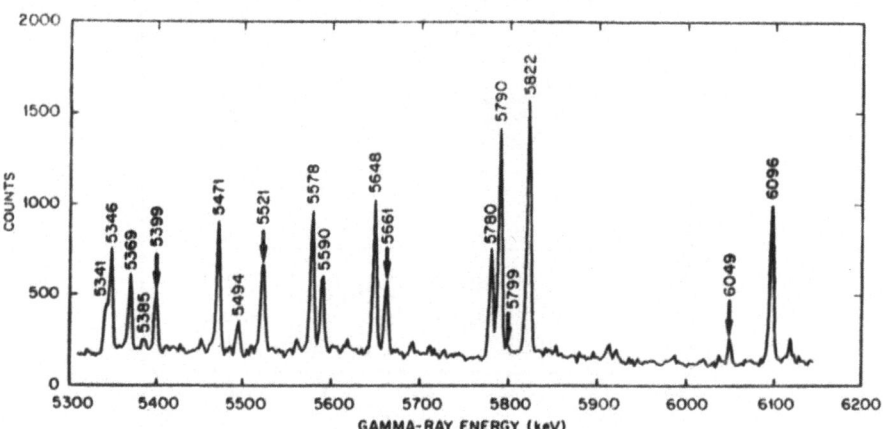

FIG. 7. A portion of the 2 keV ^{146}Nd spectrum /17/.

Fluctuations

The relative dispersions (σ/m) for resonance-averaged cross section may be directly evaluated from the Monte Carlo code RACA and compared to data. Table 1 summarizes such a calculation for the ^{168}Er case.

210

TABLE 1. Relative RMS fluctuations, $\tilde{\sigma}/m$, for ^{168}Er

J_f	2 keV				24 keV			
	s waves		s+p waves		s waves		s+p waves	
	Calc.	$2/\sqrt{\nu}$	Calc.	$2/\sqrt{\nu}$	Calc.	$2/\sqrt{\nu}$	Calc.	$2/\sqrt{\nu}$
2^-	0.194	0.200	0.197	0.141	0.119	0.133	0.109	0.080
3^-	0.135	0.137	0.135	0.098	0.081	0.092	0.077	0.054
4^-	0.133	0.137	0.133	0.098	0.078	0.092	0.073	0.054
5^-	0.188	0.189	0.187	0.134	0.109	0.127	0.101	0.073
2^+	0.194	0.203	0.171	0.141	0.119	0.133	0.077	0.080
3^+	0.135	0.133	0.116	0.098	0.082	0.092	0.056	0.054
4^+	0.133	0.139	0.120	0.098	0.080	0.092	0.053	0.054
5^+	0.188	0.188	0.161	0.134	0.109	0.127	0.074	0.073

Qualitatively the calculated dispersions may be understood in terms of the assumed Porter–Thomas dispersions of neutron and radiative models. Neglecting the correlations arising in the neutron and total widths, the expected relative dispersion becomes $2/\sqrt{\nu}$, which is what is obtained from the product of two random variables, each obeying x-square distribution with ν degrees of freedom; ν in this case represents the number of re- sonances in the averaging interval. Table 1 compares the Monte Carlo calculation with the quantity $2/\sqrt{\nu}$, and shows that the naive expectation is remarkably close to the calculated result.

Experimental observation in the case of ^{167}Er(n,γ)^{168}Er at 2 keV, for 3^- and 4^- capturing state gives a relative standard deviation of 0.137, in good agreement with the table. These fluctuations, when compared to the mean intensity for populating states of various spin and parity (Fig. 8) determine the sensi- tivity of the method. In the case of 2 keV ^{167}Er capture, the sensitivity allows us to distinguish parities, and the ($3^\pm$,$4^\pm$) spin groups from the ($2^\pm$,$5^\pm$) spin groups, over 2 MeV range of excitation.

Entrance Channel Effects

Now we are in a position to turn the method around and inquire:

are the data indicative of any violations of the assumptions
a) to f) above?

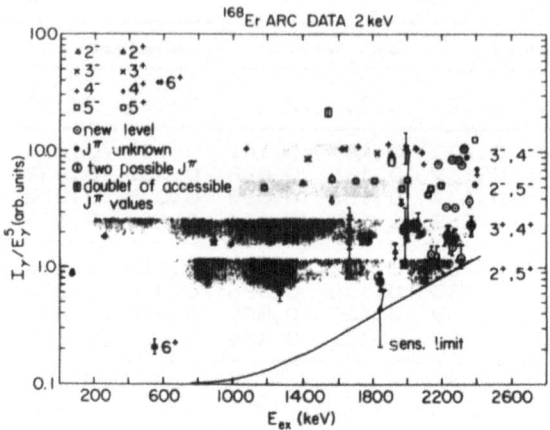

FIG. 8. Resonance-averaged spectrum for ^{168}Er.

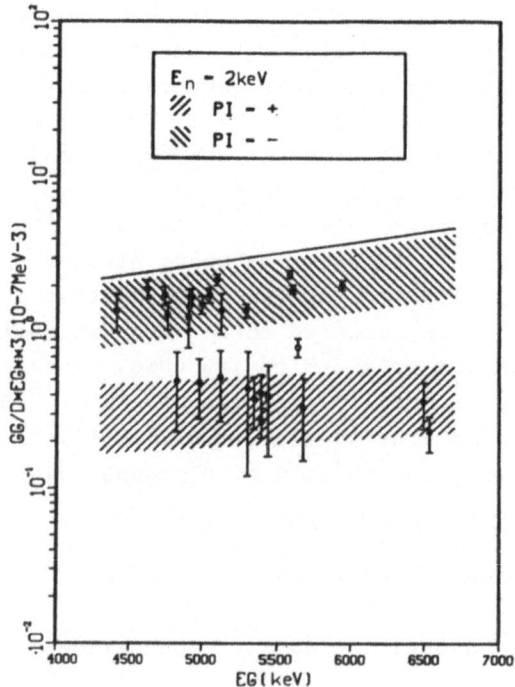

FIG. 9. Resonance-averaged spectrum for ^{240}Pu /18/.

An extreme, but highly instructive example of how the simple population patterns of Fig. 2 can be altered by entrance channel effects is afforded by the example of ^{240}Pu /18/, whose resonance-averaged 2 keV intensities are displayed in Fig. 9. A clear grouping into odd and even parity bands is clearly evident, but the expected spin grouping is completely absent. (The 0^+ and 1^+ resonances should populate final states 0^-, 1^-, 2^- in the ratios 1:2:1 by E1 transitions, and 0^+, 1^+, 2^+ states by M1 transitions in a similar pattern.)

^{240}Pu is an <u>exception</u> to the general rule because the target nucleus ^{239}Pu is fissile. It is known experimentally, and expected from the channel theory of fission, that the fission widths are strongly spin dependent. For ^{239}Pu it is found that $\langle\Gamma_F(J = 0)\rangle/\langle\Gamma_F(J = 1)\rangle \simeq 65$ /19/. The large $J = 0$ fission widths result in large total widths for $J = 0$ resonances, and since,

$$\sigma_{n,\gamma} = \sigma_c \Gamma_\gamma / \Gamma,$$

the radiative capture for $J = 0$ resonances is suppressed. The relative probability for final state population, Fig. 10, shows only two groupings for dipole transitions, with similar populations of states 0, 1 and 2.

This example is a good one for illustrating several other features of averaged spectra. Perhaps the most important is the

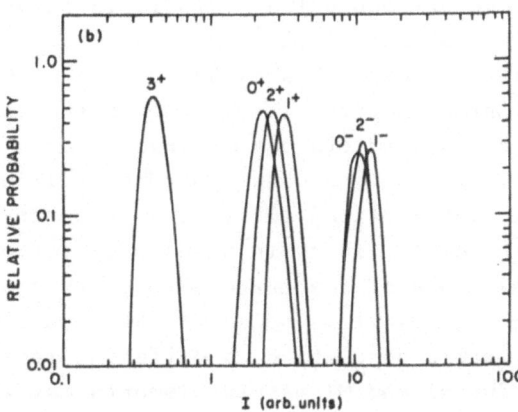

FIG. 10. Calculated intensity distributions for ^{240}Pu /18/.

apparently different energy dependences of the positive and
negative parity population. Referring first to the latter states,
it is obvious that the dominant multipole in the population is
E1 (there is a small correction for M1´s following p-wave neutron
capture). The Brink—Axel hypothesis implies that strengths of
these E1 transitions vary with excitation energy in the same way
as the photoexcitation cross section. Parametrization of the
photoexcitation cross section by the Lorentzian form

$$\sigma_a = \sigma_0 \left\{ E^2 \Gamma^2 / [(\Delta E^2)^2 + \Gamma^2 E^2)] \right\} \quad \text{where} \quad \Delta E^2 = E_0^2 - E_\gamma^2$$

and allowing for the split in the giant resonance caused by
nuclear deformation, permits us to calculate cross sections at
low nuclear excitation. The Brink—Axel hypothesis relates the
photon strength function to the photoabsorption cross section

$$\langle \Gamma_\gamma \rangle / D = (3\pi^2 \hbar^2 c^2 / E_\gamma^2) \langle \sigma_a \rangle \ ,$$

$$f_{E1} = \langle \Gamma_\gamma \rangle / DE^3 ,$$

and f_{E1} is shown plotted as the solid line in Fig. 9. Because
the intensities have been normalized to the absolutely determined photon strengths of ref. /18/, the data may be compared
absolutely. It can be seen from the figure that the giant resonance tail well reproduces the energy variation of the intensities, but the intensity observed is about 65% of that expected
from the resonance extrapolation. This is in agreement with the
observations of McCullagh et al. /20/ based on a large sample
of discrete resonance data. The recent review of Raman /10/ confirms the conclusion that the Lorentzian extrapolation of the
dipole resonance overpredicts the observed photon strengths for
(n,γ) reactions. Maureen and Donald Gardner /21/ have suggested
several possible explanations for this overprediction, and
support McCullagh et al. in their suggestion for an energy
variation in the width of the Lorentzian expression.

Returning to Fig. 9, it is apparent that the slope of the
positive parity state population is different from that of negative states. The E1 and M1 matrix elements may vary differently, but superimposed on this variation is the effect of p-wave

214

capture on the population of these positive parity states. The
effective photon strength function must be written as the
weighted average of two components: one corresponding to s-wave
capture followed by M1 radiation, the other to p-wave capture
followed by E1´s. The Monte Carlo code RACA indicates that
about 35% of the population is due to p-wave. The assumption of
an <u>energy-independent</u> $f(M1) \equiv \Gamma(M1)/DE_?^3$, leads to the calculated band shown on the figure.

The M1 variation shown here suggests an increasing divergence
between E1´s and M1 s as <u>gamma-ray energy increases</u>. Lone /22/
and, more recently, Kopecky /23/ have revived an old suggestion
that collective M1 strength may be concentrated in a giant resonance at an excitation given by the spin-orbit partner split-
tings of the shell model.

Such an M1 giant resonance is strikingly illustrated by boron
filtered beam data on ^{106}Pd (Fig. 11). Here the positive parity
state populations seem to be turning over about a broad maximum
near 8 MeV. Bohr and Mottelson /24/ give an approximate energy

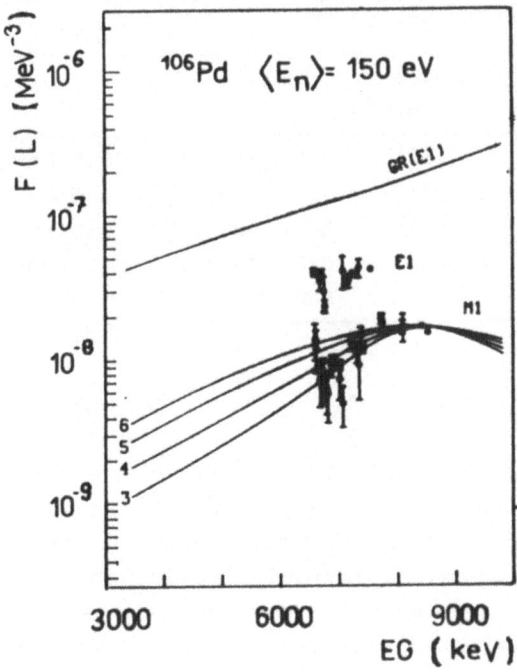

FIG. 11. Resonance-averaged spectrum for ^{106}Pd.   215

for the spin-flip strength at about one-half the E1 giant resonance energy, or $E_o \approx 41\ A^{-1/3}$. This energy would place an M1 giant resonance maximum ranging from 9 MeV for A = 100 to 6.6 MeV for A = 240.

Kopecky /23/ has collected a set of nine resonance-averaged spectra ranging from ^{96}Mo to ^{239}U, from which the exponent n in the power-law dependence $I_\gamma(M1) \approx E_\gamma^n$ may be derived. The exponent n may be related to an assumed giant resonance energy, E_o, and width Γ. The normalization of the strength function is taken from discrete resonance data.

Sufficiently precise M1 data does not yet exist to pin down these M1 giant resonance parameters with any great exactitude, but the data are certainly suggestive of a giant resonance interpretation. Table 2 shows summed M1 strength for 5 nuclides, derived from the measured M1 strength function and the giant resonance interpretation. These are in reasonable agreement with shell model calculations.

TABLE 2. M1 strengths calculated on the Giant Resonance Assumption $S(M1) = \Sigma EB(M1)$

Nucleus	$E_{\gamma M1}$ (MeV)	$f(M1)$ (10^{-8} MeV^{-3})	$S(M1)$ (μ_o^2 MeV)		
^{106}Pd	7.5	1.2(3)	220 (55)	$E_o \approx 8.5$ MeV, $\Gamma \approx 4$	MeV
^{168}Er	6.2	2.7(5)	780(144)	$E_o \approx 8.5$ MeV, $\Gamma \approx 6$	MeV
^{176}Lu	5.8	1.8(5)	580(160)	$E_o \approx 8.5$ MeV, $\Gamma \approx 6$	MeV
^{233}Th	4.2	4.0(8)	690(140)	$E_o \approx 6$ MeV, $\Gamma \approx 6{-}8$	MeV
^{239}U	4.1	3.0(6)	520(105)	$E_o \approx 6$ MeV, $\Gamma \approx 6{-}8$	MeV

Non-statistical Effects

As has been previously stated, the basis for the application of the averaging method to nuclear structure problems is the statistical model for resonance decay. Are the present data sufficient to test this assumption? Clearly, the success obtained for spin and parity determinations by this method implies that initial and final state correlations cannot be large.

Table 1 shows that a departure of an intensity fluctuation from the average value by two or three standard deviations will mask non-statistical effects which are as large as 60% of the mean — a very large effect. It follows then that averaging can only exclude the largest effects and will not be sensitive to effects at, say, the 30% level.

Nevertheless, there have been very large effects reported. One such effect is the report of Stefanon et al. /25, 26/ of the possible dependence of primary capture gamma-ray strength on the K quantum number of the final state in ^{174}Yb. Stefanon found that for resonances over the lowest 200 eV over threshold, the ratio $I(K = 2^+)/I(K = 0) \geq 3.5$, when the intensities were corrected by and E_γ^5 reduction factor.

Recently, ^{174}Yb has been re-examined with the 2 keV scandium filter at the Brookhaven High Flux Beam Reactor with the specific intention of examining such non-statistical effects. At the present time, only preliminary data are available from this experiment. Even preliminary data, however, are quite sufficient to show that the original claim is not supportable.

Transitions to $K = 0^+$ final states at 1561, 1715, 1958, 2123, 2171, and 2336 keV were compared to $K = 2^+$ final states at 1634 and 1805 keV. Only well-separated transitions are included in these sets. A ratio $I(2^+)/I(0^+) = 1.39 \pm 0.20$ was found, which is within averaging statistics.

Thus we can conclude that K-dependent effects, if present, must be at a 20% level or smaller.

Using resonance-averaging to confirm initial state correlations such as, for example, correlations between neutron and radiative widths would appear not to be possible. There is, however, an interesting way to modify the incident neutron spectrum in the filter. To introduce this idea, let me first quote from an exchange between Lowell Bollinger and myself at the 1966 Argonne Slow Neutron Capture Gamma-ray Conference (p. 530), in reference to boron filtering.

Chrien: "... there may be also an off-resonant component. If so, it would influence your results ..."

Bollinger: "What you say in undoubtedly true. In fact it may turn out to be one of the more interesting things to try in

the future. It may be possible to look for a difference ...
when capture at the resonances is somehow inhibited."

Off-resonance capture has recently been demonstrated at 2 keV
at Brookhaven /27/. For samples of relatively small mass number,
the level spacing can be large enough so that no resonances
appear in the averaging window. Such a case is $^{59}(n,\gamma)^{60}Co$ which
has no s- or p-wave resonance near 2 keV. Despite the unfavor-
able ratio of $\sigma(n,\gamma)/\sigma(n,n) = 130/2000$, the spectral quality is
remarkably good (see Fig. 12) from a sample of 9.9 gms/cm^2. It
is interesting to note that the intensities differ significantly

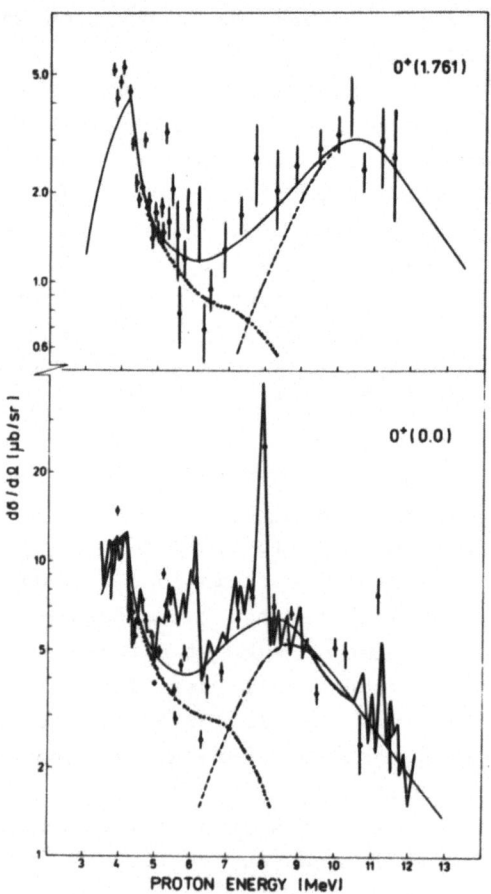

FIG. 12. Off-resonance (2 keV) spectrum for ^{60}Co (above),
compared to thermal capture (below).

from the thermal capture, even though the same resonance tails contribute.

Having made the point that off-resonance capture can be measured at 2 keV, it is evident that for most samples, the spectrum is dominated by the resonances. Let us suppose, however, that a _resonance filter of the same composition as the target_ is placed in the beam. To a first approximation, the resonance contribution is thus removed and we are left _only_ with the off-resonance contribution. We are then free to examine final state correlations and expect to see the effects of direct neutron capture enhanced to those states with significant single-particle character.

Furthermore, by varying the thickness of the resonance absorber, it may be possible to alter the effects of the Porter—Thomas averaging so that, for example, a thin absorber removes mainly the large resonances, selectively.

An experiment to test the correlations observed by Becvar /11/ in ^{173}Yb(n,γ)^{174}Yb using this variation of the "self-indication" method is in progress at the BNL HFBR. A wide range of thicknesses will be studied, and the effects on the averaging spectrum will be observed. It will be interesting to see if this method is useful for direct capture studies.

REFERENCES

1. Proceedings of the Conference on Slow-Neutron-Capture Gamma-ray Spectroscopy, ANL-7282, 1966, p. 523.
2. L. M. Bollinger and G. E. Thomas, Phys. Rev. C2 (1970) 1951.
3. R. G. Greenwood and R. E. Chrien, Nucl. Inst. Meth. 138 (1976) 125.
4. R. E. Chrien, in "Neutron Capture Gamma-ray Spectroscopy and Related Topics 1981", Conf. Ser. 62, Inst. of Physics, Bristol - London, 1982, p. 342; R. E. Chrien, Trans. N.Y. Acad. Sci. 44 (1980) 40.
5. R. F. Casten, D. D. Warner, M. L. Stelts, and W. F. Davidson, Phys. Rev. Lett. 45 (1981) 1077.
6. R. C. Block and R. M. Brugger, in "Neutron Sources for Basic Physics and Applications", Pergamon Press, Oxford, 1983, p. 177.
7. R. C. Greenwood and A. J. Caffrey, in "NEANDC Specialists Meeting on Yields and Decay Data of Fission Products Nuclides", Brookhaven National Laboratory Report 51778, 1983, p. 356.

8. R. E. Chrien, in "Neutron Capture Gamma-ray Spectroscopy, Reactor Centrum Nederlands, Petten, 1975; p. 247.
9. S. F. Mughabghab and R. E. Chrien, in "Neutron Capture Gamma-ray Spectroscopy", Plenum Press, New York, 1979, p. 265.
10. S. Raman, in ref. /4/, 1982, p. 357.
11. F. Becvar, in "Capture Gamma-ray Spectroscopy and Related Topics 1984", Amer. Inst. of Physics, New York, 1985, p. 345.
12. O. Sahal, S. Raman, G. G. Slaughter, C. Coceva, and M. Stefanon, Phys. Rev. C25 (1982) 1283.
13. P. Axel, Phys. Rev. 126 (1962) 671.
14. J. Kopecky, in ref. /11/, p. 318.
15. W. D. Hamilton, S. J. Robinson, and D. M. Snelling, J. Phys. G: Nucl. Phys. 9 (1983) L13.
16. D. L. Bushnell, G. R. Tassotto, and R. K. Smither, Phys. Rev. C14 (1976) 75.
17. S. Raman, O. Sahal, M. J. Kenny, and R. E. Chrien, J. Phys. G: Nucl. Phys. 9 (1983) L137.
18. R. E. Chrien, J. Kopecky, H. I. Liou, O. A. Wasson, J. B. Garg, and S. Dritsa, Nucl. Phys. A436 (1985) 205.
19. S. F. Mughabghab, "Neutron Cross Sections", Part B, Academic Press, New York, 1984.
20. C. McCullagh, M. L. Stelts, and R. E. Chrien, Phys. Rev. C23 (1981) 1394.
21. M. A. Gardner and D. G. Gardner, in ref. /4/, 1982, p. 319.
22. M. A. Lone and F. C. Khanna, Proc. Int. Conf. on Nuclear Physics with Electromagnetic Interactions, Sect. 1, p. 24, Mainz, 1979; M. A. Lone, personal communication.
23. J. Kopecky, contributed paper submitted to Int. Conf. on Nuclear Data for Basic and Applied Science, Santa Fe, 1985.
24. A. G. Bohr and B. R. Mottelson, "Nuclear Structure", Vol. II, W. A. Benjamin, London, 1975, p. 636.
25. M. Stefanon, in ref. /11/, p. 335.
26. S. Raman, O. Sahal, and B. G. Slaughter, Phys. Rev. C23 (1981) 1794; see ref. /12/.
27. J. Kopecky, unpublished data from BNL-ECN, Petten collaboration, 1984.

220

EXPERIMENTAL STUDY OF FAST NEUTRON CAPTURE

F. Cvelbar and A. Likar

J. Stefan Institute and Faculty of Natural Sciences
and Technology, E. Kardelj University, Ljubljana,
Yugoslavia

ABSTRACT. The status of the present-day technique of the energetic neutron capture study is presented. The experimental data
are compared with the results of the direct-semidirect capture
model.

1. INTRODUCTION

As the electromagnetic interaction is very well known, one
would expect that the fast nucleon radiative capture is one of
the most established reactions. There are, however, the collective effects - giant multipole modes (e.g. /1/), that make the details of this reaction mechanism rather unknown. Since the cross-
section for (n,γ) reaction is low, it was not studied until the
last 15 years, using projectiles of the energy of the order of
10 MeV to 20 MeV. Interest in this reaction grew up in parallel
with the improvement of the experimental technique, and especially when higher than E1 multipole giant resonance was observed in inelastic scattering of nucleons and electrons. Nowa-
days the study of the radiative capture of fast nucleons is an
interesting and promising field of research.

In this report the experimental study of fast neutron capture involves especially the in-beam spectroscopy of fast capture gamma rays at different angles and the excitation of the
angular differential (n,γ) cross-section ($\sigma(\theta) = d\sigma/d\Omega$) as a
function of neutron energy 5 MeV $< E_n < $ 30 MeV for capture transitions to the particular (selected) states of the final nucleus (excitation functions). Though yielding interesting data,
however, not directly concerning the problems considered, the

in-beam (n,γ) studies using neutrons of the energy of few MeV
(e.g. ref. /2/) are not considered here. The same is true for
the off-beam (activation) measurements of the 14 MeV neutrons
(e.g. /3/).

Advantage of the study of (n,γ) over the (p,γ) reaction,
which is to be explained later in the text, should be mentioned
already at the beginning.

As usually, the angular distribution of gamma ray is presented with the coefficients of the Legendre expansion (a_i)

$$\sigma(\theta) = A_o(1 + \Sigma a_i P_i(\cos \theta)).$$

The majority of fast neutron capture data were obtained
using the unpolarized neutrons. Experimental study using polarized neutrons was successfully introduced recently /4, 5/
(the corresponding proton measurements appeared a decade ago
/5/). In our case besides the $\sigma(\theta)$ also the asymmetry of the
product $\sigma(\theta)A(\theta)$ is studied; $A(\theta)$ means the analysing power
defined as

$$A(\theta) = 1/P\left[(N^+ - N^-)/(N^+ + N^-)\right],$$

where P is the (neutron) beam polarization, and N^+ and N^- are
the number of counts obtained for spin-up and spin-down events,
respectively. The product $\sigma(\theta)A(\theta)$ is expanded over the associated Legendre polynomials P_i^1

$$\sigma(\theta)A(\theta) = A_o \Sigma b_i P_i^1(\cos \theta).$$

Gamma rays are measured in the presence of background,
caused by direct and scattered neutrons. As, in addition, the
cross-section does not practically exceed 10 μb/sr in light
nuclei, and 100 μb/sr in heavier ones, measurements are tedious
and optimal selection of the target nuclei is unavoidable.

Experimental data were successfully reproduced with the
results of the so-called direct-semidirect (DS /6, 7/) capture
model which takes into account, besides the direct capture
process, also the effect of the giant multipole modes. Till
now it has mostly been used in the description of the giant
electric dipole and giant electric quadrupole mode contributions.

222

Parameters of the giant dipole resonance (GEDR) (L = 1, T = 1, S = 0, $E_{1R} = 77A^{-1/3}$ MeV, 4 MeV $< \Gamma_{1R} < 8$ MeV) that were studied most intensively in the past are well known now (e.g./1, 8/).

Isoscalar giant quadrupole resonance (ISGQR) (L = 2, T = 0, S = 0, $E_{2SR} = 65A^{-1/3}$ MeV, $\Gamma_{2SR} = 90A^{-2/3}$ MeV, $\eta_{2SR} \approx 100\%$ for A $\succ$ 100) (e.g. /1, 8/) was identified from the modulation of the excitation function of inelastic scattering of hadrons (e.g. alpha particles).

Isovector giant quadrupole resonance (IVGQR) (L = 2, T = 1, S = 0, $E_{2VR} = 130A^{-1/3}$ MeV, $\Gamma_{2VR} = 6$ MeV, $\eta_{2VR} \approx 100\%$ for A $\succ$ 50) (e.g. /1, 8/) cannot be intensively excited by the hadron scattering. The source of the IVGQR data is the electron scattering process (here η represents the relative part of the energy weighted sum rule, exhausted by the observed resonance; other symbols are self-explanatory).

In the capture study, the spectral intensity of the quadrupole mode contribution is usually hidden by about 10 times more intense dipole transitions. However, it can be efficiently studied through the fore-aft asymmetry of the gamma-ray angular distribution resulting from the interference between the dipole and the quadrupole radiations.

A number of excellent review papers and conference discourses devoted to the nucleon capture have been published recently /5, 9—11/.

In this report the status of the present-day experimental technique in the fast neutron capture study is presented. After a brief outline of the direct-semidirect capture model, its results are compared with the experimental data.

2. EXPERIMENTAL TECHNIQUE

2.1. Gamma-ray Spectrometers

An increasing interest in the experimental nucleon capture study brought about in the last decade a rather intensive development of the NaI(Tl) gamma-ray spectrometers for energies of about 20 MeV. The aim was to access the energy resolution as good as possible (e.g. 500 keV) in order to select the capture transitions for the individual low excited states and to measure as efficiently as possible the rather low capture cross-sections.

Since 98% of 20 MeV gamma rays are absorbed in the 25 cm thick layer or NaI(Tl), it is natural that one needs a huge crystal with dimensions of about Ø 25 cm x 25 cm to orient the development towards the goals set before. Using such a crystal, the energy resolution[*] (FWHM) of the spectrometer from 6% /12/ to 8% /13/ was obtained and, though not satisfactory, it allowed relatively reliable measurement of the (n,γ) excitation functions to the separate final states. Typical example are the results of the Uppsala group /14/.

Essential improvement of the resolution, followed from the introduction of the well type anticoincidence shield around the NaI(Tl) crystal (e.g. ref. /15/ and references therein). Events in which the radiation escaped from the main crystal and was absorbed in the shield, were not accepted in the gamma-ray spectrum measured. Combined with the selected NaI(Tl) crystal (Ø 25 cm x 25 cm) this method yielded the spectrometers, having the (FWHM) resolution of 4.6% /13/, 3.8% /16/, 3.4% (/17/ and references therein) and 3—4% (2.4% in ideal working conditions, TUNL /18, 19/). Selected crystal which had the dimensions of Ø 24 cm x 36 cm allowed the resolution of 2.2% (BNL /20/) and 1.5% (Uppsala /21/; Fig. 1) and also very low width of the peak at one tenth of the maximum FWTM, i.e. 6.5% and 7.4%, respectively. In the BNL spectrometer 56% of detected pulses were rejected due to the anticoincidence shield.

The main reason for the excellent resolution of the last two spectrometers lies in the selection of the crystals and PM tubes (RCA 4900, RCA S83021E, Ampex 2202) in working-out the crystal surface to get (within 0.6%) the same gamma-ray scintillation sensitivity over the whole volume. It was found that for this treatment gamma rays of the energy of 6.13 MeV from

---

[*]In most response function measurements, presented here, the $^{11}B(p,\gamma)^{11}C$ reaction at E_p = 7.25 MeV, yielding 22.6 MeV gamma rays, was used.

^{244}Cm^{13}C source are much more appropriate than the usually used
0.661 MeV gamma rays from the ^{137}Cs source. Some contribution
to the outstanding FWTM resolution of the BNL spectrometer results also from the design of the anticoincidence shield, of
which the front plastic scintillator is separated from the
anulus composed of 6 optically isolated sections. The scintillation efficiency of these separate sections was much easier to
equalize in comparison with the compact anulus. It was also
found that the light collection is more efficient if these scintillators are polished and wrapped in Al foil, rather than
painted with the diffuse paint TiO$_2$.

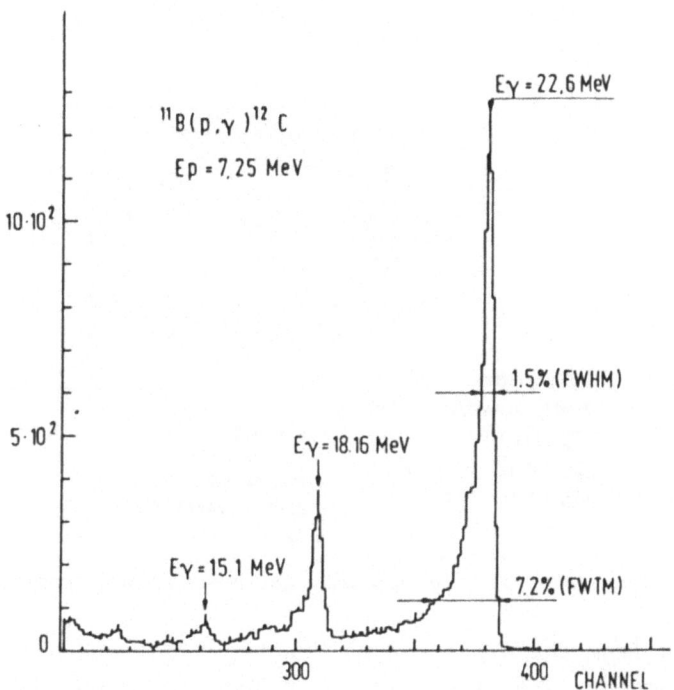

FIG. 1. Response of the Uppsala gamma-ray spectrometer /21/.

As far as the energy and time resolution is concerned it
is definitely better to attach several 7.5 cm Ø photomulti-
pliers, instead of the larger one, to the crystal. The reason
for it lies in the fact that the cathode of a large PM is rather
inhomogeneous and the first dynodes are very different.

225

The Uppsala spectrometer is presently used to study the $^{89}Y(n,\gamma)^{90}Y$ reaction separately for the transition to the $2d_{5/2}$ g.s. and $3s_{1/2}$ (1.2 MeV) excited states.

The BNL spectrometer has recently been applied to measure the reaction $^{27}Al(p,\gamma)^{28}Si$ /20/ at 11 MeV < Ep < 39 MeV, with the aim to study systematically (for the first time) the giant resonances built on the excited states.

The Uppsala spectrometer constructed for neutron capture measurements is shown in Fig. 2. A 10 cm thick layer of lead shields the apparatus against the gamma rays and cosmic rays.

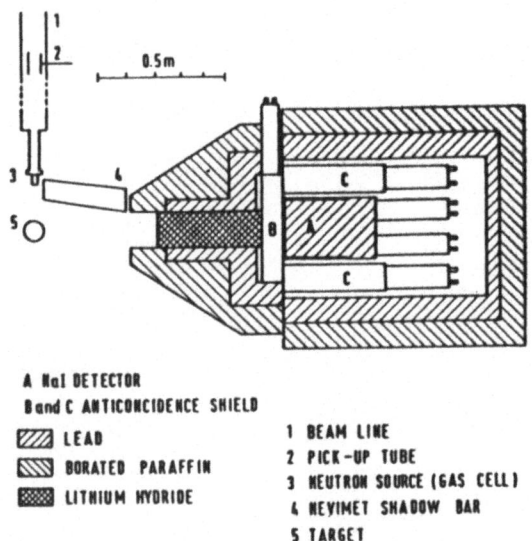

A NaI DETECTOR
B and C ANTICONCIDENCE SHIELD

▨ LEAD
▨ BORATED PARAFFIN
▨ LITHIUM HYDRIDE

1 BEAM LINE
2 PICK-UP TUBE
3 NEUTRON SOURCE (GAS CELL)
4 HEVIMET SHADOW BAR
5 TARGET

FIG. 2. Construction of the Uppsala gamma-ray spectrometer /20/.

Neutrons from the experimental chamber, scattered towards the spectrometer, are attenuated with borated 10 cm thick paraffin shield. The spectrometer entrant hole is filled with a 40 cm long cylinder of lithium hydride. The front face of the spectrometer appears at a distance of 97 cm from the centre of the (n,γ) reaction target. A hevimet shadow bar protects the spectrometer against the direct neutron flux.

In connection with the insertion of the neutron attenuator into the hole of a lead collimator, it is worth mentioning (BNL result /20/) that the 50 cm long paraffin attenuator

deteriorated the spectrometer resolution from 2.2% to 2.4%. In
the spectral region below 12 MeV, the area under the tail of
the response function increased significantly, depending also
on the effective energy threshold of the front anticoincidence
shield.

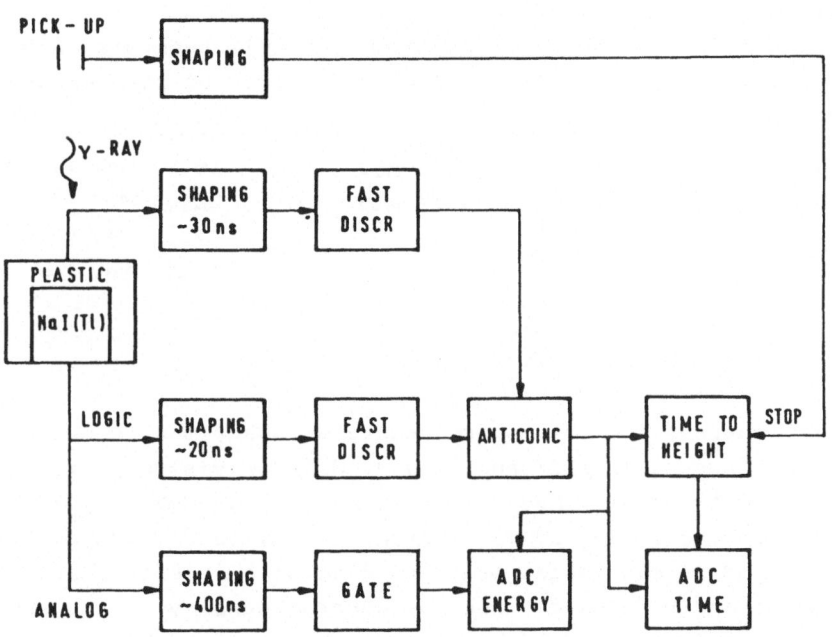

FIG. 3. Essential elements of the typical electronic chain
of the anticoincidence shielded NaI(Tl) spectrometer.

   Electronic chain of the spectrometer (Fig. 3) has to be
adapted to two experimental facts: (i) Pulses from NaI(Tl)
system cannot be clipped below ~ 400 ns in order not to lose
the energy resolution of the spectrometer /20/ and (ii) the
rate of pulses in the spectrometer is about 10^5 s^{-1}, which
means a rather high probability for pile-up events. To avoid
their registration, the output from the PM is split into a
slow (~ 400 ns) and a fast (20 ns) branch (for details see
refs. /13, 17, 19, 20/). Output pulses from the pile up reject-
ing block in the fast branch, which pass the anticoincidence
unit, vetoed by the plastic shield pulses operate the ADC of
voltage analyser and very often /12, 19/ also the ADC of the

time analyser. Time analysis, relative to the moment of the pulse of the neutron source, provides the possibility of the time-of-flight discrimination between the proper gamma-ray events and the events belonging to the delayed background neutrons /32/. The BNL spectrometer /20/ is additionally supplied with the LED system to control and stabilize the amplification factor of PM-s.

Nuclear (p,γ) reactions, yielding the well separated high-energy gamma rays, usually used for the study of the response functions are (e.g. ref./23/):

$$^{12}C(p,\gamma_o)^{13}N \qquad E_p = 14.3 \quad MeV \qquad E_\gamma = 15.07 \; MeV \; (resonance)$$
$$^{11}B(p,\gamma_o)^{12}C \qquad E_p = 7.25 \; MeV \qquad E_\gamma = 22.6 \quad MeV$$
$$T(p,\gamma)^4He \qquad E_p = 10 \; MeV \qquad E_\gamma = 27.3 \quad MeV$$
$$^{10}B(^3He,p,\gamma)^{12}C \qquad\qquad\qquad E_\gamma = 15.1 \quad MeV \; /24/$$

2.2. Neutron Sources

Neutron sources for the neutron capture measurement considered here are $^2H(d,n)^3He$ /25/ and $^3H(d,n)^4He$ /26/ yielding, for instance, neutrons of 7 MeV $< E_n <$ 14 MeV and 20 MeV $< E_n <$ 28 MeV energy, respectively, if for instance, 7 MV tandem Van de Graaff accelerator is available. Neutrons from both reactions are peaked strongly forward. Thus the neutron density at the sample position is favourably increased. In the gammay-ray angular distribution measurements this results in the spectrometer background, which is strongly dependent on the detecting angle. In the same time also the (very high) rate of low-energy pulses changes. This usually causes the change of the amplification factor and the resolution of the spectrometer. All these facts require additional care of the experimentalists.

2.5 cm long gas targets and a gas pressure from 3 b to 6 b are typical. Windows of the cells are made of nickel /12/ (5 mg/cm^2) or molybdenum /5/ (4 mg/cm^2) foils. To pulse the neutron source, the accelerator is clipped (up to 3 times) and bunched. Typically, the pulse width and the repetition rate are some ns and some MHz, respectively.

In the recent measurements of the capture of polarized fast

neutrons /19/, these were produced using the ^{2}H(^{2}H, n)^{3}He reaction. Bombarding the gas deuterium target with the beam of polarized deuterons from the Lamb-shift ion source (polarization of 0.60, average current of 150 nA) we obtain 108 n/sr/μC in the forward direction. To avoid the corrections of the experimental data of the analysing power, due to the asymmetry and time dependence of the background, the TUNL measurement system has been applied with an additional (identical) NaI(Tl) spectrometer positioned symmetrically as to be relative to the neutron beam axis and the first NaI(Tl) detector /4, 19/.

3. DIRECT-SEMIDIRECT CAPTURE MODEL ANALYSIS OF THE DATA

Experimental data are best evaluated in the light of the so-called direct-semidirect (DSD) capture model /6, 7/. Now we shall give its brief review.

In the DSD model the transition matrix element M_{if} is the sum of the term describing the direct capture of the bombarding nucleon to the (single-particle) final state, accompanied by the emission of the gamma photon, and the resonant term, resulting from the two-step process in which the resonance mode is first excited and then, later on, de-excitated by the emission of the gammay ray

$$M_{if} = <u_b \mid d^L + h^L(r)/(E_\gamma - E_{RL} + i\Gamma_{RL}/2) \mid \chi_E^+ >.$$

Here $\mid u_b >$ is the radial wave function of the captured nucleon in the bound state, χ_E^+ is the optical model continuum wave function, d^L represents the radial part of the single-particle electromagnetic operator for radiation of multipolarity L; E_{RL} and Γ_{RL} refer to the position and the width of the (multipole L) giant resonance of the combined target plus nucleon system, respectively. The quantity $h^L(r)$ represents the radial part of the incident nucleon-target nucleus vibration coupling interaction. There were proposed several forms of $h^1(r)$. Recently, the complex volume form /7/, in which the real and imaginary strengths V_1 and W_1 (though in some way connected to the corresponding strengths of the isospin part of the optical model potential) are treated as free parameters, has been considered most adequate. Parameters are usually extracted: (i) by fitting the GDR excitation function at $\theta = 90°$ /24/, and less accurately, (ii) from the description of the (n, γ) spectra belonging to the 14 MeV neutron capture (see Fig. 4 /27/). Using the V_1 and W_1 values obtained from (i), the coefficients a_i and b_i are calculated and compared with the experimental results.

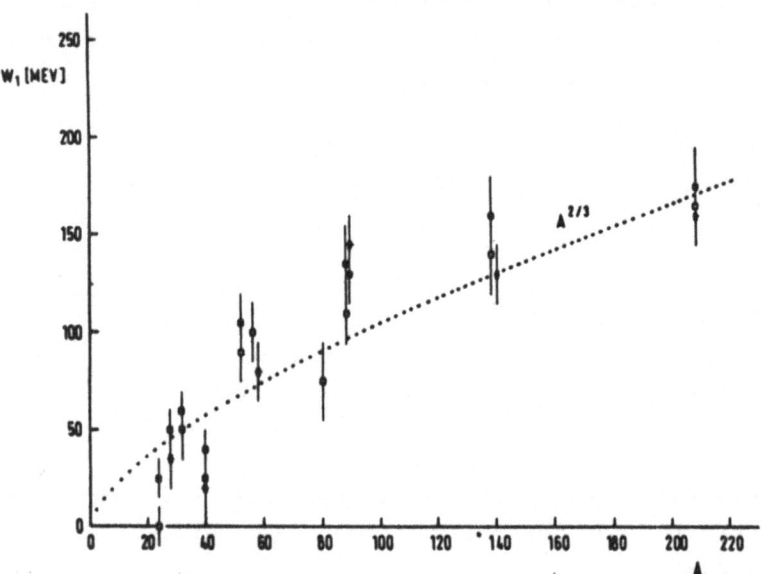

FIG. 4. Mass dependence of the strength of the imaginary part of the particle-dipole collective motion form factor W_1 extracted from the 14 MeV capture data using the DSD model /27/.

Lately a version of the DSD model, the so-called "pure resonance" (PR) model, in which the direct process was formally included into the resonant term, has been reported /28/. The model is based on the introduction of the projected optical model wave function for which the direct capture term is zero. As its results (e.g. $\sigma(90°)$) are less dependent on the selection of the optical model potential and also on the W_1 value, it was thought to replace the usually used DSD model. It seems, however, that the PRM inadequately describes the energy dependence of the expansion coefficients a_i and b_i (e.g., a_2 and b_2 - see Fig. 8), and therefore incorrectly predicts transition

230

amplitudes and phases of contributing partial waves. In this
connection see also ref. /29/.

Data from polarized capture measurements yield, in some cases,
enough coefficients: A_o, a_i and b_i (see e.g. /5/) to allow a
model-independent determination of the transition amplitudes
and their relative phases. Only the $s_{1/2}$ and $d_{3/2}$ incident channels contribute in the dipole capture of nucleon by even-even
nucleus into $p_{1/2}$ state. Unknown are therefore two transition
amplitudes and their relative phases, all these being connected
with three experimental values: A_o, a_2 and b_2 through quadratic
equations. The role of the model in this case is to select the
proper solution from the two of the quadratic equations. In less
favourable cases there are more unknown than measurable quantities. The comparison of the measured and calculated results
tells us about the adequacy of the model. The general result
is that that the DSD model fairly well describes: (i) the (n,γ)
excitation functions (Fig. 5, ref./30/, see also refs. /9, 20/)
and (ii) the energy dependence of the expansion coefficients a_2

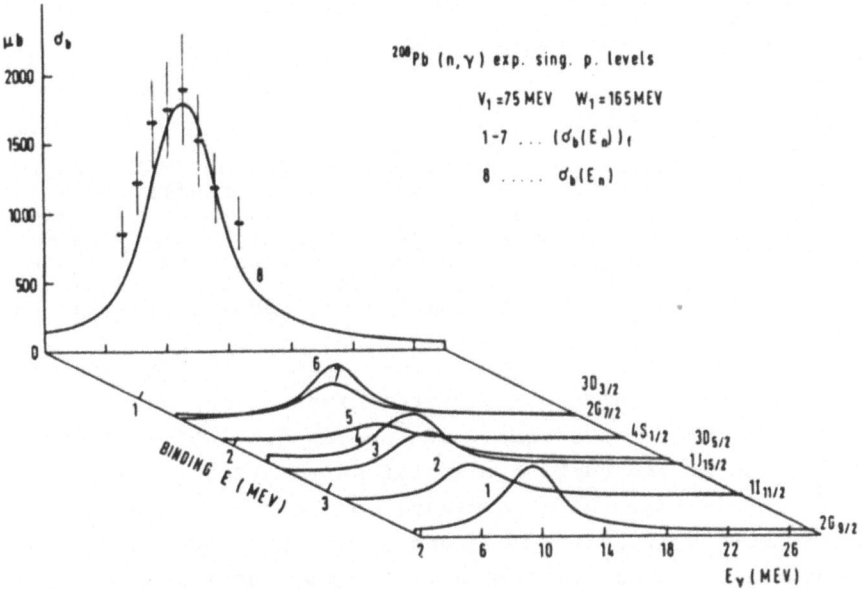

FIG. 5. The DSD model (combined with the Brink hypothesis)
calculation of the energy integrated neutron capture excitation function /30/.

and b_2 at least for not heavy nuclei (Fig. 6, see also refs. /4, 5/), the (gross) structure in $\mathcal{A}_1$ in the neutron (or excitation) energy dependence as a result of the interference between the dipole and quadrupole (Fig. 7, see also refs. /4, 5, 10, 11, 24/), and even magnetic dipole giant mode /4/.

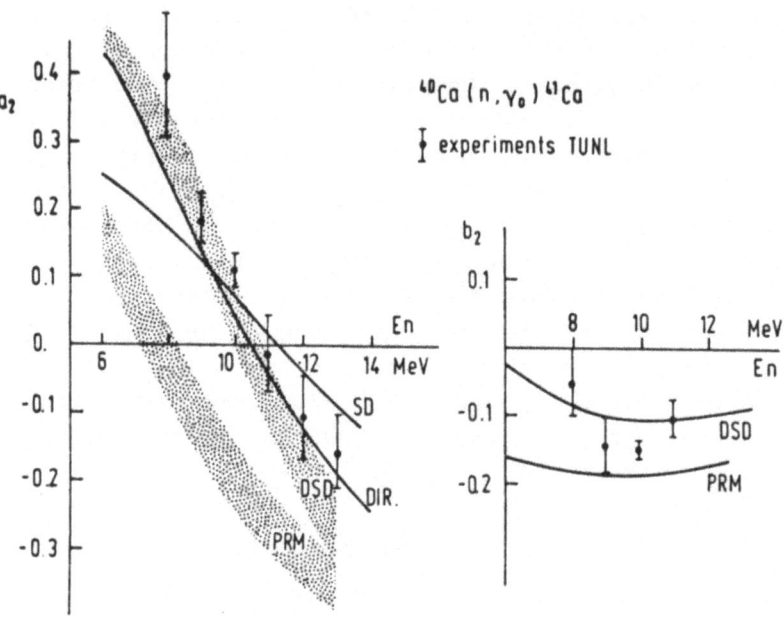

FIG. 6. Experimental /5/ and calculated (DSD and PRM) neutron energy dependence of the a_2 and b_2 expansion coefficients for the neutron capture to the $1f_{7/2}$ g.s. of ^{41}Ca. For the sake of completeness, also the separate results of the direct (DIR) and the semidirect (SD) calculations are presented. Stippled areas cover the results of the calculation using possible (and also extreme) values of the optical model potential and of the W_1 parameter.

Example given in Fig. 6 is rather typical for medium weight nuclei and shows the adequacy of the DSD model also for the description of the capture transitions to a rather high spin state. In the heavy ^{208}Pb nucleus, on the other hand, neither the DSD nor the PR model do properly describe the neutron energy dependence of the a_2 coefficient for the transition to the $2g_{9/2}$ or $1i_{11/2}$ levels (Fig. 8, ref. /4/).

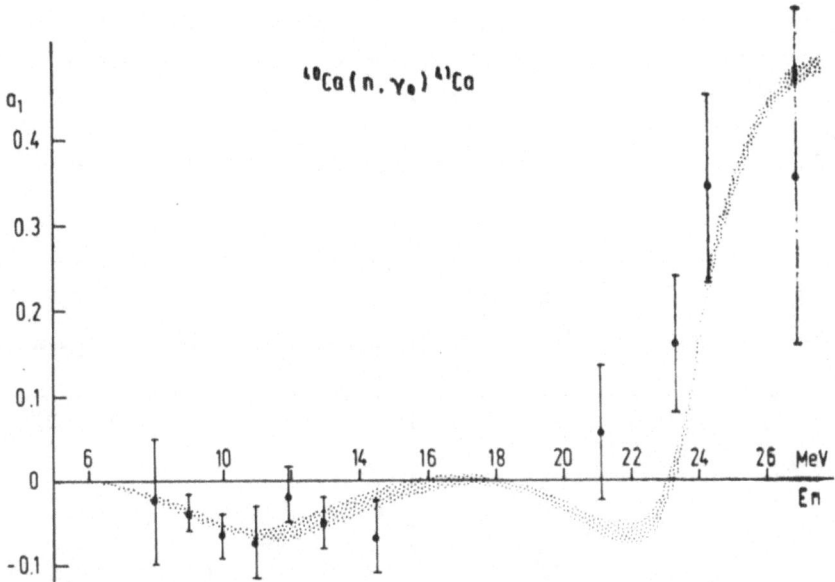

FIG. 7. Experimental /5, 24/ and DSD model calculated neutron energy dependence of the $\mathcal{A}_1$ asymmetry coefficient for neutron capture transition to $1f_{7/2}$ g.s. of ^{41}Ca. ($\mathcal{A}_1 = (I_{55}-I_{125})/(I_{55}+I_{125})$, where I_θ is the normalized yield at angle θ in degrees. Stippled area covers the results of the calculation performed with possible (and also extreme) values of the optical model parameters. In the calculation, besides the giant dipole state (at E^x = 18 MeV), also the isoscalar GQR (at E^x = 18.4 MeV) and isovector GQR (at E^x = 32 MeV) are taken into account.

4. CONCLUSION

Although the study of fast neutron radiative capture is experimentally hard, it is very important for our better understanding of the nuclear giant multipole modes. Its main power lies in the measurements of the interference between the dominating dipole and higher multipole radiations. As the sensitivity of the method is very high, we may expect it to provide us with some additional information about the resonances in low-mass region, similar to that resulting from the study of Ca(n,ᵧ) reaction (Fig. 7). The quadrupole resonances of such a light nucleus have not been observed in other reactions.

Yet we need to stress that in the higher mode (e.g. quadrupole) neutron capture, the direct (quadrupole) matrix element is practically zero and therefore the analysis of the collective effects is much easier in comparison with the results of (p,γ) reaction. In the quadrupole capture process, the neutron effective charge is exactly 4Z times /5/ smaller than that of the proton.

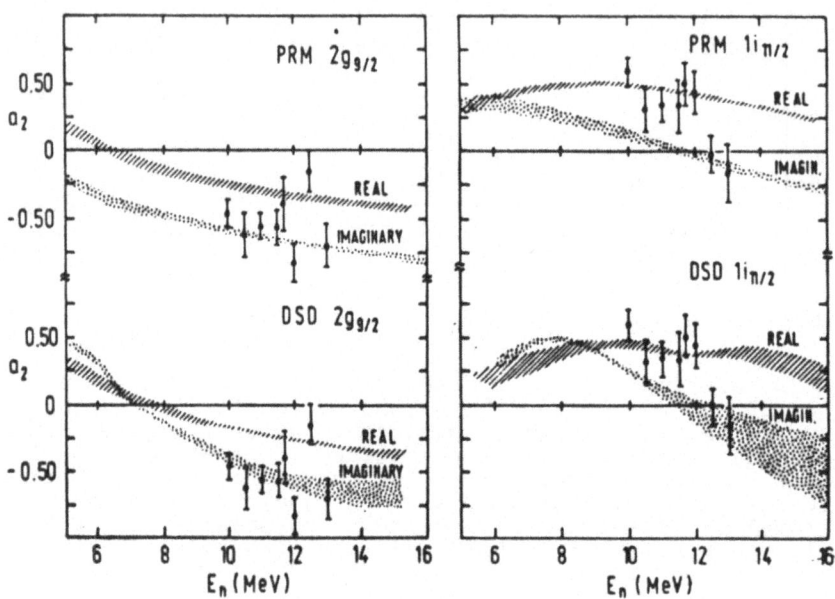

FIG. 8. Analysis of the a_2 coefficient, similar to that in Fig. 6, for neutron capture transitions to the $2g_{9/2}$ and $11_{11/2}$ states in ^{209}Pb /4/.

From what has been presented here it follows that it is but the effectiveness of the used methods that has been demonstrated in the study of the fast nucleon capture study. The main work remains to be done in the future. Let us try to list the activities expected to follow in this field:

1. Construction of an already designed large ($\emptyset$ 6 cm x 20.3 cm) germanium detector equipped with a 15 cm thick NaI(Tl) anti-coincidence shield /20/ in order to see whether the expected resolution of 0.3—0.5% (at 20 MeV) could be achieved.

2. Additional systematic examination of the giant resonance built on the excited states in order to study in detail the

Brink hypothesis stating that these resonances differ among themselves only in the (excited state) energy displacement.

3. Continuation of the higher mode giant resonance study, especially in light nuclei, where the resonances are spread over a larger energy interval than the heavier nuclei.

4. Performance of additional polarized (neutron) capture measurements in order to get the systematic information on the phase angle between the different partial waves.

5. Measurement of the high resolution (n, γ) spectra in order to see in detail whether the spectral intensity follows the single partial strength distribution of final states, as proposed in the DSD model combined with the Brink hypothesis.

6. Experimental and theoretical study of the (n, γ) reaction using neutrons of the energy between, say, 1 and 5 MeV in order to be able to examine the competition between the compound nucleus and the DSD reaction mechanism, and to continue the extraction of the gamma-ray strength function started by Joly et al. /2/.

Some of the listed activities will be efficiently worked-out using the pulsed spallation "white" neutron source in combination with the "crystal ball" gamma-ray spectrometer recently developed at the Los Alamos Meson Physics Facility (LAMPF) /31/. As it allows the time-of-flight selection of neutrons in energy channels, and the spectrometer consists of five $\emptyset$ 7.6 cm x x 7.6 cm bismuth germanate (BGO) crystals (resolution of 5% at E = 15 MeV), the excitation functions and angular distributions of capture gamma rays can be measured simultaneously.

Authors are indebted to Dr L. Nilsson who kindly supplied us with the results of the Uppsala Tandem Laboratory.

REFERENCES

1. J. Speth and A. van der Woude, Rep. Prog. Phys. __44__ (1981) 46.
2. S. Joly, D. M. Drake and L. Nilsson, Phys. Rev. __C20__ (1979) 2072.
3. P. Anderson, R. Zorro and I. Bergqvist, Proc. Int. Conf. Nuclear Data for Science and Technology, Antwerp, 1982, ed. by K. H. Bockhoff, D. Reidel, Dordrecht, 1983, p. 866.

4. H. R. Weller, Prof. Int. Symp. Neutron Capture Gamma-ray Spectroscopy, Grenoble, 1981, ed. by T. von Egidy et al., The Inst. of Phys. Conf. Ser. No. 62, 1982, p. 494.
5. H. R. Weller and N. R. Roberson, Rev. Mod. Phys. 52 (1980) 699.
6. G. E. Brown, Nucl. Phys. 57 (1964) 339.
   C. F. Clement, A. M. Lane and J. R. Rock, Nucl. Phys. 66 (1965) 273, 293.
7. M. Potokar, Phys. Lett. B46 (1973) 346.
8. F. E. Bertrand, Nucl. Phys. A354 (1981) 129c.
9. I. Bergqvist and M. Potokar, Neutron Capture Gamma-ray Spectroscopy, ed. by R. E. Chrien and W. R. Kane, Plenum Press, New York - London, 1979, p. 299.
10. L. Nilsson, Neutron Capture Gamma-ray Spectroscopy, ed. by T. von Egidy et al., The Inst. of Phys. Conf. Ser. No. 62, 1982, p. 465.
11. L. Nilsson, Int. Symp. on Capture Gamma-ray Spectroscopy, Knoxville, 1984, Amer. Int. of Phys. Conf. Proc. No. 125, 1985.
12. H. Coude, L. E. Person, L. G. Stromberg, A. Lindholm, G. Lodin, L. Nilsson and C. Nordborg, Tandem Lab. Report TLU 9/73, Uppsala, 1973.
13. E. M. Diener, J. F. Amann, S. L. Blatt and P. Paul, Nucl. Instr. Meth. 83 (1970) 115.
14. A. Likar, A. Lindholm, L. Nilsson, I. Bergqvist and B. Palson, Nucl. Phys. A298 (1978) 217.
15. P. Paul, in "Nuclear Spectroscopy and Reactions", ed. by J. Cerny, Academic Press, New York, 1974, p. 345.
16. M. Suffert and A. Debre, J. Phys. 32 (1971) C5b, 261.
17. M. D. Hasinoff, S. T. Lim, D. F. Measdy and T. J. Mulligan, Nucl. Instr. Meth. 117 (1974) 375.
18. H. R. Weller, R. A. Blue, N. R. Roberson, D. G. Rickel, C. P. Cameron, R. D. Ledford and D. R. Tilley, Phys. Rev. C13 (1976) 922.
19. H. R. Weller and N. R. Roberson, IEEE Trans. Nucl. Sci. NS28 (1981) 1268.
20. A. M. Sandorfi and M. T. Collins, Nucl. Instr. Meth. 222 (1984) 479.
21. L. Nilsson, private communication.
22. J. M. Paul, Nucl. Instr. Meth. 89 (1970) 582.
23. H. Weller, Fast Neutron Workshop, Chapel Hill, 1984, unpublished.
24. I. Bergqvist, R. Zorro, A. Hakanson, A. Lindholm, L. Nilsson, N. Olsson and A. Likar, Nucl. Phys. A419 (1984) 509.
25. H. Liskien and A. Paulsen, Nucl. Data Tables 11 (1973) 588.
26. Nucl. Data Tables 11 (1973) 603.
27. R. Martinčič, A. Likar, F. Cvelbar and M. Mikuž, Proc. Int. Symp. Neutron Capture Gamma-ray Spectroscopy, Grenoble, 1981, ed. by T. von Egidy et al., The Inst. of Phys., Conf. Ser. No. 62, 1982, p. 526.
28. F. S. Dietrich and A. K. Kerman, Phys. Rev. Lett. 43 (1979) 114.
29. M. Potokar, Phys. Lett. 92B (1980) 1.
30. F. Cvelbar, R. Martinčič and A. Likar, Atomkernenergie -Kerntechnik 45 (1984) 179.

31. S. A. Wender, G. F. Auchampaugh, J. F. Wilkerson, N. W. Hill, L. R. Nilsson and N. R. Roberson, Nucl. Instr. Meth. 220 (1984) 371.
32. R. Zorro, A. Lindholm and L. Nilsson, The Large Gamma-ray Spectrometer at the Uppsala Tandem Acc. Lab. LUNDFD6/ (NFFR-3053)/1-25, 1984.

PHOTON STRENGTH FUNCTIONS

M. A. Lone

Atomic Energy of Canada Limited, Chalk River Nuclear
Laboratories, Chalk River, Ontario, Canada

ABSTRACT. We discuss the gross features and formalism of the
distribution of reduced transition probabilities of dipole
transitions between excited states. By considering a partial
breakdown of the Brink hypothesis, we develop an analytic expression for strengths of E1 transitions between excited states. The M1 data supporting the existence of an M1 giant resonance at $\sim$ 8 MeV is surveyed.

1. INTRODUCTION

Strengths of electromagnetic transitions between nuclear states provide data /1–6/ for tests of nuclear models, radiative
capture reaction mechanisms, nuclear potentials and nucleon-
nucleon interactions. Systematic evaluation of the gamma-ray
strengths have led to simple analytic expressions that make it
possible to include electromagnetic decay channels in Hauser-
Feshbach calculations.

There is now a large body of data on gamma-ray strength
functions that describes the energy dependence of average transition strength. The majority of these data have been obtained
from the intensities of primary transitions, E1, M1 and to some
extent E2, observed following radiative neutron capture that
populates compound nuclear states. The transitions are from the
capturing state to low lying states and show statistical behaviour, i.e. Porter-Thomas distributions, expected of a compound
nuclear process.

The interpretation and compatibility of these data with the
photo-absorption data that involve transitions between the

ground state and higher excited states have been based on the assumption that the excitation function in the continuum region is independent of the initial target state. This was justified /4/ in view of the expectation that in the continuum region the compound nuclear formation would result in washing out any nuclear structure details of the initial state. The discovery of the collective E1 giant resonance cast doubt on this conjecture. But the collective nature of the giant resonance, involving core excitations, implies that such resonances could be excited starting from any state. This led to what is known as Brink hypothesis /7/, that assumes Excited State Giant Dipole Resonances (ESGDR) with identical shapes, independent of the initial excited state. This model provides reasonable, although not exact, predictions of the energy dependence and the magnitude of the E1 strengths at $E_\gamma \sim 3$ MeV.

However one does not expect the above model to hold for initial states well above the ground state. Recent $(\dot{p}, \gamma)$ studies /8/ show a partial breakdown of the model since the spreading width and even the integrated strength seem to depend on the structure of the initial state. Experimental data from $(n, \gamma\alpha)$ studies /9, 10/ also show an energy dependence of the strength function, a constant value at $E_\gamma \prec 2$ MeV, that is inconsistent with the E_γ dependence predicted by the Lorentzian tail of the GDR. Admittedly the Lorentz line shape of the ground state photo absorption cross section has never been proved at low energies because of non-statistical level distributions and other quantum mechanical effects.

Thus the strength function related to the transitions between excited states may in fact differ from the one related to the ground state photo absorption cross section. There is now enough experimental data on E1 strength functions to explore this. The small discrepancies /6, 11/ between the model predictions and the data may in fact be due to a partial breakdown of the Brink hypothesis. So far there has been very little study of this topic.

The present review follows the style and formalism of the earlier reviews by Lone /6/ and Axel /12/, with emphasis on the dipole strength functions. Sections 2–5 discuss the nomen-

clature and formalism necessary for understanding the model
predictions. Section 6 explores the consequences of a partial
breakdown of the Brink hypothesis. A simple analytical expression based on a phenomenological model is derived that provides a clear distinction between the two types of strength functions while maintaining compatibility with the ground state
photo excitation data. Sections 7 and 8 discuss the comparison
of data with model predictions of E1 and M1 strength functions
in medium and heavy nuclei.

2. TRANSITIONS BETWEEN DISCRETE LEVELS

There are several recent reviews /2/ of the strengths of individual gamma-ray transitions between discrete levels. This information is useful for recommendation of upper limits for transitions of different characters. These limits are often an in-
dispensable tool in nuclear spectroscopy for exclusion of "impossible" J^{π} values.

The data also provide information on the fragmentation of
the single particle strength as a function of the nuclear mass
and excitation energy. For this purpose one defines /2/ the
strength of a gamma-ray transition relative to the Weisskopf
estimates /2-4/ as

$$S_{XL} = \frac{\Gamma_{\gamma}(XL)}{\Gamma_W(XL)} \quad , \tag{1}$$

where XL denotes the multipole character. The Weisskopf single-
particle estimates are, with $R = 1.2\ A^{-1/3}$ fm,

$$\Gamma_W(E1) = 6.74 \times 10^{-2}\ A^{2/3}\ E_{\gamma}^3 \ , \tag{2a}$$

$$\Gamma_W(M1) = 2.07 \times 10^{-2}\ E_{\gamma}^3 \ , \tag{2b}$$

where the width Γ is in eV and E_{γ} in MeV units.

Histograms /2/ of the hindrance factors, defined as $\log(S_{XL})$,
show that in general E1 and M1 transitions are weaker than the
Weisskopf estimates whereas the E2 transitions are stronger.

240

Schumacher /12/ has recently moted that the distribution of the
E1 hindrance factors can be explained on the basis of the Porter
—Thomas distribution if one assumes an average hindrance for
$\bar{S}_{E1}$, that depends on the nuclear mass. This corroborates a recent statement by Brody et al. /13/ that properties of the statistical model are likely to be found valid at all excitation
energies.

3. TRANSITIONS FROM THE CONTINUUM REGION

The strengths of individual transitions from a region of high
density of states have limited usefulness as indicators of average strengths because of the statistical nature of the decay
process. The average partial decay width, $\bar{\Gamma}_{\lambda\gamma f}$, from an excitation energy E_λ to a final state f, averaged over the set of
levels, λ, of same spin and parity, shows a linear dependence
on the average spacing D_λ of these levels. Consequently the ra-
tio $\bar{\Gamma}_{\lambda\gamma f}(XL)/D_\lambda$ is a better indicator of the gross features of
the distribution of strength. With the assumption of identical
photo excitation functions for any target state, f, of a particular nucleus the above ratio is a function only of the multipole character and the energy of the gamma-ray.

The gamma-ray strength function is defined as /1, 5, 6/

$$f_{XL}(E_\gamma) = \frac{\bar{\Gamma}_{\lambda\gamma f}(XL)}{D_\lambda \ E_\gamma^{(2L+1)}} \ . \tag{3a}$$

With D and Γ expressed in eV and E_γ in MeV, the unit of f_{XL} is
$MeV^{-(2L+1)}$. This definition is analogous to the definition of
the neutron strength function.

The gamma-ray strength function defined in eq. (3a) is proportional to the amount of downward reduced transition probability, $B(XL)\downarrow$, per MeV excitation energy interval i.e.

$$f_{XL}(E_\gamma) \propto \frac{d}{dE} B(XL)\downarrow \ , \tag{3b}$$

where

$$B(E1)\downarrow = 0.95 \ E_\gamma^{-3} \Gamma(E1) \ \dots \ (e^2 fm^2), \tag{4a}$$

$$B(M1)\downarrow = 86 \ E_\lambda^{-3} \ \Gamma(M1) \cdots \ (\mu_0^2) \tag{4b}$$

with Γ in eV and E_λ in MeV units.

However, a strength function in $(MeV)^{-(2L+1)}$ units does not provide a meaningful assessment of the degree of collectivity or fragmentation of the strength. It is more revealing if the strength function is expressed in terms of Weisskopf units. Thus we define /6, 14/ a relative strength function, b(XL) as

$$b(XL) = \frac{1}{D_\lambda} \frac{\bar{\Gamma}_{\lambda\downarrow f}(XL)}{\Gamma_W(XL)} \tag{5}$$

$$= \frac{B(XL)\downarrow \ \text{per MeV}}{B_W(XL)\downarrow} \tag{6}$$

with D_λ in MeV units. Values of $B_W(XL)\downarrow$ can be calculated from eqs. (2a,b) and (4a,b).

The constants relating b(XL) and f_{XL} for dipole transitions are /6, 14, 15/

$$b(E1) = 1.48 \times 10^7 \ A^{-2/3} \ f_{E1} \cdots \ (W.U. \ MeV^{-1}), \tag{7a}$$

$$b(M1) = 4.82 \times 10^7 \qquad f_{M1} \cdots \ (W.U. \ MeV^{-1}). \tag{7b}$$

4. AVERAGE PHOTOABSORPTION CROSS SECTIONS AND THE GAMMA-RAY STRENGTH FUNCTIONS

Let us consider a narrow resonance of spin I_λ that is excited by the interaction of a photon with the ground state of spin I_g, then /14/

$$\int_\lambda \sigma_{XL}(E_\lambda) \ dE_\lambda = \pi^2 \chi^2 \frac{(2I_\lambda+1)}{(2I_g+1)} \ \Gamma(XL, I_\lambda - I_g) \ , \tag{8a}$$

where $\Gamma(XL, I_\lambda - I_g)$ is the partial width for the decay of the excited state λ to the ground state. Converting this width to the upward reduced transition probability,

242

$$B_\lambda(XL)\uparrow = \frac{(2I_\lambda+1)}{(2I_g+1)} B_\lambda(XL)\downarrow \tag{8b}$$

we get

$$\int_\lambda \sigma_{XL}(E_\gamma)dE_\gamma = 0.4 E_\lambda B_\lambda(E1)\uparrow \tag{8c}$$

$$= 4.4 \times 10^{-3} E_\lambda B_\lambda(M1)\uparrow \tag{8d}$$

in units of fm^2 MeV if $B(E1)$ is in e^2fm^2 and $B(M1)$ in μ_o^2. These equations provide a relationship between the energy weighted sum and the total integrated cross section.

From eq. (8a) we can also write an expression for the average photo absorption cross section from ground state of spin I_g to excited states of spin I_λ, as /14/

$$\bar{\sigma}_{XL}(E_\gamma) = 3.86 \times 10^5 \frac{1}{E_\gamma^2} \frac{\bar{\Gamma}_{\lambda\gamma}(XL)}{D_\lambda} \left(\frac{2I_\lambda+1}{2I_g+1}\right) . \tag{9}$$

If the excitations from the ground state involve core nucleons (rather than valence nucleons) then the ratio $\bar{\Gamma}_{\lambda\gamma}/D_\lambda$ is independent /14/ of spin and the average cross section summed over all possible I_λ spins is

$$\bar{\sigma}_{XL}(E_\gamma) = \frac{3.86 \times 10^5 (2L+1)}{E_\gamma^2} \frac{\bar{\Gamma}_{\lambda\gamma}(XL)}{D_\lambda} (fm^2) ; \tag{10}$$

where Γ and D are in eV and E_γ in MeV units. Now if we invoke the Brink hypothesis then from eqs. (3) and (10) we get

$$f_{XL}(E_\gamma) = \frac{2.6 \times 10^{-6}}{(2L+1)} \frac{\bar{\sigma}_{XL}(E_\gamma)}{E_\gamma^{2L-1}} , \tag{11}$$

where σ_{XL} is in fm^2 and E_γ in MeV units.

5. GIANT RESONANCES AND SUM RULES

Often the experimental value of the average (averaged over fine structure to smooth out Porter-Thomas fluctuations) photon absorption cross section, $\sigma_{XL}(E_\gamma)$, is not available. However, in

the excitation region below 20 MeV, the presence of giant resonances, E1, M1 and E2, strongly influences the strengths of
transitions observed in radiative neutron capture.

The photo neutron cross section data /16/, for nuclei with
A ≈ 60, clearly display a Lorentz line shape of the E1 GDR at
energies above 7 MeV. The data /6/ on M1 strength functions in
nuclei with A ≈ 100 also indicate the influence of an M1 GDR
peaked at ∼ 8 MeV ($E_R \simeq 44 \ A^{-1/3}$), although there are insufficient data, as yet, to map out the line shape of this resonance.

In the absence of any other guidelines and for simplicity
we assume that the average photo absorption cross section for
any multipole excitation in the continuum region is dominated
by the appropriate giant resonance with a classical Lorentz
line shape

$$\overline{\sigma}_{XL}(E_\gamma) = \overline{\sigma}_{XL}(E_R) \ \frac{\Gamma_R^2 E_\gamma^2}{(E_\gamma^2 - E_R^2)^2 + \Gamma_R^2 E_\gamma^2} \ . \tag{12}$$

For E1 GDR a compilation of the measured resonance parameters,
energy E_R, spreading width Γ_R and peak cross section $\overline{\sigma}_{E1}(E_R)$
is given in ref. /16/. In deformed nuclei the GDR splits into
two Lorentz components.

The peak cross section can also be computed from the energy-
weighted sum of the upward reduced transition probability. From
eq. (12) we can write

$$\overline{\sigma}_{XL}(E_R) = \frac{2}{\pi \Gamma_R} \int_0^\infty \overline{\sigma}_{XL}(E_\gamma) \ dE_\gamma \ . \tag{13}$$

Then from eq. (8) we get

$$\overline{\sigma}_{E1}(E_R) = \frac{0.25}{\Gamma_R} \sum E \ B(E1)\uparrow, \tag{14a}$$

$$= \frac{2.8 \times 10^{-3}}{\Gamma_R} \sum E \ B(M1)\uparrow . \tag{14b}$$

244

6. BRINK HYPOTHESIS AND EXCITED STATE GIANT RESONANCE (ESGR) MODEL

There is now considerable evidence /8, 17, 18/ that the ESGDR's do indeed exist, but as yet there is insufficient quantitative investigation of their relative strengths and spreading widths. Recent (p,γ) reaction studies /8/ reveal excitation functions that are dominated by excited state giant dipole resonances. These resonances are found to peak at about the same gamma-ray energy with respect to the initial excited state. The resonances appear to have similar line shapes but their spreading width increases progressively with the initial state excitation energy. Even the inverse (γ,p_o) cross sections are found to be proportional to the single-particle spectroscopic factors of the initial excited state.

The above observations imply that there is at least a partial breakdown of the Brink hypothesis of identical ESGDR's. This suspicion is further corroborated by the energy dependence of the low energy, $E_\gamma \prec 2$ MeV, primary gamma-ray strengths observed in the (n,$\gamma\alpha$) reaction studies /9, 10/. These transitions, mostly dipole from continuum to continuum states, yield an energy independent value of $\sim 2 \times 10^{-8}$ MeV^{-3} for the strength function compared with the energy dependent, $f_{E1} \propto E_\gamma$, predicted by the Lorentz line shape of the photo absorption cross section.

Let us examine the consequence of a partial breakdown of the Brink hypothesis with the following simplifying assumptions:

1. The spreading width, Γ_{RS}, of the ESGDR depends on the initial state excitation energy, E_S. We will adopt the Fermi liquid model /10/ prediction for the spreading width

$$\Gamma_{RS} = \Gamma_{RO} \left[1 + \frac{4\pi^2}{a} \frac{E_S}{E_{RO}^2} \right] = \Gamma_{RO} \, \beta, \tag{15a}$$

where RO refer to the ground state GDR and symbol a is the Fermi gas level density parameter (MeV^{-1}) such that the thermodynamic temperature T_S of the initial state is

$$T_S^2 = E_S/a. \tag{15b}$$

245

2. Due to the progressively increasing fragmentation, the resonance structure accounts for only a part of the total sum strength. That is, the resonance peak cross section varies as

$$\sigma_{E1}(E_{RS}) = \sigma_{E1}(E_{RO}) \frac{\Gamma_{RO}}{\Gamma_{RS}} \delta ,$$ (15c)

where $\delta = [1+E_S/E_{RO}]^{-\eta}$.

3. The remainder of the strength is uniformly distributed, as in the single particle model of Weisskopf, over an energy region E_o. Then from eq. (11) we can write

$$f_{E1}(E_\gamma,E_S) = C(E_S) + 0.87\times10^{-6} \frac{\sigma_{E1}(E_\gamma,E_S)}{E_\gamma} .$$ (15d)

The values of the constant $C(E_S)$ can be computed from the requirement of the energy weighted sum rule that implies

$$\int E_\gamma f_{E1}(E_\gamma,E_s) \, dE_\gamma = \int E_\gamma f_{E1}(E_\gamma,0) \, dE_\gamma.$$ (15e)

Then

$$C(E_S) = 0.87\times10^{-6} \frac{\pi \sigma_{E1}(E_{RO})\Gamma_{RO}}{E_0^2} [1-\delta]$$ (15f)

and

$$\sigma_{E1}(E_\gamma,E_S) = \frac{\sigma_{E1}(E_{RO})\Gamma_{RO}^2 E_\gamma^2 \delta \beta}{(E_\gamma^2-E_{RO}^2)^2 + E_\gamma^2 \Gamma_{RO}^2 \beta^2} .$$ (15g)

Using relationships in eqs. (15a—g), the strength function, $\tilde{f}_{E1}(E_\gamma,E_\lambda)$, for the decay transitions from the excited state E_λ can be expressed in terms of the strength function, $\tilde{f}_{E1}(E_\gamma,0)$, for the E1 excitations from the ground state. Thus for $E_\gamma < E_{RO}$

$$\tilde{f}_{E1}(E_\gamma,E_\lambda) = \left\{ \frac{\pi E_{RO}^4}{E_0^2 \Gamma_{RO} E_\gamma} (1-\delta) + \delta\beta \right\} \tilde{f}_{E1}(E_\gamma,0),$$ (16a)

where

$$\delta = [(E_{RO}+E_\lambda-E_\gamma)/E_{RO}]^{-\eta} \quad \text{and} \quad \beta = \left[1 + \frac{4\pi^2}{a} \frac{(E_\lambda-E_\gamma)^2}{E_{RO}^2} \right].$$ (16b)

246

When $E_\gamma = E_\lambda$

$$\overleftarrow{f}_{E1}(E_\gamma, E_\lambda) = \overrightarrow{f}_{E1}(E_\gamma, 0) \tag{16c}$$

independent of η, δ, β and E_0 as expected since the same matrix elements are involved. Also for $\eta = 0$ and $\beta \simeq 1$, at $E_\gamma \prec E_\lambda$

$$\overleftarrow{f}_{E1}(E_\gamma, E_\lambda) \simeq \overrightarrow{f}_{E1}(E_\gamma, 0). \tag{16d}$$

Thus if there is only an increase in the spreading width then at low energies the downward strength function will be bigger than the upward strength function. Also the energy dependence will be such that

$$\overleftarrow{f}(E_\gamma, E_\lambda) \propto E_\gamma \tag{16e}$$

at low E_γ as determined by eqs. (11) and (12).

If on the other hand $\eta \succ 0$ then at low energies

$$\overleftarrow{f}_{E1}(E_\gamma, E_\lambda) \prec \overrightarrow{f}_{E1}(E_\gamma, 0) \tag{16f}$$

and the first term in eq. (16a) will give an energy independent strength function at low gamma-ray energies. The experimental data strongly suggest that $\eta \succ 0$ and $C(E_S)$ is very small.

Figure 1 shows a comparison between the upward and downward strength functions. For these model calculations we have assumed $a = 20$ MeV^{-1}, a typical value for heavy nuclei; $E_0 = 140$ MeV, the (γ, π) threshold, and $\eta = 2$ as a starting guess. The GDR resonance parameters are $\Gamma_{RO} = 4$ MeV; $E_{RO} = 14$ MeV and $\sigma_{E1}(E_{RO}) \simeq 40$ fm^2, typical values for a heavy nucleus.

The solid line, curve 1, refers to $\overrightarrow{f}_{E1}(E_\gamma)$ that represent the Lorentzian tail of the ground state GDR. The dash-dot line, curve 2, is the downward strength function $\overleftarrow{f}_{E1}(E_\gamma, 8)$ that would represent gamma-ray transitions from the decay of compound states at $E_\lambda = 8$ MeV. The dashed curve 3 represents the downward strength function for transitions from $E_\lambda = 10$ MeV. It is interesting to note that at $E_\gamma \sim 5$ MeV the downward strength function has almost the same energy dependence as that of the upward strength function but the absolute magnitude is lower. Also at $E_\gamma \prec 2$ MeV the downward strength functions are nearly constant.

247

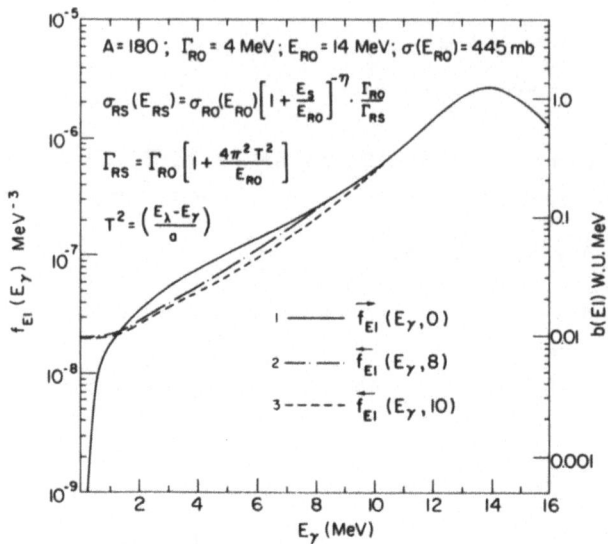

FIG. 1. Comparison of upward and downward strength functions calculated by considering a partial breakdown of the Brink hypothesis.

7. E1-DATA

There are several recent surveys /5, 6, 11, 14, 19—21/ of the dipole radiation strength functions obtained from radiative neutron capture data. A global description of the data $E_\gamma \geq 3$ MeV is rather satisfactorily given by the Lorentzian extrapolation of the E1 GDR. The data also show local perturbations from the smooth Lorentzian description. These could be caused by non-statistical reaction mechanisms such as direct or valence capture /22—24/ that can lead to the enhancement of E1 strength. Another cause may be the decoupled E1 strength /1, 25, 26/ that lies at energies well below the giant resonance energy.

The measured strength functions at $E_\gamma < 8$ MeV range in value from about 0.01 to 0.15 WU MeV^{-1} whereas the peak value at the resonance energy is 1 WU MeV^{-1} in ^{182}Ta, see Fig. 2. It is not clear to what extent the theory can predict such small values of the strength function so far away in the tail region. They seem to be "in the noise" of the existing theoretical calculations.

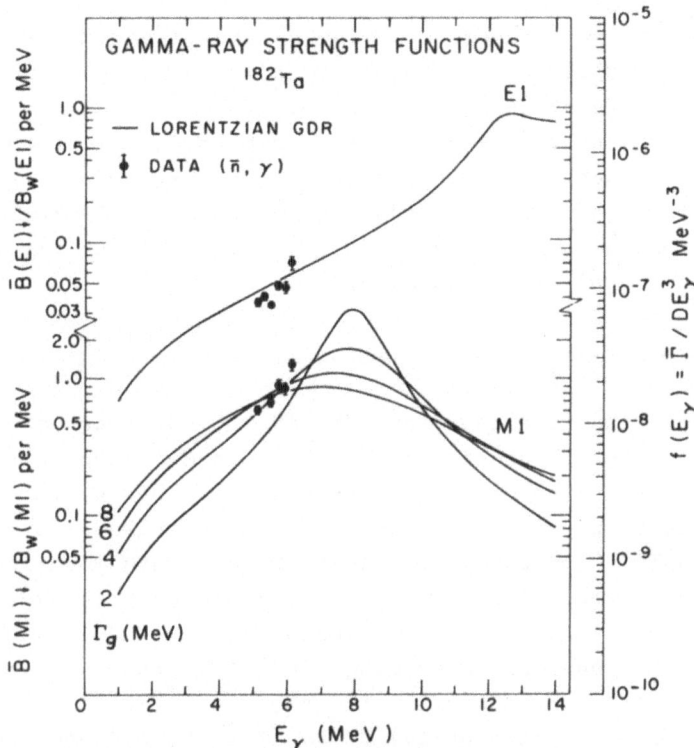

FIG. 2. Strength functions for E1 and M1 transitions in ^{182}Ta. Resonance parameters for GED2 are from ref. /16/. The data points are from ref. /35/. For GMDR the energy weighted sum is assumed to be μ_o^2 MeV.

Comparison of the E1 strength function from decay of highly excited states with the Lorentz tail of the GDR does show a significant discrepancy in the absolute magnitude, particularly for spherical nuclei. The ratio

$$b_{(E1)}\text{measured}/b_{(E1)}\text{Lorentz tail} \simeq 0.7 \qquad (17a)$$

at $E_\gamma \simeq 6$ MeV. This has been interpreted /19, 28/ as evidence for the energy dependent spreading width of the GDR. An energy dependence /17/ of

$$\Gamma_R(E_\gamma) = \Gamma_R(E_R) \left(\frac{E_\gamma}{E_R} \right)^{0.5} \qquad (17b)$$

yields a better fit /19/ to the data. However this will not explain the energy independent value of the strength function at $E_\gamma < 2$ MeV implied by the $(n,\gamma\alpha)$ data /9, 10/.

The discrepancy between the absolute magnitudes of the decay strength function and the photo excitation strength function may in fact be due to the partial breakdown of the Brink hypothesis. Figure 1 shows that $\bar{f}$, calculated with the model parameters discussed in sect. 6, provides a better fit to the (n,γ) and $(n,\gamma\alpha)$ experimental data.

8. M1-DATA

One of the recurring questions /6, 15, 30—33/ about the M1 radiation is whether or not there exists an M1 resonance in heavy nuclei in the range of energy accessible to radiative neutron capture gamma rays. The presence of such a collective resonance would result in an energy dependent M1 strength function.

In general the M1 photon absorption strength is expected /14, 15/ to be proportional to the number of filled $j = 1+1/2$ orbitals that have corresponding $j = 1-1/2$ vacant orbitals. Bohr and Mottelson /15/ estimated that in ^{208}Pb, the spin flip transitions will yield B(M1)↑ of $36\,\mu_o^2$ at 8 MeV resulting in an energy weighted sum of $300\,\mu_o^2$ MeV. However, no such localized M1 strength has been found thus far.

There is now increasing evidence /31—34/ that in heavy nuclei the M1 strength is fragmented and spread over a wider region. Recent measurements of Leszewski et al. /32/ in ^{206}Pb show a resonance structure centred at about 7.5 MeV with a total B(M1)↑ of $19\,\mu_o^2$ at excitation energies between 6.7 and 8.1 MeV. They conclude that there probably is no "missing" M1 strength in ^{208}Pb but rather the discrepancy between the current theory and experiment can be attributed to local fragmentation of the strength into states in the vicinity of 7.5 MeV, that are individually too weak to be identified.

In nuclei with $A > 100$, the M1 strength functions obtained from (n,γ) studies /6, 21, 30/ for $E_\gamma \approx 3$ to 8 MeV show a well pronounced energy dependence. At $E_\gamma \approx 6$ MeV, the majority of the data show that

$$\bar{\Gamma}_{M1} \propto E_\gamma^5 , \qquad\qquad\qquad\qquad (18a)$$

$$\bar{\Gamma}_{E1}/\bar{\Gamma}_{M1} \simeq 7 , \qquad\qquad\qquad\qquad (18b)$$

$$b(M1) \propto E_\gamma^2 , \qquad\qquad\qquad\qquad (18c)$$

$$b(M1) \approx 0.9 \text{ W.U. MeV}^{-1} . \qquad\qquad\qquad (18d)$$

This behaviour of the M1 strength function strongly indicates the influence of an underlying M1 collective resonance nearby.

Figure 2 shows a comparison of the E1 and M1 strength functions /35/ in ^{182}Ta with predictions of the E1 and M1 giant resonance model. In the case of E1, measured resonance parameters /16/ are used whereas for M1 we assume

$$E_R \simeq 8 \text{ MeV} ,$$

$$\Gamma_R = 4 \text{ MeV} ,$$

$$\sum EB(M1)\!\uparrow = 450\,\mu_0^2 \text{ MeV.}$$

In fact the majority of the data in A > 100 nuclei can be explained with a range of $\Gamma_R \approx 4$ to 6 MeV and energy weighted sum of 300 to 600 μ_0^2 MeV. Thus in heavy nuclei the M1 GDR model would be a better recipe for the M1 strength function than the Weisskopf single particle model.

REFERENCES

1. G. A. Bartholomew, E. D. Earle, A. J. Ferguson, J. W. Knowles and M. A. Lone, Adv. Nucl. Phys. 7 (1973) 229.
2. P. M. Endt, At. Data Nucl. Data Tables 23 (1979) 3, 547; 26 (1981) 47.
3. D. H. Wilkinson, "Nuclear Spectroscopy B", edited by F. Ajzenber-Selove, Academic Press, New York, 1960, 859 p.
4. J. M. Blatt and V. F. Weisskopf, "Theoretical Nuclear Physics", John Wiley and Sons, New York, 1952, 644 p.
5. B. J. Allan, I. Bergqvist, R. E. Chrien, D. Gardner and W. P. Poenitz, "Neutron Radiative Capture", Pergamon Press, Oxford, 1984.
6. M. A. Lone, "Neutron Capture Gamma-ray Spectroscopy", edited by R. E. Chrien and W. R. Kane, Plenum Press, New York, 1979, p. 161.
7. D. M. Brink, Doctoral Thesis, Oxford University, 1955.

B. B. Kinsey, "Handbuch der Physik XL", Springer-Verlag, Berlin, 1957, p. 317.
P. Axel, Phys. Rev. $\underline{126}$ (1962) 671.

8. D. H. Dowell, Capture Gamma-ray Spectroscopy and Related Topics, AIP Conference Proceedings No. 125, American Institute of Physics, New York, 1984, p. 597.
9. V. A. Vtyurin and Yu. P. Popov, Report JINR RY-10775, Dubna, 1977.
10. S. G. Kadmenskii, V. P. Markushev and V. I. Furman, Sov. J. Nucl. Phys. $\underline{37}$ (1983) 165.
11. R. E. Chrien, Trans. N.Y. Acad. Sci., Ser. II, $\underline{40}$ (1980) 40.
12. M. Schumacher, in ref. /8/, p. 166.
13. T. A. Brody, J. Flores, J. B. French, P. A. Mello, A. Pandey and S. S. M. Wong, Rev. Mod. Phys. $\underline{53}$ (1981) 385.
14. P. Axel, in ref. /6/, p. 825; Trans. N.Y. Acad. Sci., Ser. II, $\underline{40}$ (1980) 20.
15. A. Bohr and B. R. Mottelson, "Nuclear Structure", Vol. 1, W. A. Benjamin, New York, 1969; Vol. 2, 1975.
16. B. L. Berman and S. C. Fultz, Rev. Mod. Phys. $\underline{47}$ (1975) 713.
17. S. L. Blatt, in ref. /8/, p. 570.
18. K. A. Snover, in ref. /8/, p. 660.
19. C. M. Cullagh, M. L. Stelts and R. E. Chrien, Phys. Rev. $\underline{C23}$ (1981) 1394.
20. S. Raman, Neutron Capture Gamma-ray Spectroscopy and Related Topics 1981, edited by T. V. Egidy, F. Gonnenwein and B. Maier, Conference Series 62, Institute of Physics, Bristol, 1982, 357.
21. J. Kopecky, in ref. /20/, p. 423; Nucl. Data Conference, Santa Fé, 1985.
22. A. M. Lane and J. E. Lynn, Nucl. Phys. $\underline{17}$ (1960) 563.
23. S. F. Mughabghab, M. A. Lone and B. C. Robertson, Phys. Rev. $\underline{C26}$ (1982) 2698.
24. Y. K. Ho and M. A. Lone, Nucl. Phys. $\underline{A406}$ (1983) 1.
25. D. S. Armstrong, S. K. Saha, C. W. Cheng, E. Adamides, A. Henrikson, M. A. Lone and B. C. Robertson, Nucl. Phys. (1985).
26. A. M. Lane, Ann. Phys. $\underline{63}$ (1971) 171.
27. M. Harvey and F. C. Khanna, Nucl. Phys. $\underline{A221}$ (1947) 77.
28. M. Gardner, in ref. /5/, p. 85.
29. F. K. Thielemann and M. Arnold, Int. Conf. on Nuclear Data for Science and Technology, Antwerp, 1982.
30. L. M. Bollinger, Proc. Int. Conf. Photonuclear Reactions and Applications, edited by B. L. Berman, USAEC Report Conf.-73-301, $\underline{11}$ (1973) 783.
31. G. E. Brown and S. Raman, Comments Nucl. Part. Phys. $\underline{9}$ (1980) 79.'
32. L. M. Laszewski, P. Rullhusen, S. D. Hoblit and S. F. LeBrun, Phys. Rev. Lett. $\underline{54}$ (1985) 530.
33. S. Müller, G. Kuchler, R. A. Richter, H. B. Block, H. Block, G. W. deJager, H. deVries and J. Wamback, Phys. Rev. Lett. $\underline{54}$ (1985) 293.
34. D. Cha, B. Schwesing, J. Wamback and J. Speth, Nucl. Phys. $\underline{A430}$ (1984) 321.
35. H. E. Jackson, Japan Atomic Energy Research Institute Report JAE2I-M-5984 (1975) 119.

APPLICATIONS AND MISAPPLICATIONS OF THE CHANNEL-CAPTURE FORMALISM OF DIRECT NEUTRON CAPTURE *

S. Raman and * *J. E. Lynn[+]

Oak Ridge National Laboratory, Oak Ridge, Tennessee, USA
* *Atomic Energy Research Establishment, Harwell, England

ABSTRACT. We discuss the channel-capture approximation of slow neutron direct-capture theory. We show that this approximation gives a generally good representation of the neutron capture cross sections for several electric dipole transitions in a broad range of nuclides from A = 9 to A = 136; these are mostly near-spherical nuclei. Despite this body of agreement, we examine the accuracy we can expect from the simple channel-capture theory. Comparison with calculations of the potential-capture cross section from physically more realistic optical model calculations shows that, in general, the channel-capture cross section can be up to $\approx 40\%$ in error. In cases where the expected channel-capture cross section is much smaller than the "hard-sphere" capture cross-section estimate, the disagreement with potential capture can be much worse than this. Also, in these cases, compound-nucleus capture can be of comparable or greater magnitude. These effects have been shown to completely undermine recent attempts to determine nuclear interaction radii for targets, such as ^{12}C and ^{9}Be, by application of the channel-capture formula to capture cross-section data.

1. INTRODUCTION

The low-energy neutron radiative capture reaction has been a rich source of data on nuclear spectroscopy for several decades, but, surprisingly perhaps, considerable uncertainty still exists

---
*Research sponsored in part by the U.S. Department of Energy under Contract No. DE-AC05-840R21400 with the Martin Marietta Energy Systems, Inc.

[+]Eric Lynn is grateful to ORNL for its hospitality during April-May, 1985.

in our understanding of the detailed mechanism of the reaction itself. For most heavy nuclides not in proximity to a closed shell, development of statistical estimates of radiative transition strengths, based on the spreading of the electric-dipole giant resonance into a broad range of compound nucleus states, seems to be the best that we can do. Demonstrations also have shown, however, that for a large number of light nuclides and near-closed-shell nuclides, a direct-capture mechanism dominated by single-particle transitions from the entrance channel offers a good explanation of off-resonance-capture cross sections for primary E1 transitions to final states with a certain degree of single-particle character.

The success of the channel-capture estimates of the cross sections in many such cases is now leading to its use (or misuse) as a tool for determining other nuclear quantities of interest, such as spectroscopic factors of final states, total thermal absorption cross sections, scattering lengths, and nuclear potential radii. It is therefore pertinent to ask the question of how exact we can expect the formula for the channel-capture cross section to be even in relatively ideal situations. In this study we address this question, first drawing attention to the nature of the physical approximations inherent in the derivation of the simple channel-capture expression and then to some of the numerical approximations incorporated in the present commonly used formula. Finally, we look at a few of the recent applications of the formula and draw attention to the limitations of the methodology.

2. A SIMPLE CHANNEL-CAPTURE FORMULA

In 1960, Lane and Lynn /1/ derived a simple expression for slow-neutron direct capture in the channel-capture approximation (which retains only the extranuclear contribution to the matrix element) for the cross sections of E1 transitions in terms of a cross section given by the hard-sphere approximation (extranuclear capture associated with the limiting case of hard-sphere scattering) to the direct-capture theory. This expres-

sion has been used by several authors and, in particular, by
Mughabghab, who gives the following version /2/

$$\sigma_{\gamma f}(\text{channel}) = \sigma_{\gamma f}(\text{hard sphere}) \left[1 + \frac{R-a_s}{R} Y_f \frac{Y_f+2}{Y_f+3}\right]^2, \quad (1)$$

where

$$\sigma_{\gamma f}(\text{hard sphere}) = \frac{0.062}{R\sqrt{E_n}} \left[\frac{Z}{A}\right]^2 \mu \frac{2J_f+1}{6(2I+1)} S_{dp}\left[\frac{Y_f+3}{Y_f+1}\right]^2 Y_f^2 \quad (2)$$

and

$$Y_f^2 = \frac{2mE_\gamma R^2}{\hbar^2}, \quad (3)$$

where R is the interaction radius ($1.35 \times A^{1/3}$ fm), a_s is the
coherent scattering length for channel spin s, J_f is the total
spin of the final state, I is the spin of the target nucleus,
S_{dp} is the (d,p) spectroscopic factor, E_γ is the energy of gamma ray due to capture of an s-wave neutron feeding a final p-
state, E_n is the energy of the incident neutron = 0.0253 eV
for 2200 m/s neutrons, μ is the unity in all cases (see below).

To apply the previous expression, one needs to know the interaction radius R, the scattering length a_s, and the (d,p)
spectroscopic factors. Without the spectroscopic factors, the
comparison with data would be jejune, but with these factors
in hand, R and a_s can still be left as adjustable parameters.
Tables 1 and 2 show the application of the previous expression
in analyzing thermal capture data /3—16/. The overall agreement between calculation and experimental data is good. The
agreement can be improved (for example by using R = 5.1 fm instead of R = 4.24 fm as in ref. /16/ for ^{31}P), but we emphasize that when R departs significantly from either ($1.16\ A^{1/3}$
+0.6) fm or from ($1.35 \times A^{1/3}$) fm, such an agreement is somewhat
contrived. So it is also when the scattering length is not
known from experiment.

The quantity μ in eq. (2) was supposed to be unity or two
if I $\neq$ 0, depending on whether J_f = I±3/2 or J_f = I±1/2, respectively /2/. It follows that if the theory is valid, appli-

TABLE 1. Comparison between calculated and measured partial
cross sections from thermal neutron capture on even targets

Final nucleus	E(level) (keV)	(d,p) $(2J+1)S$	Primary E_γ(keV)	Calc. σ_γ(mb)	Exp. σ_γ(mb)
(i) ^{12}C(n,γ) /3/: Assumed R = 2.96 fm; known a_S = 6.15 fm					
^{13}C	0	2.2	4945	2.4	2.4
	3684	0.4	1262	1.3	1.1
(ii) ^{32}S(n,γ) /4/: Assumed R = 4.32 fm; known a_S = 2.74 fm					
^{33}S	3221	1.90	5421	266	302
	4211	0.30	4431	33	25
	4918	0.09	3724	8	13
	5711	1.06	2931	74	87
	5889	0.44	2753	29	29
	6425	0.34	2217	18	13
	7188	0.18	1454	6	3
(iii) ^{34}S(n,γ) /5/: Assumed R = 4.32 fm; known a_S = 3.40 fm					
^{35}S	2348	2.04	4638	150	163
	3802	0.36	3184	18	18
	4189	0.28	2797	13	16
	4903	1.55	2083	52	46
	4963	0.87	2023	29	34
(iv) ^{36}S(n,γ) /6/: Assumed R = 4.50 fm; assumed a_S = 3.20 fm					
^{37}S	646	2.71	3657	172	161
	1992	0.22	2312	9	9
	2638	1.60	1666	46	52
	3262	0.58	1042	11	8
	3493	0.31	811	4	2
(v) ^{40}Ca(n,γ)/7/: Assumed R = 4.62 fm; known a_S = 4.90 fm					
^{41}Ca	1943	2.60	6420	119	184
	2462	0.90	5901	39	33
	3613	0.19	4750	7	9
	3730	0.02	4633	1	2
	3944	1.17	4419	43	90
	4603	0.16	3760	5	10
	4753	0.37	3610	12	33
(vi) ^{42}Ca(n,γ)/8/: Assumed R = 4.69 fm; known a_S = 3.10 fm					
^{43}Ca	593	0.16	7340	30	39
	2046	2.80	5886	406	360
	2611	0.27	5322	35	28
	2878	0.18	5054	22	16
	2943	0.19	4989	23	24
	3286	0.12	4646	13	16
	3572	0.19	4360	19	20
	4207	0.84	3725	72	56

TABLE 1 (Continued)

Final nucleus	E(level) (keV)	(d,p) (2J+1)S	Primary E_γ(keV)	Calc. σ_γ(mb)	Exp. σ_γ(mb)
(vii) ^{44}Ca(n,γ) /9/: Assumed R = 4.77 fm; known a_s = 1.79 fm					
^{45}Ca	1435	0.40	5980	100	94
	1900	2.20	5515	493	480
	2249	0.30	5166	62	85
	2842	0.34	4573	59	36
	3242	0.14	4173	22	21
	3419	0.49	3996	72	90
	3783	0.08	3632	10	9
	3838	0.19	3577	24	11
	4616	0.34	2799	32	30
	4999	0.36	2415	28	19
(viii) ^{128}Te(n,γ)/10/: Assumed R = 6.80 fm; assumed a_s = 5.20 fm					
^{129}Te	2040	(0.04)	4043	3	4
	2267	0.06	3815	5	5
	2361	0.57	3722	42	43
	2379	0.30	3703	22	23
	2705	0.33	3377	22	22
	3502	0.07	2579	3	3
	3558	0.05	2525	2	12
	3792	0.21	2290	9	8
(ix) ^{130}Te(n,γ)/11/: Assumed R = 6.84 fm; known a_s = 5.53 fm					
^{131}Te	2511	0.25	3418	14	13
	2582	0.87	3347	49	57
	3002	0.38	2928	19	26
	3690	0.11	2240	4	3
(x) ^{136}Xe(n,γ)/12/: Assumed R = 6.94 fm; assumed a_s = 5.75 fm					
^{137}Xe	601	1.96	3425	106	102
	986	0.68	3039	33	33
	1841	0.72	2184	25	26
	1936	0.40	2089	13	12
	2106	0.12	1829	3	3
	2490	0.60	1535	15	15
	3609	0.16	416	1	2
(xi) ^{136}Ba(n,γ)/13/: Assumed R = 6.94 fm; assumed a_s = 3.00 fm					
^{137}Ba	2182	1.06	4723	267	291
	2663	0.36	4243	80	91
	3316	0.15	3590	26	12
	3403	0.19	3503	31	10
	3680	0.09	3226	13	9
	3778	(0.10)	3127	14	10
	3799	0.06	3106	8	4

TABLE 2. Comparison between calculated and measured partial
cross sections from thermal neutron capture on odd targets

Final nucleus	E(level) (keV)	(d,p) $(2J+1)S$	Primary E_γ (keV)	Calc. σ_γ(mb)	Exp. σ_γ(mb)
(xii) ^{9}Be(n,γ)/14/: Assumed R = 3.00 fm; known a_s = 7.01 fm					
^{10}Be	0	2.1	6809	12.8	4.9
	3368	1.75	3444	0.41	0.86
	5959	(3.94)	853	2.0	2.0
	6180	(0.03)	632	0.016	0.018
(xiii) ^{13}C(n,γ) /3/: Assumed R = 3.10 fm; known a_s = 5.47 fm					
^{14}C	0	2.09	8174	1.16	1.15
	6589	(0.06)	1587	0.14	0.12
(xiv) ^{27}Al(n,γ)/15/: Assumed R = 4.08 fm; known a_s = 3.34 fm					
^{28}Al	3464	0.78	4262	8.2	14.2
	3592	0.90	4134	9.2	17.3
	4690	1.2	3036	9.2	16.7
	4765	1.0	2961	7.5	20.7
(xv) ^{31}P(n,γ) /16/: Assumed R = 4.24 fm; known a_s = 4.99 fm					
^{32}P	3264	1.10	4670	12.3	22.9
	4007	0.33	3927	3.5	6.5
	4036	1.30	3898	13.7	28.6
	4663	(0.40)	3271	3.9	8.9
	4878	(0.42)	3056	4.0	10.0
	5350	0.72	2584	6.4	8.8
	5509	0.24	2425	2.1	3.8
	5776	0.74	2158	6.1	11.7
	6060	(0.26)	1874	2.0	3.4
	6195	(0.20)	1739	1.5	4.3
(xvi) ^{33}S(n,γ) /17/: Assumed R = 4.32 fm; known a_s = 4.68 fm					
^{34}S	4624	0.84	6792	8.4	24.2
	5680	2.70	5737	24.7	43
	5756	1.68	5661	15.2	18.4
	6169	1.00	5248	8.7	11.8
	6342	0.80	5075	6.8	0.4
	6479	3.64	4938	30.6	22.2
	6685	1.28	4731	10.5	1.6
	6954	0.84	4462	6.7	7.9
	7110	0.52	4306	4.1	8.3
	7630	3.92	3787	28.4	26.5
	7781	1.08	3636	7.6	5.2
	8138	1.04	3279	6.9	3.2

cation of eq. (1) could lead to a spin assignment for a level
as was done in refs. /12/ and /15/. Such a conclusion is un-
physical and never intended in the formulation of the theory.
The quantity μ appearing in the eq. (2) should be removed.

3. PHYSICAL APPROXIMATIONS IN THE CHANNEL-CAPTURE FORMULA

The electric dipole radiative transition matrix element between initial i and final state f contains the factor

$$\int_0^\infty dr\ u_i(r)rw_f(r)\ ,$$ (4)

if the motion of the system can be approximated by single-particle motion in a potential well. The radial wave function of
this motion is denoted by $u_i(r)$ for the initial state and by
$w_f(r)$ for the final state. In general, the internal behavior
of the compound nucleus at several MeV of excitation cannot be
approximated as single-particle motion, but beyond a certain
radial separation of neutron and residual nucleus, denoted as
the channel radius R, residual forces that cannot be described
by a simple potential-energy function become negligible and
the single-particle approximation is then very good. If beyond
the channel radius the potential energy is also negligible,
the radial wave functions assume simple analytical forms, related to spherical Bessel and Neumann functions for the initial state (describing the scattering of a neutron by the nucleus) and to spherical Hankel functions for the final state.
In this case the capture cross section for (very) low-energy
neutron capture can be estimated by replacing the integral (4)
by its component in the channel

$$\int_R^\infty dr\ u_i(r)rw_f(r)\ .$$ (5)

The resulting expression is /17/

$$\sigma_\gamma(i\to f)CH = \frac{\overline{e}^2}{\hbar v}\ \frac{32\pi}{3}\ R^5 k_\gamma^3 \sum_{J=|I-1/2|}^{I+1/2} g_J\ \frac{1}{|1-iP_0\mathcal{R}_J|^2}\ \cdot$$

$$\cdot\ \frac{f(y)}{y^4}\ w_{J_f jJ}\theta_f^2\ ,$$ (6)

where

$$f(y) = \left[\frac{y+3}{y+1}\right]^2 + 2\mathrm{Re}\,\mathcal{R}_J\,\frac{y(y+3)(y+2)}{(y+1)^2} +$$

$$+ \left[(\mathrm{Re}\,\mathcal{R}_J)^2 + (\mathrm{Im}\,\mathcal{R}_J)^2\right]\,y^2\,\left[\frac{y+2}{y+1}\right]^2. \tag{7}$$

In these equations, $\bar{e}$ is the effective charge of the neutron ($\bar{e} = -eZ/A$, where e is the proton charge, Z is the proton number, and A is the mass number of the target nucleus), $\hbar$ is Planck's constant divided by 2π, v is the relative velocity of neutron and target nucleus, k_γ is the wave number of the photon emitted in the transition, J is the total angular momentum of the initial scattering system, g_J is the statistical weight for this angular momentum selection [$g_J = (2J+1)/2(2I+1)$] for unpolarized particles, I is the spin of the target nucleus, P_o is the penetration factor for s-wave neutrons (this can be expressed in terms of the wave number k for the relative motion of neutron and target nucleus as $P_o = kR$), $\mathcal{R}_J$ is the reduced $\mathcal{R}$-function that is central to the description of the scattering function amplitude when the total angular momentum of the scattering function is J, and $y = \kappa R$, κ being the reciprocal attenuation length of the channel wave function of the final state. In terms of the binding energy E_f of the final state and the reduced mass μ of the (neutron + target nucleus) system, $\kappa = (2\mu E_f)^{1/2}/\hbar$. If the neutron scattering length a_{sJ} of the target nucleus is known at the incident neutron energy, the real part of $\mathcal{R}_J$ is determined (in the absence of strong neutron capture) by the simple expression $\mathrm{Re}\,\mathcal{R}_J = 1-(a_{sJ}/R)$. (Note that this substitution then closely relates eqs. (6) and (7) with eq. (1) if $\mathrm{Im}\,\mathcal{R}_J = 0$ and the scattering lengths are not J-dependent.) The quantity $W_{J_f j J}$ is a spin-coupling factor depending on the angular momentum J_f of the final state and on j, the coupled spin and orbital momentum of the single-particle motion of the neutron in the final state as well as on J. The value $W_{J_f j J}$ can be written in terms of the Wigner W-coefficients and vector-coupling coefficients, and numerical values of $W_{J_f j J}$ have been tabulated in ref. /17/. Finally, the quantity θ_f^2 is the spectroscopic factor for the fractionation of the single-particle bound state into the final state f.

The ideal situation in which eq. (6) would be a very accurate representation of the cross section for the transition i→f would be a) if the nucleus were very sharp-edged, having very little interaction of any kind with the incident neutron outside the channel radius R and showing full effect of the overall potential field just within it; b) if the neutron energy were strongly off-resonance; and c) if the nucleus were very large and the nuclear potential deep (compared to the final-state binding energy) so that the radial wave function of the single-particle final state within the nucleus had many nodes. Condition a) would guarantee that the analytic forms of the radial wave functions used in deriving eq. (6) are fully accurate and that the integrand of eq. (4) becomes highly oscillatory and therefore largely self-cancelling for r ≺ R (or alternatively strongly damped). Condition b) would imply that in any case the contribution to the integral (eq. (4)) from the region r = 0 to r = R would be of small amplitude. Condition c) would guarantee that the approximate value of the final-state wave function at the channel radius, namely $w_f^2(R)$ = 2/R, employed in deriving eq. (6) is accurate.

4. CENTER-OF-MASS FACTORS IN CHANNEL-CAPTURE FORMULAS AND COMPUTATIONS

If eq. (1) is going to be employed willy-nilly in analyzing experimental results, it is important to consider center-of-mass factors in greater detail. We need to begin with the expression for hard-sphere capture (eq. (9) of ref. /17/).

$$\sigma_{HS} = \frac{\bar{e}^2}{\hbar v} \frac{32\pi}{3} \frac{R^5 k_\lambda^3}{y^4} \left[\frac{y+3}{y+1} \right]^2 , \qquad (8a)$$

where

$$y = \kappa R \quad \text{and} \quad \kappa = \sqrt{\frac{2\mu E_f}{\hbar^2}} = \sqrt{\frac{M}{M+m}} \sqrt{\frac{2m E_f}{\hbar^2}} . \qquad (8b)$$

Here μ is reduced mass, M target mass, m neutron mass, E_f binding energy of the final state, and v relative velocity of incident neutron and target. Suppose y is expressed without its $(M/M+m)^{1/2}$ factor. The error introduced in the $[(y+3)/(y+1)]^2$ term is small. But the y^{-4} term will require a factor

$$\left(\frac{M+m}{M}\right)^2 . \tag{8c}$$

Eq. (8a) is usually developed for numerical work by recognizing that at low neutron energy $E_\gamma \approx E_f$. Hence $k_\gamma = E_\gamma/\hbar c = E_f/\hbar c$, and eq. (8a) becomes

$$\sigma_{HS} = \frac{\bar{e}^2 \hbar^2}{c^2} \frac{4\pi}{3} \frac{1}{v} \frac{1}{\mu^3} \frac{1}{R} \; y^2 \left[\frac{y+3}{y+1}\right]^2 . \tag{9a}$$

The above universal formula (eq. (9)) will now contain the factor

$$\underbrace{\left(\frac{M+m}{M}\right)^3}_{\text{from } \mu^{-3}} \quad \underbrace{\left(\frac{M}{M+m}\right)^2}_{\text{from } y^2} = \left(\frac{M+m}{M}\right)^2 \quad \text{as before.} \tag{9b}$$

In the formula for channel capture, additional terms in y will appear with different powers. These terms often cancel each other and hence depend critically on the precise value of y. It is important therefore to use the exact value of y in this formula, that is,

$$y = R \sqrt{\frac{M}{M+m}} \sqrt{\frac{2mE_f}{\hbar^2}} . \tag{10}$$

Eq. 17(a) of ref. /17/ is then, in terms of laboratory neutron energy,

$$\sigma_{\gamma(i \to f)(CH)} = \frac{\bar{e}^2}{\hbar} \frac{32\pi}{3} \frac{R^5 E_f^{\,3}}{\hbar^3 c^3} \sqrt{\frac{m}{2E_{lab}}} \frac{1}{y^4} \cdot$$

$$\cdot \sum_{J=|I-1/2|}^{I+1/2} g_J \frac{1}{|1-iP_o \mathcal{R}_J|^2} \left[\left(\frac{y+3}{y+1}\right)^2 + \right.$$

$$+ 2\text{Re}\,\mathcal{R}_J \, \frac{y(y+3)(y+2)}{(y+1)^2} + |\mathcal{R}_J|^2 y^2 \left(\frac{y+2}{y+1}\right)^2 \Bigg] W_{J_f j J} \theta_f^2 \, . \qquad (11)$$

If E_f^3 is written in terms of κ, i.e. $E_f = \hbar^2 \kappa^2 / 2\mu$
$= ((\hbar^2 \kappa^2)/2m)((M+m)/M)$, we obtain eq. 17(b) of ref. /17/:

$$\sigma_{\gamma(i \to f)(CH)} = \frac{0.06137}{R \sqrt{E_{lab}}} \left(\frac{Z}{A}\right)^2 \left(\frac{M+m}{M}\right)^3 \sum_J g_J \frac{1}{|1-iP_o \mathcal{R}_J|^2} \cdot$$

$$\cdot \left[y^2 \left(\frac{y+3}{y+1}\right)^2 + 2\text{Re}\,\mathcal{R}_J \, \frac{y^3(y+1)(y+2)}{(y+1)^2} + \right.$$

$$\left. + |\mathcal{R}_J| \, y^2 \left(\frac{y+2}{y+1}\right)^2 \right] W_{J_f j J} \theta_f^2 \, . \qquad (12)$$

We emphasize that the factor in square bracket contains the exact value of y given by eq. (10) above.

5. OPTICAL MODEL COMPARISONS WITH THE CHANNEL-CAPTURE FORMULA

The limitations of the channel-capture formula in its application to neutron capture by real nuclei can be demonstrated at
one level by comparison with the study of neutron capture within the framework of the optical model. This was briefly discussed in ref. /1/ and more extensively in ref. /18/. Recent explicit work on the comparison of channel capture with the fuller version of slow-neutron direct capture (which is known as
potential capture) as calculated in the optical model framework, with special reference to the sulfur isotopes, is given
in ref. /17/. Briefly, within the optical model the radial dipole integral (eq. (4)) is calculated numerically for the full
range of the radial variable by integrating the Schrödinger
equation for initial (scattering) and final (bound) states
using a realistic form (such as Woods—Saxon) for the potential-
energy function; within the integral region of the nucleus,
the effect of residual interactions that remove strength from
the incident nucleon channel is represented crudely by the in-

clusion of an imaginary potential-energy term. Thus, the de-
limiting conditions a), b) and c) in the last paragraph of
section 3 for ensuring the accuracy of the channel-capture formula are largely relaxed by explicitly calculating the potential-capture cross section. Direct comparisons in the A = 30
—70 range of the potential-capture cross section calculated in
this manner to the channel-capture cross section have been
given in ref. /17/.

The ratio of potential-capture cross section to channel-
capture cross section for a given set of conditions (final sin-
gle-particle-state binding energy E_f, channel radius identi-
fieo with the potential-well radius R for the purpose of calculating the channel-capture cross section in conjunction with
the potential scattering radius R' substituted for the scattering length a_s) is termed C_{opt}. The behavior of C_{opt} as a
function of final-state binding energy for various parameteri-
zations of the Woods-Saxon surface diffuseness d and the imaginary component of the optical model potential is shown in
Figs. 1 and 2, reproduced from ref. /17/. It should be noted

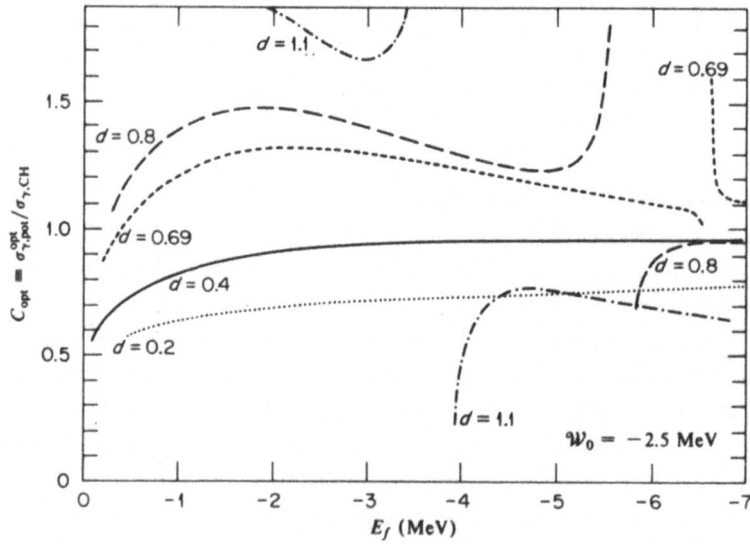

FIG. 1. Dependence of the ratio of potential-capture cross section to channel-capture cross section on the surface diffuseness
parameter d. Calculations have been made for j_f = 3/2 and the
volume absorption model. (Reproduced from ref. /17/.)

that a change in the final-state binding energy implicitly governs a change also in the potential-radius R and hence in the scattering length. For fairly realistic values of the optical model parameters, C_{opt} <u>usually</u> differs from unity by less than ~40%. This is a measure, therefore, of the accuracy we can reasonably expect from the channel-capture formula in the absence of other capture mechanisms and is a quite remarkable vindication of its considering the nature of the approximations inherent in it. We note, however, that some of the approximations leading to the channel-capture expression are clearly partially cancelling. Thus, on taking small values of the surface diffuseness parameter, the channel-capture formula underestimates the potential-capture cross section, contrary to condition a). This can be explained as a result of the approximation $w_f^2(R) = 2/R$ being a poor one for a p-wave state in a square potential-well with small radius.

Although Figs. 1 and 2 demonstrate the general usefulness (within ~40%) of the channel-capture expression it is also

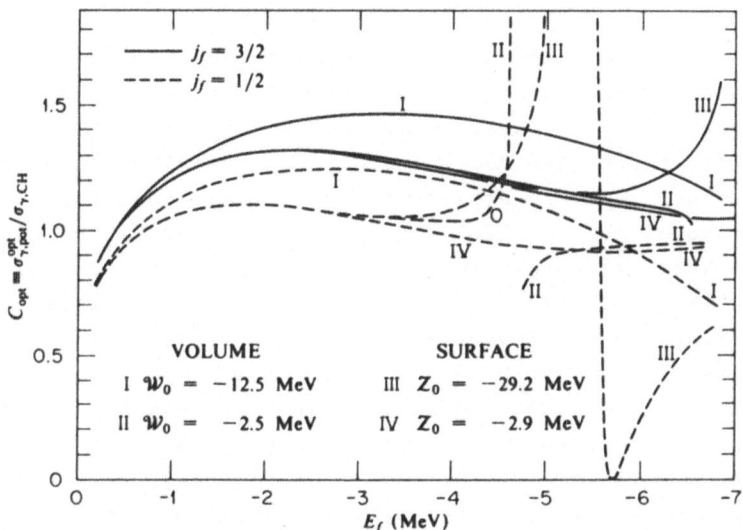

FIG. 2. Dependence of the ratio of potential-capture cross section to channel-capture cross section on the magnitude of the imaginary potential. Calculations have been made for surface and volume absorption and for both values of spin-orbit coupling. The surface diffuseness parameter is fixed at d = 0.69 fm. (Reproduced from ref. /17/.)

265

clear that in certain situations this expression can be highly inaccurate, with C_{opt} going through a major fluctuation from very large values to very small values for certain narrow ranges of the final-state binding energy. This fluctuation is associated with destructive interference of the terms in $f(y)$ of eq. (7) (occurring when Re $\mathcal{R}_J$ is negative, i.e., $a_{sJ} \sim R$) giving rise to a very small value of the channel-capture cross section. In this case the individual terms of $f(y)$ will not be estimated accurately enough from the assumed simple analytic forms of the radial wave functions in the external region; furthermore, the ignored contribution to the radial dipole integral from the internal region becomes of comparable importance. Thus, it is dangerous to use the channel-capture formula as a precise estimate when this formula gives rise to a value of the capture cross section that is very much smaller than the estimate of the hard-sphere capture cross section of eq. (9).

6. COMPOUND-NUCLEUS CONTRIBUTIONS

The previously discussed limitations of the channel-capture formula apply even when no other mechanisms contributing to the capture process exist. However, in the neutron-nucleus interaction, the target nucleus cannot be described as a totally inert core, simply providing a potential field for the scattering of the incident nucleon. One aspect of the residual nucleon interactions in the system is implied by the use of the spectroscopic factor θ_f^2 to give the single-particle purity of the final state. The remaining configurations of the final states can be connected in the E1 transition strength to configurations mixed into the initial state by excitation of highly complicated states of the compound nucleus. Even at off-resonance energies of the incident neutron, small components of such compound-nucleus states will be mixed into the initial state; and therefore, it is necessary to make some estimate of the interfering compound-nucleus contribution to the radiative capture cross section.

    An overall survey of the radiation widths for individual

transitions from low-energy compound nucleus resonances was carried out by Cameron /19/ with the overall result

$$\Gamma_{\gamma,CN}(\text{in meV}) = 0.33 \times 10^{-6} E_\gamma^3 (\text{in MeV}) A^{2/3} D(\text{in eV}) \, , \qquad (13)$$

where D is the mean resonance spacing (for resonances of single spin and parity). An alternative estimate of the compound-nucleus radiation width can be derived using Brink's method from the giant dipole resonance model. This gives

$$\Gamma_{\gamma,CN} = \frac{4}{3\pi} \frac{NZ}{A} \frac{e^2}{\hbar c} \frac{(1+0.8x)}{mc^2} \frac{\Gamma_G E_\gamma^4}{(E_\gamma^2 - E_G^2)^2 + (\Gamma_G E_\gamma)^2} D, \qquad (14)$$

where E_G is the energy of the electric dipole giant resonance, Γ_G is its width, and x is the fraction of exchange force present in the nuclear force. Equation (14) probably gives a more nearly accurate representation of the gamma-ray energy dependence of the partial radiative width, but it has been established from the study of total radiation widths that it overestimates the bulk of radiative transitions width energies in the range 2 to 3 MeV from neutron resonances by a factor of the order of 2 to 4. On the other hand, eq. (13) underestimates the average strength of high-energy transitions ($E_\gamma \gtrsim 5$ MeV). Hence, this equation can be taken as giving a conservative estimate of the average width for the typical compound-nucleus component of radiative transitions. An estimate of the compound-nucleus contribution to the cross section may be obtained from eq. (13) by assuming that for each channel spin J a single nearby compound nucleus level of energy $E_{\lambda J}$ is responsible for the major part. Then,

$$\sigma_{\gamma(i \to f)CN} \approx \pi \lambda^2 \sum_J g_J \frac{\Gamma_{\lambda}(n) \Gamma_{\gamma(i \to f)CN}}{(E_{\lambda J} - E)^2}$$

$$\approx \pi \lambda^2 \sum_J g_J \frac{2P_o Re \, \mathcal{R}_J}{E_{\lambda J} - E} \Gamma_{\gamma(i \to f)CN} \, , \qquad (15)$$

where P_o is the s-wave neutron penetration factor, $P_o = R/\lambda$. The amplitude of this contribution to the radiative transition can be taken to interfere either constructively or destructively with the channel-capture contribution, giving a final estimate for the overall capture cross section of

$$\sigma_{\gamma(i \to f)} \approx \left| \sigma_{\gamma(i \to f)CH}^{1/2} \pm \sigma_{\gamma(i \to f)CN}^{1/2} \right|^2 . \tag{16}$$

The actual value of $\sigma_{\gamma(i \to f)CN}$ is subject to Porter-Thomas fluctuations inherent in the partial radiative width $\Gamma_{\gamma(i \to f)CN}$, so the two extreme values of $\sigma_{\gamma(i \to f)}$ given by eq. (16) can be taken as a measure of the range of variation that can be expected in the observed value of the cross section. Numerical examples of the possible effect of the compound-nucleus contributions are given in sections 7 and 8 on the application of the channel-capture theory to some specific data.

7. CAPTURE BY ^{12}C

In low-energy neutron capture by ^{12}C, significant primary transitions to the ground state and 3.686 MeV excited state of ^{13}C (gamma-ray energies of 4.945 MeV and 1.262 MeV, respectively) occur; the cross sections for neutrons of energy 0.0253 eV are 2.38±0.05 mb and 1.14±0.02 mb, respectively. The spectroscopic factors of the final state are 1.1 and 0.1, respectively. The scattering length of thermal neutrons interacting with ^{12}C is 6.15 fm. The channel-capture cross section calculated from eq. (12) for these two transitions is shown as a function of potential radius R in Fig. 3. Note that these calculations are 27% higher than those given in ref. /13/ because of the $[(M+m)/M]^3$ factor discussed in section 4.

The functional behavior of the channel-capture cross section has been used in ref. /3/ to determine the potential radius from the measured cross sections; this is quoted as 2.91±0.03 fm. It is seen from Fig. 3 that application of this method using the corrected formula would give R = 2.96 and 2.8 fm from the two cross-section values.

When the capture cross section has such a small value (only 2.38 mb for the ground-state transition), it is important, in general, that a possible compound-nucleus contribution should be considered because such a contribution could distort the calculated cross section very considerably. Using eq. (13) to estimate the compound-nucleus radiation width for the ground-state transition, we find $\Gamma_\gamma \approx 0.3$ eV (using a conservative estimate for the s-wave resonance spacing of $D \approx 1.5$ MeV). Substitution of this in eq. (16) for the cross sections gives the two shaded regions as very rough estimates for the cross sections for these two transitions. It can be seen that inclusion of this amount of compound-nucleus effect would result in an uncertainty in deducing the potential radius of the order of ± 0.2 fm while employing the channel-capture formula.

It is possible to argue, however, that any compound-nucleus contribution to the radiative transition to the ground state should be extremely small, because the very high spectroscopic factor (S = 1.1) for the ground state implies very little ad-

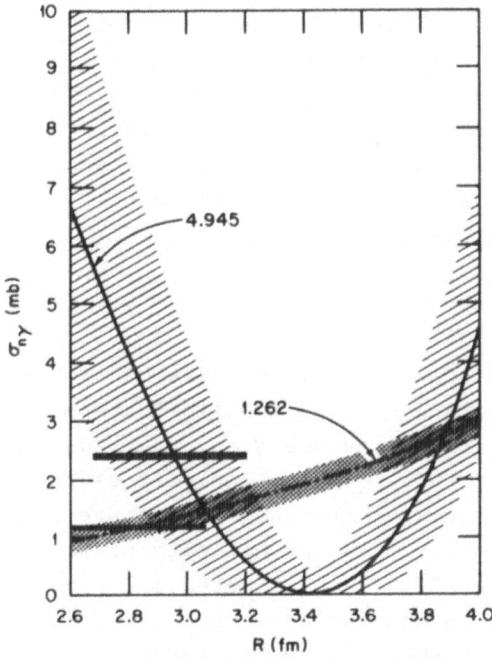

FIG. 3. Calculated channel-capture cross section as a function of the nuclear potential radius for the 4.945 MeV and 1.262 MeV transitions from the $^{12}C(n,\gamma)$ reaction. Shaded regions show the effect of including a small compound-nucleus contribution. Measured cross sections are shown by horizontal bars.

mixture of configurations other than the single-particle p-
wave component. Hence the only important contribution to the
cross section would be from the direct and valency mechanisms.
If we dismiss the possibility of any significant compound-nucleus component, we must then consider these two mechanisms
much more carefully.

Although the channel-capture cross section appears particularly sensitive to the potential radius when there is large
destructive interference between the principal terms of eq.
(12), this is just the situation in which the channel-capture
formula can be expected to give unreliable results. Ho and
Lone /20/ have shown, in fact, that a form of the channel-capture cross section, modified numerically to account for the
departure of the channel wave functions from the simple analytical forms assumed in the absence of any potential energy tail
in the channel, is quite insensitive to the values of the potential radius, provided that the well depth is adjusted to
reproduce the binding energy of the final state. Allowing for
interference with a valency component from the bound state at
-2.02 MeV, these authors obtain a realistic estimate of the
modified channel-capture cross section which is close to the
measured value; and again their estimate is quite insensitive
to the value of the potential radius.

8. CAPTURE BY ^{9}Be

Four electric dipole transitions have been observed following
slow neutron capture by ^{9}Be. The final state, their spectroscopic factors, gamma-ray energies, and measured capture cross
sections used in ref. /14/ for analysis of these transitions
within the framework of the channel-capture theory are shown
in Table 2, together with the estimate of the capture cross
section from eq. (12), using a potential radius of R = 3.00 fm
and a slow neutron scattering length a = 7.01 fm (essentially
independent of channel spin). The theoretical estimates differ
from the experimental values of the cross sections by a factor
of up to about 2.5 in either direction. These discrepancies
have been overcome in ref. /14/ by adjustment of the nuclear

radius R, <u>separately for each incident channel spin J</u>, i.e.,
eq. (7) is modified by defining $y_J = \kappa R_J$. In addition, allow-
ance is made for mixing of the two different spin-orbit configurations $j = 1/2$ and $3/2$ in the final state f by using the
factor $|\Sigma_j w_{JfjJ} \theta_{f,j}|^2$, the phase in the product of Wigner 6-j
symbols and vector coupling coefficient being implied in the
process of taking the square root. The available data and calculations on the spin-orbit mixing of the 3.368 and 5.959 MeV
state of ^{10}Be imply that the major contributions to the channel-capture cross section come from opposite incident channel
spins in the two cases and allow an apparently clean separation of the "spin-dependent potential radii", R_J. The analysis
of the data in this way leads to a quoted difference $R_{J=1} -$
$R_{J=2} = 1.12$ fm. From this difference, a spin-dependent term
$V_{\vec{1}.\vec{\sigma}}$ (where $\vec{\sigma}$ denotes the neutron spin in the optical potential for ^{9}Be+n) is deduced.

From our discussion in sect. 6, it is clear that the development given in ref. /14/ is unnecessary in order to explain
the experimental capture cross section data and, indeed, that
the deduction of a channel-spin-dependent nuclear potential
radius, if that development is followed, must be a very impre-
cise procedure. The reason is that conditions are such that
large interference terms are again present (the hard-sphere
capture cross sections calculated for the first three transitions of Table 2 are 25, 13, and 10 mb, respectively, i.e., 5
to 15 times greater than the experimental quantities). Thus,
the C_{opt} factor resulting from a more realistic optical model
calculation of the direct-capture cross section could differ
very considerably from unity, and the factor of up to 2.5 noted
in the preceding paragraph is not an unreasonable expectation
for C_{opt}. Furthermore, the interfering effect of the compound-
nucleus mechanism, as described in some detail for ^{12}C(n,γ)
in section 7, could also be substantial here. A compound-nucleus contribution based on the Cameron estimate (eq. (8a)), a
bound s-wave level at $E_\lambda = -0.85$ MeV, and a level spacing con-
servatively estimated at $D \sim 1$ MeV, would give a range of cross
sections of 3.8 to 26 mb, 0.005 to 1.4 mb, and 1.8 to 2.2 mb,
respectively, for the first three transitions in ^{10}Be of

Table 2, thus easily encompassing the experimental data.

9. CONCLUSION

We have pointed out in this paper that a commonly used formulation of the theory of slow neutron direct capture, namely
the channel-capture formula of Lane and Lynn /1/, although appearing surprisingly accurate in reproducing a large body of
experimental data, is based on an idealization of the scattering function that can lead to an analytically tractable and
simple expression. The degree of accuracy that can be expected
from the channel-capture formula has been estimated elsewhere
/17/ by comparisons with numerical computations of the potential-capture cross section from physically realistic optical
models of neutron scattering. We have pointed out that, generally speaking, the channel-capture formula is accurate to better than 40% in the mass number range $A \approx 30$ to 70 (but usually underestimating the direct-capture cross section) if the
spectroscopic factors are known reliably and accurately. It is
worthwhile to stress that these factors (derived from comparisons of experimental results with DWBA predictions) generally carry an irreducible overall uncertainty of 20%. We have
also shown that for light nuclei certain mass-dependent factors not normally included in the channel-capture formula can
be numerically significant.

In certain situations where there is large destructive interference among the terms entering the channel-capture formula, the inaccuracy inherent in the formula can be much greater
than 40%. Furthermore, in such situations the contribution of
the compound-nucleus mechanism to the capture amplitude from
the tails of neutron resonance states can be comparable to or
greater than the direct component.

It follows that the application of the channel-capture formula can be very unreliable in such cases. In particular, we
have examined its application to the capture cross-section
data of ^{12}C in order to determine the nuclear potential radius
/3/ and to data of ^{9}Be in order to determine channel-spin-dependent potential radii /14/ and hence the magnitude of a pos-

272

sible spin-spin term in the optical potential. We find in both cases that the possible magnitude of compound-nucleus contributions is sufficient to prevent the determination of such quantities with any worthwhile degree of precision. Because both cases also involve large destructive interference among the terms entering the channel-capture formula, the very application of this formula to treat these data is questionable.

Added in proof. Contrary to what it states in the text, the calculated values in Tables 1 and 2 are based not only on eq. (1) but also on the correct center-of-mass factors (see section 4).

REFERENCES

1. A. M. Lane and J. E. Lynn, Nucl. Phys. 17 (1960) 563; 17 (1960) 686.
2. S. F. Mughabghab, M. Divadeenam and N. E. Holden, "Neutron Cross Sections", Vol. 1, Part A, Academic Press, New York, 1981, p. 12.
3. S. F. Mughabghab, M. A. Lone and B. C. Robertson, Phys. Rev. C26 (1982) 2698.
4. S. Raman, in "Neutron-capture Gamma-ray Spectroscopy and Related Topics 1981", ed. by T. von Egidy, F. Gönnenwein and B. Maier, Institute of Physics, Bristol, 1982, p. 357.
5. R. F. Carlton, S. Raman and E. T. Jurney, in "Neutron-capture Gamma-ray Spectroscopy and Related Topics 1981", ibid., p. 375.
6. C. E. Thorn, J. W. Olness, E. K. Warburton and S. Raman, Phys. Rev. C30 (1984) 1442.
7. Data from P. M. Endt and C. Van der Leun, Nucl. Phys. A310 (1978) 1.
8. S. Mughabghab, in "Proceedings of the Specialists' Meeting on Neutron Cross Sections of Fission Product Nuclei", Bologna, 1979, ed. by C. Coceva and G. C. Panini, Report NEANDC(E)209 "L", p. 179.
9. S. F. Mughabghab, in "Proceedings of the Conference on Nuclear Data Evaluation Methods and Procedures, BNL, Upton, N.Y., 1981, ed. by B. A. Magurno and S. Pearlstein, Report BNL-NCS-51363, Vol. 1, p. 339.
10. J. Honzátko, K. Konečný, F. Bečvář, E. A. Eissa and M. Králik, Z. Phys. A299 (1981) 183.
11. J. Honzátko, K. Konečný, F. Bečvář and E. A. Eissa, Czech. J. Phys. B30 (1980) 763.
12. S. F. Mughabghab, Phys. Lett. 81B (1979) 93.
13. J. Honzátko, K. Konečný, F. Bečvář, Z. Kosina and M. Králik, in "Neutron Induced Reactions", Proc. of the Europhysics Topical Conference, Smolenice, 1982, ed. by P. Obložinský, Institute of Physics, Bratislava, 1982, p. 249.
14. S. F. Mughabghab, Phys. Rev. Lett. 54 (1985) 986.

15. Shi Zongren, Zeng Xiantang and Guo Taichang, Chin. J. Nucl. Phys. 4 (1982) 88.
16. R. L. Macklin and S. F. Mughabghab, Phys. Rev. C, to be published.
17. S. Raman, R. F. Carlton, J. C. Wells, E. T. Jurney and J. E. Lynn, Phys. Rev. C, to be published.
18. J. Cugnon and C. Mahaux, Ann. Phys. (N.Y.) 94 (1975) 128.
19. A. G. W. Cameron, Can. J. Phys. 37 (1959) 322.
20. Y. K. Ho and M. A. Lone, Nucl. Phys. A406 (1983) 18.

VERIFICATION OF THE BRINK HYPOTHESIS - GIANT DIPOLE RESONANCES BUILT ON EXCITED STATES

Z. Szefliński

Institute of Experimental Physics, Warsaw University,
Warsaw, Poland

ABSTRACT. The proton capture excitation functions for the reactions $^{63}Cu(p,\gamma_i)^{64}Zn$ and $^{89}Y(p,\gamma_i)^{90}Zr$, leading to several resolved states in the final nuclei, have been measured in the GDR excitation energy region. They display distinct resonance shape. The energy dependence of the gamma-ray strength function built on an excited state in ^{90}Zr was determined. The results support the Brink hypothesis. The excitation functions have been well reproduced by the combined CN and DSD calculations. The extracted spectroscopic factors compared with those derived from the stripping reactions show a distinct correlation.

1. INTRODUCTION

The probability of gamma-ray transitions between highly excited states and low-lying states has been the subject of investigations since a long time. It has been shown that the gamma-ray emission process is governed by the properties of the electric giant dipole resonance (GDR). The study of the GDR has been most extensively performed by gamma-ray absorption experiments in which the resonance built on the ground state is excited. These experiments showed that the gamma-ray strength function for E1 absorption has a Lorentzian shape and a magnitude described by the dipole sum rule. The photonuclear absorption can be described by formula

$$\sigma_{abs}(E_\gamma) = \frac{2}{\pi} \frac{60NZ}{A} \frac{\Gamma E_\gamma^2}{(E_\gamma^2 - E_G^2)^2 + E_\gamma^2 \Gamma^2} \quad (MeV\ mb), \quad (1)$$

where E_G and Γ are the energy and width of the GDR. By analogy with the ground state photoabsorption it is easy to introduce the gamma-ray strength function /1/

$$f(E_\gamma) = \langle \Gamma^J \rangle \, \rho_J(E_i + E_\gamma)/E_\gamma^3 =$$

$$= 8.7 \times 10^{-8} \, \sigma_{abs}(E_\gamma) E_\gamma^{-1} \qquad (MeV^{-3}), \qquad (2)$$

where $\langle \Gamma^J \rangle$ is the average radiative width for a transition J $\rightarrow$ J' (J' = J or J+1), $\rho_J(E_i + E_\gamma)$ is the density of states with given spin J. However, when considering other reactions involving high-energy photons, e.g. capture reactions and heavy ion induced fusion reactions, one has to determine the gamma-ray strength function built on excited states.

In order to describe the strength function built on excited states, it is conventional to use the so-called Brink hypothesis /2/. This hypothesis states that identical giant resonances are built upon each excited state, just shifted upwards by the excitation energy of the corresponding state. The validity of the Brink hypothesis is of great importance for the description of the gamma-emission and absorption processes as well as for the analysis of the gamma-ray spectra emitted from highly excited compound nuclei formed in heavy ion induced reactions. The assumption that emission processes are governed by the excited states giant dipole resonances (ESGDR) and hence by the Brink hypothesis allows us to determine the shape of the average ESGDR, as well as the average nuclear deformation of highly excited states /3, 4/.

These wide applications of the Brink hypothesis have motivated us to investigate the energy dependence of gamma-ray strength function built on excited state. In order to test the Brink hypothesis one has to prove that the strength function built on an excited state has the same functional form as that built on the ground state /1/. There were some previous attempts to verify the Brink hypothesis, but they concerned gamma-ray strength function measurements for just a single excitation energy /5, 6/. It was also demonstrated that a set of strength functions built on successive excited states, populated in resonance-averaged nucleon capture reactions /5, 6/, describes quite well the behaviour of the low-energy GDR tail. However, in none of the papers mentioned above, the energy dependence of the gamma-ray strength function for a definite excited state in a final nucleus was determined.

It is necessary to notice that the recently studied radiative capture reactions at intermediate proton energies on light nuclei /7/ are dominated by ESGDRs. The measured radiative capture excitation functions related to the dipole strength are obviously folded by the reaction mechanism. These folded strength functions show peaks near gamma-ray energy equal to the position of the GDR built on the ground state. The width of the observed peaks increases monotonically with the final state excitation energy. The last observation can be treated as a deviation from the Brink hypothesis.

The radiative capture reaction offers a unique possibility to study the photonuclear properties of excited states. Up to now, such studies have been restricted to transitions to the ground state and the first excited state. The noticeable increase of interest in the analysis of transitions to the excited states encourages us in contributing to some investigations of the structure of these states as well as their coupling to the GDR.

Among the motives which actuated us to undertake the experimental studies on this subject was also the desire to examine the validity of a very simple approach to the description of the proton capture to ESGDRs, which has recently been demonstrated by Dowell et al. /7/. This interesting approach leads to the proportionality between the integrated resonance strengths derived from (γ_i, p_o) cross-sections, to the proton stripping spectroscopic factors. Unfortunately, such a reaction cannot be directly used for investigation of the ESGDR, because the target nuclei in excited states are not available. This difficulty can be overcome by measuring the cross-sections of the (p, γ_i) reactions from which the cross-sections of the (γ_i, p_o) reactions can be obtained by detailed balance.

Another aim of the studies we have initiated has been to check the usability of the direct-semidirect (DSD) model in the investigations of the ESGDR and to test the model parameters. The ^{64}Zn and ^{90}Zr nuclei - the objects of our studies - are very interesting because of a large variety of spins of the states which can be investigated (e.g., $p_{1/2}$ and $g_{9/2}$).

2. EXPERIMENT

The measurements were performed using the Warsaw Van de Graaf generator and the Heidelberg EN Tandem accelerator. The excitation functions of the $^{63}Cu(p,\gamma)^{64}Zn$ and $^{89}Y(p,\gamma)^{90}Zr$ reactions were measured in the proton energy range of 6.5–11.0 MeV and 3.7–12.0 MeV, respectively. Self-supporting targets of ^{63}Cu (99.7% purity) and ^{89}Y were used. The thicknesses of the targets were approximately 800 $\mu g/cm^2$. These thicknesses were determined using alpha-particle energy loss measurements.

In order to reduce the background, only one diaphragm was placed between the switching magnet and the target. The Faraday

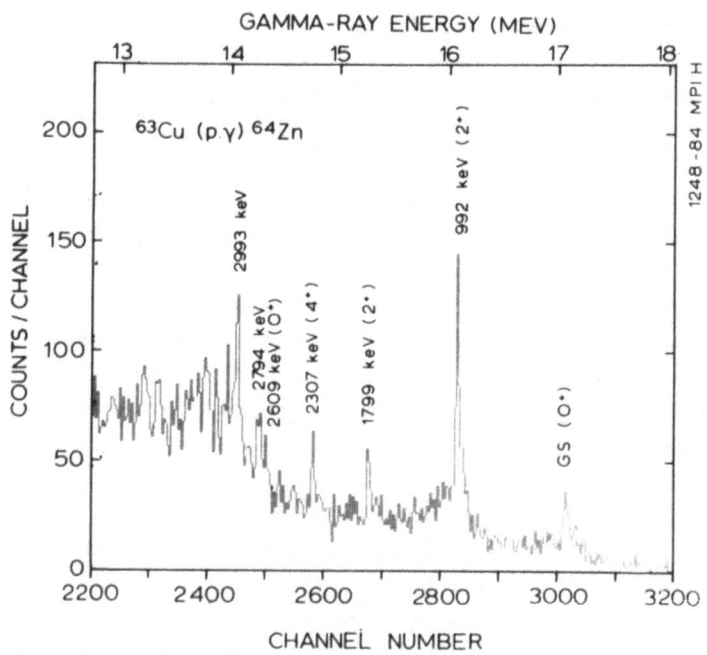

FIG. 1. An example of the measured high-energy primary gamma-ray spectra from the $^{63}Cu(p,\gamma)^{64}Zn$ reaction for $E_p = 9.0$ MeV, $\theta = 90^\circ$.

cup was surrounded by paraffin, cadmium and lead shields located in 2.3 m distance from the target.

The gamma-ray spectra for the case of the $^{63}Cu(p,\gamma)^{64}Zn$ reaction were measured using a central Ge(Li) detector surrounded by a big NaI(Tl) 30 cm x 30 cm crystal, optically divided into two parts, and for the case of the $^{89}Y(p,\gamma)^{90}Zr$ reaction – using a simple high-resolution Ge(Li) detector. The former system worked in the pair spectrometer mode. The efficiency of this pair spectrometer was approximately 30% of that for the double escape peak in the primary Ge(Li) spectrum. In both cases the Ge(Li) detector was placed at 90° with respect to the proton beam. The detectors' arrangement was shielded with cadmium, lead and borated paraffin.

An example of a typical proton capture gamma-ray spectrum of the $^{63}Cu(p,\gamma)^{64}Zn$ reaction is given in Fig. 1. The primary gamma transitions to the ground and excited states in the ^{64}Zn nucleus are labelled. The cross-section uncertainty was about 20%, mainly due to the uncertainty in the determination of the detector efficiency as well as the counting statistics.

3. ANALYSIS

3.1. Test of the Brink Hypothesis

In order to test the Brink hypothesis we have concentrated our interest on the first excited state in ^{90}Zr because it has the same J^{π} value (0^{+}) as the ground state. The analysis of proton radiative capture reactions shows that this process when induced by protons of low energy is dominated by statistical mechanism. At higher energy collective capture prevails. In the statistical model, the Hauser—Feshbach formula connects the proton capture cross-section with the average radiation width. In order to describe the dipole radiative strength function for gamma decay to a state of the same spin and parity as the ground state we use the expression /5, 8/

$$\mathcal{F}_{k\lambda}(E_{\gamma k}) = \mathcal{F}_{o\lambda}(E_{\gamma o}) \, [\sigma(p,\gamma_k)/\sigma(p,\gamma_o)] \left(\frac{E_{\gamma o}}{E_{\gamma k}}\right)^3 . \qquad (3)$$

We use here $\vec{f}$ and $\bar{f}$ to differentiate between the corresponding photoexcitation and gamma-decay strength functions; $\mathfrak{S}(p,\gamma_k)/\mathfrak{S}(p,\gamma_o)$ is the ratio of the measured cross-sections leading to the population of the excited and the ground states; $\vec{f}_{o\lambda}(E\gamma_o)$ is the ground state photoabsorption cross-section.

In ^{90}Zr the configurations of the ground state and the first excited state are, respectively /9/

$$\Psi(0) = a\,\Psi(p_{1/2})^2 - b\,\Psi(g_{9/2})^2,$$

$$\Psi(1) = b\,\Psi(p_{1/2})^2 + a\,\Psi(g_{9/2})^2, \tag{4}$$

where a and b are related to the spectroscopic factors S_o and S_1, respectively ($S_o = a^2 = 0.72$, $S_1 = b^2 = 0.28$).

The ground state of the target nucleus ^{89}Y is a pure $p_{1/2}$ state /10/.

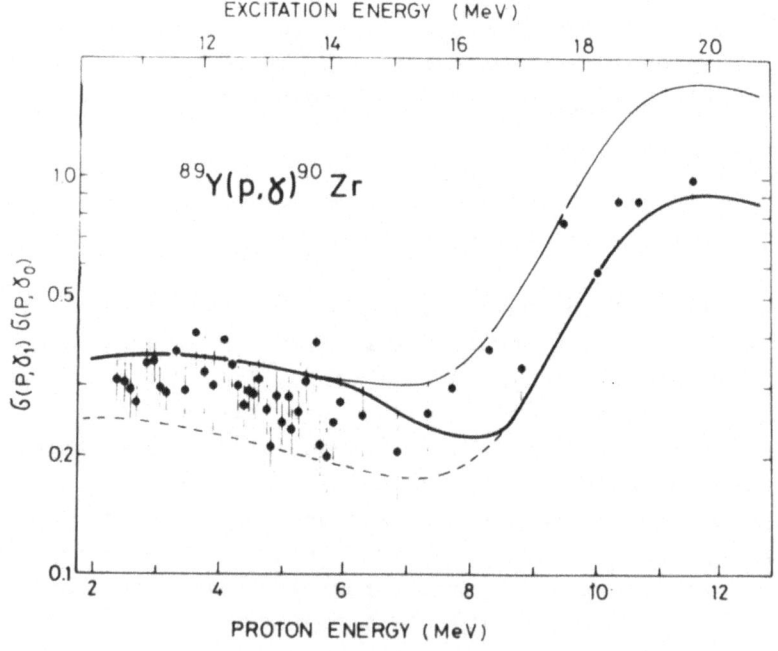

FIG. 2. The $\mathfrak{S}(p,\gamma_1)/\mathfrak{S}(p,\gamma_o)$ measured as a function of excitation energy, E_x. The dashed and dotted curves correspond to the calculations based on statistical and collective models, respectively. The solid line represents the combined calculations.

Making use of the pure resonance model of Dietrich and Kerman /11/ we obtained

$$\overline{f}_{1\lambda}(E_{\gamma_1}) = \overline{f}_{0\lambda}(E_{\gamma_0})[6(p,\gamma_1)/6(p,\gamma_0)]S_0/S_1. \qquad (5)$$

In order to determine the contributions of the statistical (CN) and the collective capture mechanisms to the measured cross-sections, the ratio of the experimental excitation curves $6(p,\gamma_1)/6(p,\gamma_0)$ was studied (Fig. 2). The predictions of CN and collective calculations are denoted by dashed and dotted curves, respectively. As seen from Fig. 2 the statistical model dominates up to proton energy of 6 MeV while the collective capture prevails above 9 MeV. Therefore we have used eq. (3) for determination of the strength function at excitation energies up to 14 MeV, eq. (5) at energies above 17 MeV and a linear interpolation at energies E = 14—17 MeV. As we determine in

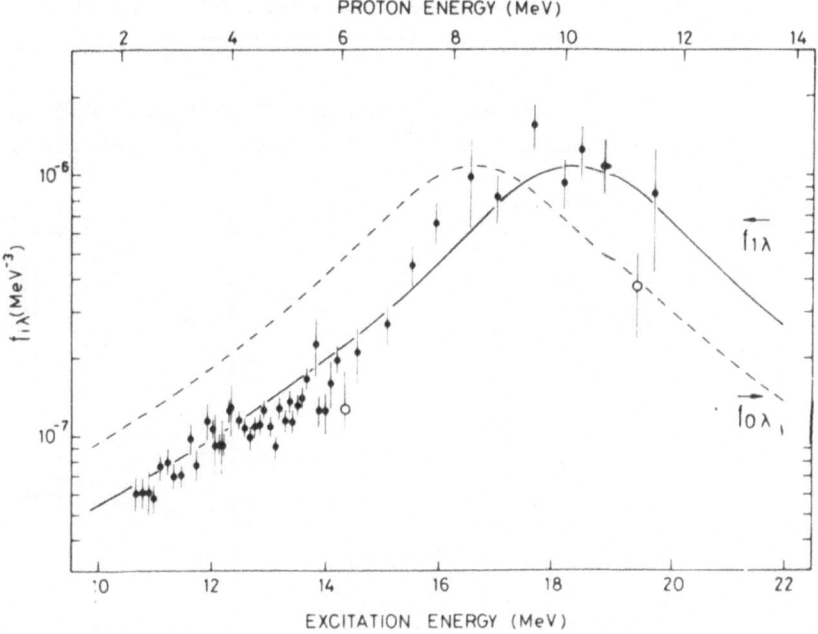

FIG. 3. The radiative strength function for the first excited state in ^{90}Zr (0^+, 1.762 MeV). Open circles denote isobaric analogue states. The dashed line is the photoabsorption strength function. The solid line labelled as $\overline{f}_1$ was obtained by shifting $\overline{f}_0$ by 1.762 MeV, according to the Brink hypothesis.

our work the ratio of strength functions, the model parameters have practically no influence on the results.

The first excited state strength function determined from the measurements is given in Fig. 3. The ground state strength function, $\bar{f}_0$ (dashed curve) was taken in the form given by formula (2) with parameters of GDR obtained in photoabsorption experiment on ^{90}Zr by Lepretre et al. /12/. The solid line denoted as $\bar{f}_1$ was obtained by shifting $\bar{f}_0$ over the excitation energy of the first excited state, i.e. 1.762 MeV, according to the Brink hypothesis. One can consider the obtained result as a positive, quantitative test of the Brink hypothesis.

3.2. Giant Resonances Built on Excited States

In order to study ESGDRs the ^{90}Zr(γ_i,p$_0$)^{89}Y excitation functions were determined from the measured (p,γ_i) excitation functions using detailed balance. The obtained photoproton excitation functions display a distinct resonance shape (Figs. 4, 5).

It is well known, that in nuclei with $T_3 = T_0$ the GDR splits into two components with $T_<$ and $T_>$ resonances, respectively.

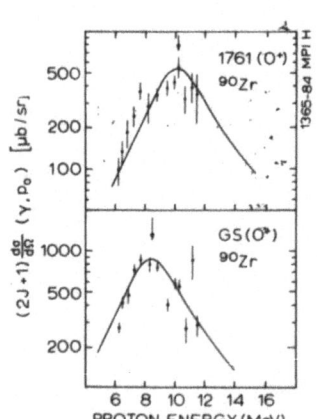 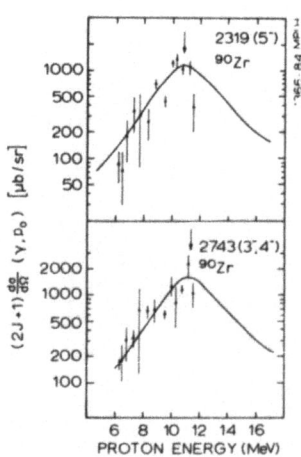

FIGS. 4, 5. A comparison of the differential cross-sections of the (γ_i,p$_0$) reactions at 90° with the Lorentzian curve fitted to the experimental data. The arrows show the positions of the fitted peaks.

The $T_<$ states can decay by both neutron and proton emission according to the isospin selection rules. Neutron decay will dominate because proton emission is strongly suppressed by the Coulomb barrier. As the neutron decay of $T_>$ states to $T_0 - 1/2$ states is forbidden, the decay of $T_>$ states will predominantly occur in the proton channel.

According to theoretical calculations of Akyüz and Fallieros /13/ the isospin splitting energy is

$$E_> - E_< = 60 \frac{T_0 + 1}{A} \quad \text{(MeV)}. \tag{6}$$

This isospin splitting in ^{90}Zr amounts to 4 MeV. This is in a fairly good agreement with the results by Lepretre et al. /12/ and Shoda et al. /14/, who observed two isospin components of GDR located at $E_< = 16.7$ MeV and $E_> = 20.7$ MeV, in photonuclear reactions.

The calculated ratio of the sum rule limits for the $T_>$ to the $T_<$ components is /15/

$$\frac{S_>}{S_<} = \frac{1}{T_0} \frac{1 - 3/2T_0A^{-2/3}}{1 + 3/2A^{-2/3}} , \tag{7}$$

where

$$S = \int E^{-1} 6(E) \, dE. \tag{8}$$

For ^{90}Zr nucleus the value $S_>/S_< = 0.12$. On the basis of the above estimate we have concluded that in the excitation energy region covered by our experiment, i.e. $E_x = 12-20$ MeV, the contribution of the $T_>$ resonance is negligible, so only one peak corresponding to $T_<$ component of the GDR is expected in our (γ_i, p_0) excitation functions.

As seen from Figs. 4 and 5, the extracted excitation functions (γ_i, p_0) show resonance-like structure centred at excitation energies approximately equal to

$$E_x = E_< + E_i, \tag{9}$$

where $E_<$ is the position of the $T_<$ component of the GDR and E_i is the excitation energy of the final state. The arrows in the figures show the positions of ESGDRs predicted according to formula (9).

It is worth mentioning that the good agreement of those predictions with our experimental data supports the Brink hypothesis.

The solid lines in the figures denote the results of fitting the Lorentzian curve to the experimental data. Taking the same width $\Gamma = 4$ MeV for both excited states and the ground state one evidently receives good results.

This is not contradictory to the earlier observations /7/ of broadening the GDRs built on high-lying states, because in our case we are dealing with relatively low-lying states only.

In the schematic model proposed by Dowell et al. /7/, mentioned in the Introduction, the integrated resonance strength, $\int \sigma(\gamma_i, p_o)$ dE, has quite simple quantitative relation to the proton stripping spectroscopic factors C^2S. This relation is

$$\int \sigma(\gamma_i, p_o) \ dE = K_{n,1} C^2 S(n,1). \tag{10}$$

The $K_{n,1}$ factor can be calculated using harmonic oscillator matrix elements in direct emission of photoproton /7/

$$K_{n,1} = 39.5 \left(\frac{N}{A}\right)^2 \left(n + \frac{1}{2} + \frac{1}{2}\right) \quad (\text{MeV mb}), \tag{11}$$

where n and 1 are the principal and orbital quantum numbers of the ejected nucleon. For the subshells engaged in the excited states of ^{90}Zr, i.e. $2p_{1/2}$ and $1g_{9/2}$, the values of $K_{n,1}$ are 37 MeV mb and 43 MeV mb, respectively.

TABLE 1. Comparison of theoretical and experimental $K_{n,1}$ values

Level energy (MeV)	J^{π}	1	$(2J+1)C^2S$	$\int \sigma(\gamma,p) \ dE$ (MeV mb)	$K_{n,1}$ (MeV mb) exp.	$K_{n,1}$ (MeV mb) theor.
0.0	0^+	1	1.31	52.5	40	37
1.762	0^+	1	0.52	36.6	70	37
2.186	2^+			17.0		
2.374	5^-	4	12.43	78.2	6.3	43
2.764	$4^-, 3^-$	4	9.99	110.2	11	43

The experimental $K_{n,1}$ values determined from eq. (10) are compared to the theoretical ones in Table 1. The results show that the simple relation (eq. (10)) between integrated (γ_i, P_o) cross-sections and stripping spectroscopic strength, which has been proved to be valid in the case of light nuclei /7/, does not apply to the medium heavy nucleus ^{90}Zr. Particularly large differences between theoretical and experimental $K_{n,1}$ values are observed for the ejection of protons with high angular momentum (l = 4). In this case drastic reduction of the measured cross-section (and hence $K_{n,1}$) is probably due to the high centrifugal barrier. The other possible source of the observed discrepancy is connected with the contribution of the compound nucleus (CN) process (not taken into account in this simple approach).

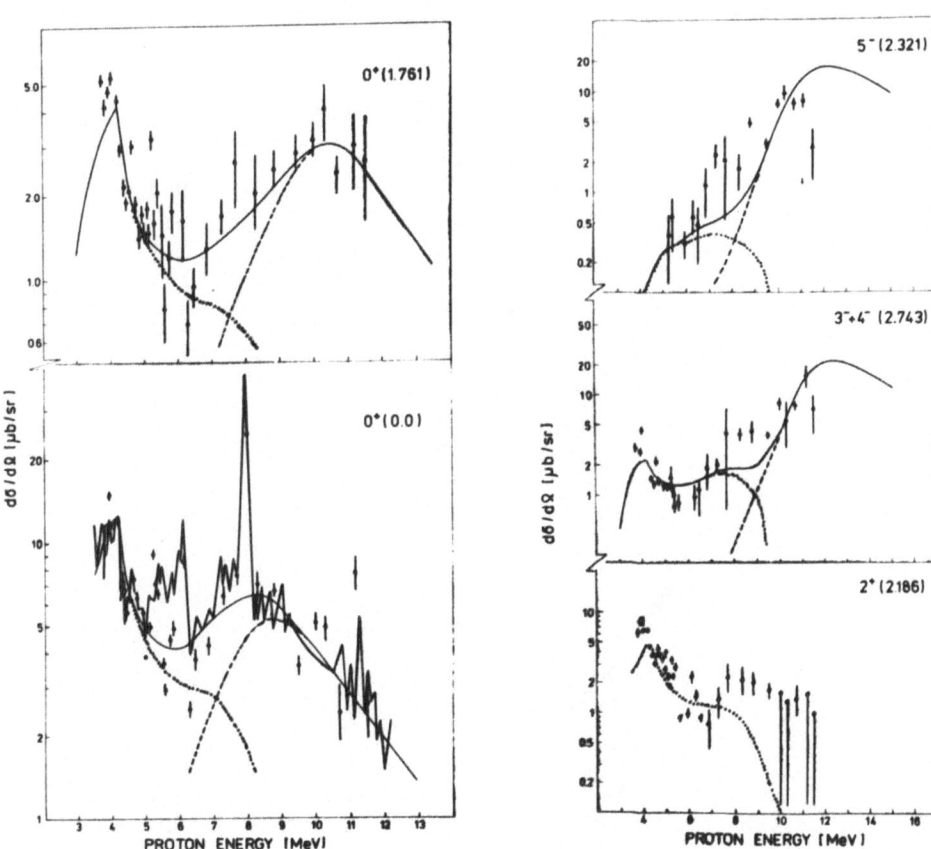

FIGS. 6, 7. The differential cross-sections of the ^{89}Y(p,γ_i)^{90}Zr reactions at 90°. The experimental data are taken from ref. /24/ (thin line) and from the present experiment (closed circles). The CN (dotted line) and DSD (dashed line) curves were obtained from the calculations described in the text. The full curve is the sum of CN and DSD.

In order to explain the disagreement between the simple
theory and our experimental data detailed calculations were
performed.

3.3. Calculations and Results

It is well known that the nucleon capture cross-sections at
low energies are mainly due to the CN process, but the direct
and semidirect DSD mechanism should be introduced in order to
describe the excitation functions in the GDR region. Since our

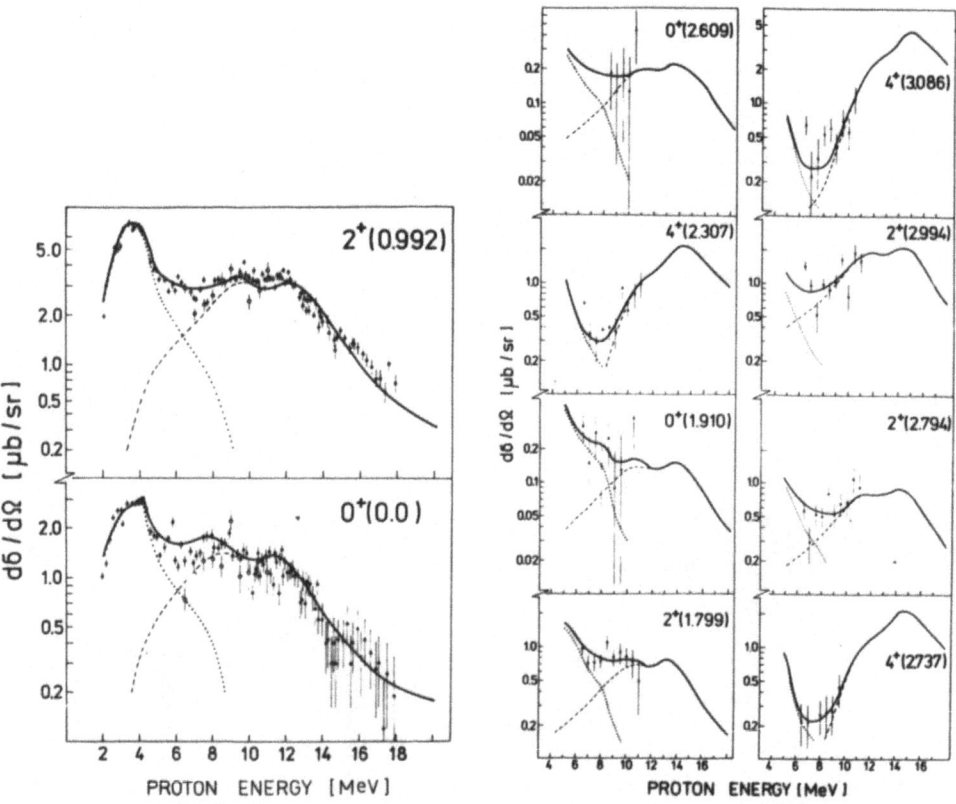

FIGS. 8, 9. The differential cross-sections for the ^{63}Cu(p, γ_i)
^{64}Zn reactions at 90°. The experimental data are taken from
ref. /19/ (solid circles on Fig. 8) and from the present experiment. The CN (dotted line) and DSD (dashed line) and combined
(full line) curves were obtained from the calculations described in the text.

measurements were done upon the proton energy region where both processes contribute (see Figs. 6—9), both of them had to be accounted for in the calculations.

In the cross-section predictions the standard Hauser—Feshbach formula was used with transmission coefficients using Perey /16/ optical potential parameters and Gilbert—Cameron /17/ level densities. The GDR parameters were taken from photoabsorption reactions /12, 18/ and the (p, γ) experiment /19/. The calculations have been performed without any free parameters.

The DSD calculations were performed using the formalism proposed by Clement et al. /20/. The incident wave functions were calculated using Perey /16/ and Becchetti—Greenlees /21/ optical model parameters for the $^{63}Cu(p, \gamma)^{64}Zn$ and $^{89}Y(p, \gamma)^{90}Zr$, respectively. In these calculations the strength of the real part of the symmetry optical potential, V_1, in form of surface peaked function proposed by Clement et al. /20/, was kept constant for all states. V_1 was determined by fitting procedure of the ground state, the spectroscopic factor of which was taken from $(^3He,d)$ data /22, 23/. The values $V_1 = 50$ MeV and 140 MeV for $^{63}Cu(p, \gamma)^{64}Zn$ and $^{89}Y(p, \gamma)$ reactions, respectively, were adjusted in that way. The fitting of the excited states made it possible for us to determine the spectroscopic factors of the final states under consideration.

The results of the fit of the cross-sections are shown in Figs. 6—9. The excellent fit of the predicted cross-sections to the experimental data is very remarkable. It can also be seen that both processes mentioned above contribute to the measured cross-sections of the investigated reactions, except that leading to the 2^+ state at 2.186 MeV in ^{90}Zr nucleus. This interesting exception is connected with the particular structure of this state, which is dominated by a $(g_{9/2})^2$ configuration. The excess of the measured cross-section over CN prediction (Fig. 7), observed above 9 MeV, can probably be explained by a contribution of the precompound process. The fitting procedure has delivered the spectroscopic factors of the excited states.

Figures 10 and 11 show the comparison of the factors obtained by us with those taken from (^{3}He,d) reactions /22, 23/. As the ground state spectroscopic factors were taken from other paper, all spectroscopic factors for the excited states are in a sense "normalized" to that for the ground state, to which the particle vibration coupling strength, V_1, was adjusted.

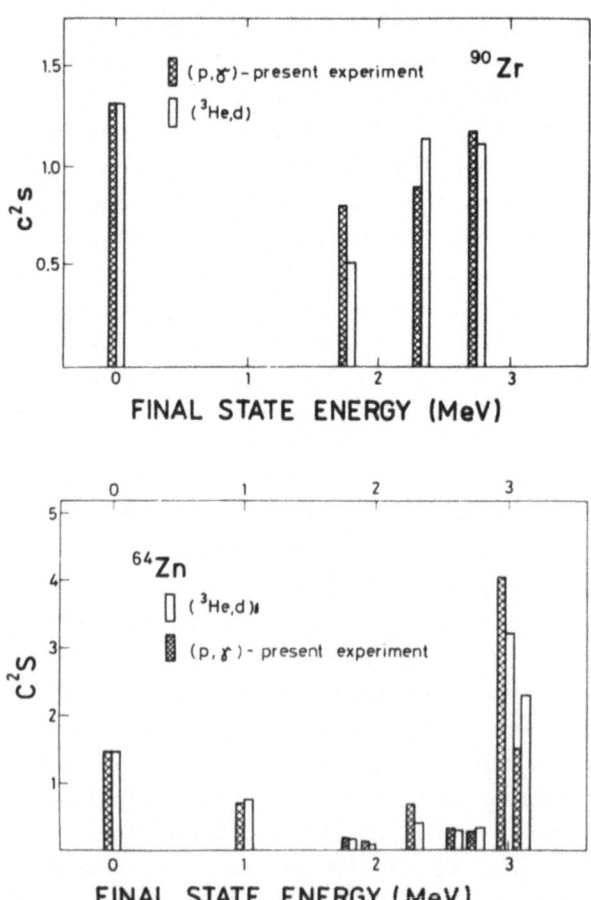

FIGS. 10, 11. Comparison of the spectroscopic factors obtained from the present experiment with those from the (^{3}He,d) reaction.

4. CONCLUSIONS

It can be seen from Figs. 10 and 11 that the agreement of the C^2S values obtained from (p,γ) and $(^3He,d)$ reactions is quite good. This fact, together with the good reproduction of the experimental excitation functions by the model calculations, demonstrates that the contributions of various reaction mechanisms were properly accounted for.

At the same time, this fact can be considered as an indication of the validity of our assumption that the strength of the real part of the symmetry optical potential, V_1, remains constant for all investigated states. The procedure applied to the considered medium mass nuclei allows us to conclude that one can use radiative capture reaction described by the DSD model in order to determine spectroscopic factors of low-lying levels. This means a further development of an idea by Rolfs /25/ to apply the direct reaction in order to determine the spectroscopic factors. The brilliant reproduction of the experimental data by the model calculations gives also support to the Brink hypothesis, since in the DSD model this hypothesis is automatically included.

I am indebted to my colleagues who have shared in this work, including Z. Wilhelmi, H. V. Klapdor, G. Szeflińska, T. Rząca-Urban, M. Kicińska, K. Grotz, J. Metzinger and J. J. Damaschke.

REFERENCES

1. G. A. Bartholomew, E. D. Earle, A. J. Fergusson, J. W. Knowles and A. M. Lone, Adv. Nucl. Phys. 7 (1973) 229.
2. D. M. Brink, Ph.D. thesis, Oxford University 1955.
   B. B. Kinsey, in "Handbuch der Physik", Vol XL, p. 314, Springer-Verlag, Berlin, 1957.
3. J. O. Newton, B. Herskind, R. M. Diamond, E. L. Diens, J. E. Draper, K. H. Lindenberger, C. Schuck, S. Shih and F. S. Stephens, Phys. Rev. Lett. 46 (1981) 1383.
4. C. A. Gossett, K. A. Snover, J. A. Behr, G. Feldman, J. L. Osborne, Phys. Rev. Lett. 54 (1985) 1486.
5. G. Szeflińska, Z. Szefliński and Z. Wilhelmi, Nucl. Phys. A223 (1979) 253.
6. S. Raman, O. Shahal and G. G. Slaughter, Phys. Rev. C23 (1981) 2794.

7. D. H. Dowell, G. Feldman, K. A. Snover and A. M. Sandorfi, Phys. Rev. Lett. 50 (1983) 1191.
   D. H. Dowell, Proc. Int. Symp. on Capture Gamma-ray Spectroscopy, ed. by S. Raman, New York, 1985.
8. Z. Szefliński, G. Szeflińska, Z. Wilhelmi, T. Rzaca-Urban, H. V. Klapdor, E. Andersson, K. Grotz and J. Metzinger, Phys. Lett. 126B (1983) 159.
9. A. Adam, D. Adam, O. Bersillon and S. Joly, Nuovo Cimento 33A (1976) 171.
10. G. Vourvopoulos, Nucl. Phys. A174 (1971) 581.
11. F. S. Dietrich and A. K. Kerman, Phys. Rev. Lett. 43 (1978) 114.
12. A. Lepretre, H. Beil, R. Bergere, P. Carlos, A. Veyssiere and M. Sugawara, Nucl. Phys. A175 (1971) 609.
13. R. O. Akyüz and S. Fallieros, Phys. Rev. Lett. 27 (1971) 1016.
14. K. Shoda, H. Miyase, M. Sugawara, T. Saito, S. Oikawa, A. Suzuki and J. Uegaki, Nucl. Phys. A239 (1975) 397.
15. S. Fallieros and B. Goulard, Nucl. Phys. A147 (1970) 593.
16. F. G. Perey, Phys. Rev. 131 (1963) 745.
17. A. Gilbert and A. G. W. Cameron, Can. J. Phys. 43 (1965) 1446.
18. P. Carlos, H. Beil, R. Bergere, J. Fagot, A. Lepretre, A. Veyssiere and G. V. Solodukhov, Nucl. Phys. A258 (1976) 365.
19. P. Paul, J. F. Amann and K. A. Snover, Phys. Rev. Lett. 27 (1971) 1013.
20. C. F. Clement, A. M. Lane and J. R. Rook, Nucl. Phys. 66 (1965) 273.
21. F. D. Becchetti and G. W. Greenlees, Phys. Rev. C2 (1970) 639.
22. J. L. C. Ford, Jr., K. L. Warsh, R. L. Robinson and C. D. Moak, Nucl. Phys. A103 (1967) 525.
23. G. Vourvopoulos and J. D. Fox, Phys. Rev. 177 (1969) 1558.
24. M. Hasinoff, G. A. Fisher and S. S. Hanna, Nucl. Phys. A126 (1973) 221.
25. C. Rolfs, Nucl. Phys. A217 (1973) 29.

GAMMA-RAY COMPETITION IN CONTINUUM

R. Antalík

Institute of Physics, Electro-Physical Research Centre of
the Slovak Academy of Sciences, Bratislava, Czechoslovakia

ABSTRACT. Recently Stengl et al. /1/ have shown that a statistical cascade model with a realistic description of the strength
functions for particles and gamma-ray channels cannot describe
the gamma-ray competition in the continuum of ^{56}Fe. In this
work we have shown that inclusion of the unequal distribution
of continuum states according to their parity gives the quantitative agreement with the experimental data. The parity distribution function we have used was calculated in a frame of a
thermodynamic model of independent quasiparticles /2/, with a
realistic Woods—Saxon single particle basis.

The inelastic scattering of 14 MeV neutrons with the excitation of continuum states of ^{56}Fe proceeds mainly through a compound mechanism, and this is a basis for our belief that deexcitation of these continuum states is dominated by the compound
mechanism. This, however, means that the Hauser-Feshbach theory
has to work very well.

It was very surprising, when Stengl et al. /1/ observed
that the statistical model with a realistic parametrization
and with good account of excitation functions and spectra of
emitted particles and gamma rays cannot describe the gamma-ray
competition in the continuum. They were able to interpret their
data only with normalization of the total gamma decay width to
Γ_γ = 20 eV which is one order of magnitude higher than a realistic value.

So, our observations have shown that the compound mechanism
is well justified in this case, that particles and gamma-ray
strengths are realistically described, and also that the energy
and spin dependences of the level densities are realistic·

However, we also known that in all even-even fp-shell nuclei the ratio

$$f_{\pi}(E, \pi = -1) = \frac{\varrho(E, \pi = -1)}{\varrho(E, \pi = -1) + \varrho(E, \pi = +1)}$$

which we call a parity distribution function, reaches the equi-probable value 0.5 at excitation energies well above a neutron binding energy (B_n). This prevailing of the positive parity states in the continuum of ^{56}Fe hinders the neutron decay to ^{55}Fe where all accessible states have the negative parity.

Results of our calculations including the parity distribution function from ref. /2/ and standard assumption that $f_{\pi} = 0.5$ are in Fig. 1, in which we can see two lines that des-

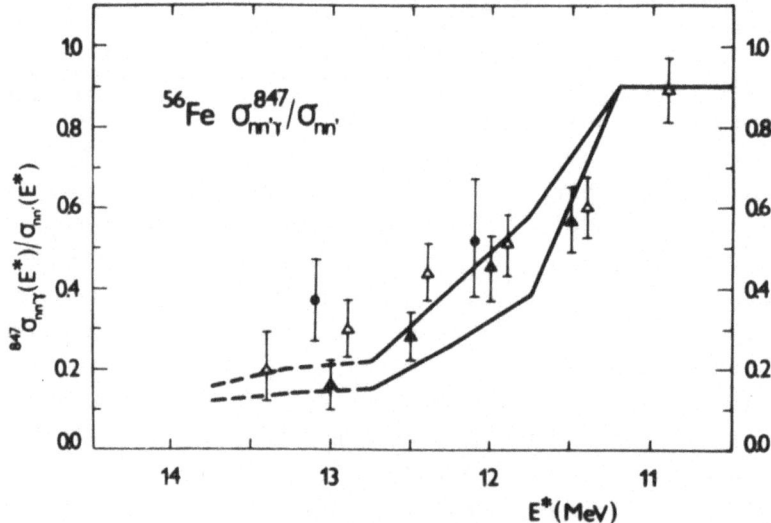

FIG. 1. The ratio of the cross-section in the (n,n'γ - 847 keV) channel to the (n,n') channel as a function of the excitation energy of ^{56}Fe. The upper (lower) line represents the statistical model prediction based on the unequal (equal) distribution of continuum states according to their parity. Experimental data are from ref. /1/ ($\blacktriangle$) and from ref. /3/ ($\bullet$).

cribe the ratio of cross-sections in (n,n'γ-decaying through the 847 keV line) and (n,n') channels as a function of the excitation energy of ^{56}Fe. The lower line represents results of the statistical reaction model calculations with the $f_\gamma(E)$ = 0.5. The upper line represents the result with the $f_\gamma(E)$ $\neq$ 0.5 /2/, and as we can see from Fig. 1, only this line is in agreement with experimental data.

Consequently we may conclude that to understand the gamma competition in the continuum of ^{56}Fe, it is necessary to assume the unequal parity distribution and not the unrealistic value of the total gamma width Γ_γ.

REFERENCES

1. M. Stengl et al., Nucl. Phys. A290 (1977) 109.
2. R. Antalík, Proc. Europhys. Top. Conference on Neutron Induced Reactions, Smolenice, 1982, edited by P. Obložinský, Institute of Physics, Bratislava, 1982, p. 357.
3. S. Hlaváč, Thesis, Bratislava, 1982.

NEUTRON-GAMMA COINCIDENT CALCULATIONS WITHIN THE PRE-EQUILIBRIUM MODEL

E. Běták and *J. Dobeš

Institute of Physics, Electro-Physical Research Centre of the Slovak Academy of Sciences, Bratislava, Czechoslovakia

*Institute of Nuclear Physics, Czechoslovak Academy of Sciences, Řež, Czechoslovakia

ABSTRACT. The differential gamma multiplicities for 14 MeV neutron induced reactions are calculated fully within the master-equation formulation of the pre-equilibrium exciton model. The results obtained correspond to the data; however, due to the lack of spin and parity in the model the present approach cannot be used to extract more physical information from the data.

The gamma emission has been incorporated into the pre-equilibrium model less than a decade ago. Though seriously handicapped by the absence of spin in the formalism used, rather reasonable results were obtained for continuous energy spectra and relative gamma widths, as well as for the exclusive nucleon spectra.

In the present paper we aim to test the ability of the model to give a more refined correlation quantity, namely the differential gamma multiplicities (those associated with the emission of a nucleon with given energy). The calculations are performed within the master-equation formulation of the exciton model. The set of master equations reads /1/

$$\frac{dP(n,t,E,i)}{dt} = P(n-2,t,E,i)\,\lambda^+(n-2,E,i) +$$

$$+ P(n+2,t,E,i)\,\lambda^-(n+2,E,i) - P(n,t,E,i)\,.$$

$$.[\lambda^+(n,E,i) + \lambda^-(n,E,i) + L(n,E,i)] +$$

$$+ \sum_{j,m,x} \int P(m,t,E',j)\lambda_x^c(m,E',j,\varepsilon) \, d\varepsilon. \qquad (1)$$

Here, $P(n,t,E,i)$ is the occupation probability of the state of i-th nucleus with n excitons and the excitation energy E at time t, $\lambda^{\pm}$ are the transition rates, λ_x^c are the emission rates and

$$L(n,E,i) = \sum_x \int \lambda_x^c(n,E,i,\varepsilon) \, d\varepsilon \qquad (2)$$

is the total emission rate from given state. The summation in (2) goes over all possible ejectiles; especially, the gamma emission has also to be included. Two different models for the pre-equilibrium gamma emission are used /2, 3/. Both of them share some common features and can be expressed as

$$\lambda_\gamma^c(n,E,\varepsilon_\gamma) = \frac{\varepsilon_\gamma^2 \, \sigma_a(\varepsilon_\gamma)}{\pi^2 \, \hbar^3 c^2} \, \frac{\displaystyle\sum_{m=n,n-2} b(m,\varepsilon_\gamma)\,\omega\,(m,E-\varepsilon_\gamma)}{\omega(n,E)}. \qquad (3)$$

The difference between them lies in the interpretation of the Brink–Axel hypothesis and in the equilibrium limit for the gamma emission. The former one /2/ couples the photon-absorption cross-section entirely to the $(n-2) \rightarrow n$ exciton-state transition (by the analogy of the ground state excitation, taken to be 0p0h $\rightarrow$ 1p1h nature /4/). Consequently,

$$b_{BD}(n-2,\varepsilon_\gamma) = 1 \ ,$$
$$b_{BD}(n,\varepsilon_\gamma) = \frac{gn}{\omega(1,1,\varepsilon_\gamma)} \qquad (4)$$

On the other hand, the recent version /3/ requires the photon absorption cross-section to be split between both the processes, what results in proper normalization at equilibrium and consistency with the standard (i.e. compound) model. In this approach,

$$b_{AG}(n-2,\varepsilon_\gamma) = \frac{\omega(1,1,\varepsilon_\gamma)}{g(n-2) + \omega(1,1,\varepsilon_\gamma)} \qquad (5)$$

$$b_{AG}(n, \varepsilon_\gamma) = \frac{gn}{gn + \omega\,(1,1,\varepsilon_\gamma)} \,. \tag{5}$$

Effectively, the near-equilibrium gamma emission in /3/ is approximately one half of that of ref. /2/; whereas the emission far from the equilibrium remains essentially unchanged.

The calculations were performed using somewhat modified PEQGM code /1/ and compared to the data of $^{56}Fe(n,xn\gamma)$ and $^{52}Cr(n,xn\gamma)$ reactions at 14 MeV, measured by Hlaváč and Obložinský /5, 6/ and are presented in Fig. 1. Two values for the

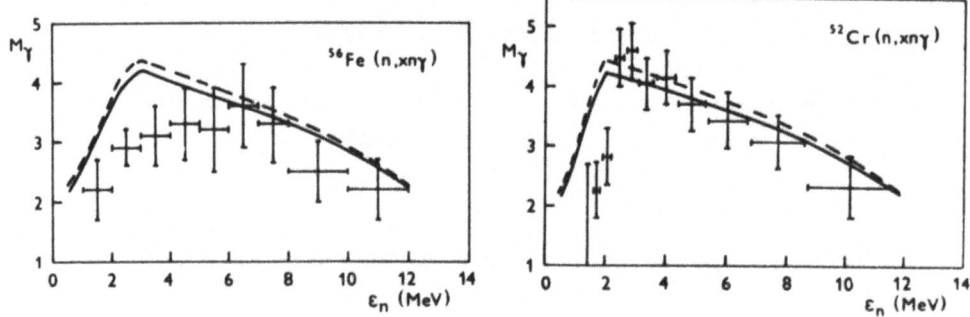

FIG. 1. A. Differential gamma multiplicities versus the energy of emitted neutron in $^{56}Fe(n,xn\gamma)$ reaction at 14.6 MeV. The data are of ref. /5/, the lines represent calculations performed for different models of the pre-equilibrium gamma emission: the dashed line stands for that of ref. /2/, the full line for that of ref. /3/.
    B. The same as in A but for $^{52}Cr(n,xn\gamma)$ reaction. The data as of ref. /6/.

matrix element constant (of the form suggested by Kalbach /7/) were taken, namely $K' = 100$ MeV^3 and $K' = 500$ MeV^3. The energy step of 0.5 MeV was used in the calculations. The multiplicities above the second neutron threshold are nearly identical for both the values of K'. Below it, the typical difference is about 1% (and therefore the two calculations are indistinguishable in our picture). Somewhat surprising is, how close the multiplicities calculated within both the models of gamma emission come out to be, especially when one remembers the factor of about 2 in the ratios of their (near-equilibrium) emission rates. The nucleus is really very near to the equilibrium after

296

the first neutron emission in 14 MeV reactions, but it has to decay somehow and the only way below the nucleon threshold is the gamma emission, so that the total intensity (integral over time) should be nearly the same, independently of the model used for the gamma emission.

The overall agreement between the calculations and the experimental data is rather encouraging, even if our model lacks completely both spin and parity (their influence on the gamma multiplicities calculated within the standard equilibrium model is remarkable, cf. ref. /8/).

Generally, we have tried to extend the abilities of the pre-equilibrium (exciton) model to the calculations of differential gamma multiplicities. Doing that, two different assumptions for the pre-equilibrium gamma emission were used. The results for both of them are rather close each to the other and do not contradict the data. However, the absence of spin and parity variables in our model disables one to extract more information and lowers the meaning of the analysis. The further development of the model requires extensions mainly in this direction.

The authors thank R. Antalík, S. Hlaváč and P. Obložinský for valuable discussions.

REFERENCES

1. E. Běták and J. Dobeš, Report IP EPRC SAS, No. 43/1983, Bratislava, 1983.
2. E. Běták and J. Dobeš, Phys. Lett. 84B (1979) 368.
3. J. M. Akkermans and H. Gruppelaar, Phys. Lett. 157B (1985) 95.
4. V. K. Lukyanov et al., Yad. Fiz. 21 (1975) 992 (Sov. J. Nucl. Phys. 21 (1975) 508).
5. S. Hlaváč et al., Izv. AN SSSR, Ser. Fiz. 46 (1982) 903.
6. S. Hlaváč et al., this volume.
7. C. Kalbach, Z. Phys. A287 (1978) 319.
8. R. Antalík, to be published.

THE ^{52}Cr(n,xnγ) REACTIONS AT 14.6 MeV STUDIED BY COINCIDENT IN-BEAM TECHNIQUES *

S. Hlaváč, P. Obložinský and J. Pivarč

Institute of Physics, Electro-Physical Research Centre of
the Slovak Academy of Sciences, Bratislava, Czechoslovakia

ABSTRACT. Reactions (n,xnγ) were studied by bombarding highly enriched ^{52}Cr sample with 14.6 MeV neutrons, using coincident in-beam techniques. Bulk of new experimental data includes average gamma multiplicities related to 8 discrete gamma transitions, average gamma multiplicities following emission of 1.3 —11.6 MeV neutrons, average energies of these gamma rays and exclusive neutron spectrum. Preliminary analysis indicates that gamma competition is higher than should be expected by GDR gamma ray strength function and standard statistical nuclear reaction model calculations.

1. INTRODUCTION

Chromium belongs among basic elements of constructional materials and its interactions with neutrons are of interest for both fusion and fission reactor technology. Inelastically scattered 14.6 MeV neutrons can effectively populate unbound states of target nucleus and can be exploited in studies of their decay modes, specifically by gamma emission, via (n,n'γ) reactions. In order to probe such states more directly we have used coincident γ-γ and n-γ in-beam techniques.

In this report we present a complete set of our experimental data for ^{52}Cr sample obtained by the coincident techniques. Besides the average gamma-ray multiplicities we report for the first time on exclusive neutron spectrum which unlike total spectrum contains pure inelastic component only.

---

*Work performed under the IAEA Coordinated Research Programme, Research Agreement No. 3436/CF.

2. EXPERIMENTAL PROCEDURES

Measurements were carried out with the multidetector set-up developed for combined in-beam neutron and gamma spectroscopy experiments using the associate particle method /1/. Geometrical arrangement of the present run is shown in Fig. 1.

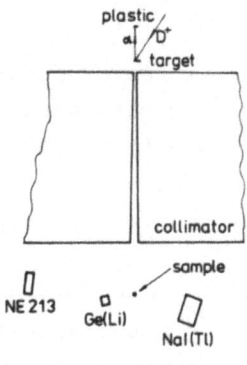

FIG. 1. Experimental arrangement. Size of spectrometers, their orientation towards the incident beam and the sample-to-detector front face distances were $\emptyset$ 16 cm x 10 cm, 70°, 30 cm for the NaI(Tl), $\emptyset$ 12 cm x 4 cm, 100°, 60 cm for NE 213 and 70 cm³, 70°, 15 cm for Ge(Li), respectively. The NaI(Tl) was shielded by an additional lead collimator with an apperture 11 cm in diameter.

A thin polyethylene bag (1.5 g) of cylindrical shape was filled with enriched sample (99.8±0.1%, weight 119.8 g) in powder form. The average sample thickness was 0.0609 at/b.

Singles spectra from all the spectrometers and 2 coincident spectra were measured. Spectrum of discrete gamma rays observed by the Ge(Li) spectrometer in coincidence with another gamma ray detected by the NaI(Tl) detector was measured as first, followed by two-parametric spectrum comprising neutron tof events as observed by the NE 213 in coincidence with energy of gamma ray events from the NaI(Tl).

The spectra were processed and analysed by standard procedures /1, 2/. Average gamma-ray multiplicities were obtained by comparing coincident to singles Ge(Li) yields and by comparing

coincident to singles neutron tof spectra. Neutron fluence was
determined from the counts of the associated alpha-particle
detector and the stilbene monitor. Corrections for sample-to-
detector geometry and self-absorption were calculated by the
Monte Carlo procedure, no correction for multiple scattering
was applied.

3. RESULTS

Average gamma-ray multiplicities of cascades passing through a
specific low-lying transition were observed in 8 instances in
the $(n,n'\gamma)$ channel. The results are summarized in Table 1.

TABLE 1. Observed gamma multiplicities including specific discrete transition in ^{52}Cr

Transition (keV)	Multiplicity
647.4, $4^+ \rightarrow 4^+$	5.6 (0.6)
704.6, $(3^+) \rightarrow 4^+$	6.8 (0.7)
744.2, $6^+ \rightarrow 4^+$	5.8 (0.6)
848.2, $5^+ \rightarrow 4^+$	4.8 (1.5)
935.5, $4^+ \rightarrow 2^+$	3.7 (0.3)
1246.2, $5^+ \rightarrow 4^+$	4.4 (1.4)
1333.6, $4^+ \rightarrow 2^+$	3.7 (0.4)
1434.1, $2^+ \rightarrow 0^+$	3.7 (0.2)

Uncertanties are given in brackets.

Average gamma-ray multiplicities of cascades following emission of a neutron with specified energy are summarized in Fig.
2. Multiplicities were extracted from ratios of given bins of
coincident and singles neutron tof spectra. Background was subtracted from the raw spectra by the procedure devised by Klein
et al. /3/.

Two-parametric spectrum n tof x gamma energy contains information about gamma-ray spectrum as a function of energy of emitted neutron. First moment of the spectrum represents the average

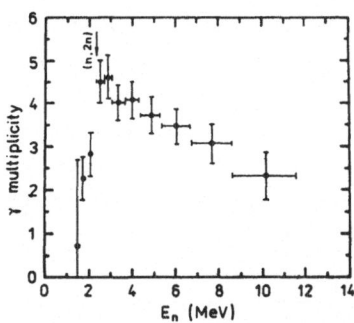

FIG. 2. Average gamma-ray multiplicity as a function of observed energy of emitted neutron in the laboratory frame. The
(n,2n) threshold is given by an arrow.

TABLE 2. Average gamma-ray energy as a function of observed
energy of emitted neutrons

E_n (MeV)	$\overline{E}_\gamma$ (MeV)
1.0 — 1.2	1.71 (0.25)
1.2 — 1.6	2.50 (0.36)
1.6 — 1.9	2.33 (0.22)
1.9 — 2.2	2.30 (0.16)
2.2 — 2.6	2.39 (0.12)
2.6 — 3.1	2.43 (0.11)
3.1 — 3.5	2.40 (0.15)
3.5 — 3.9	2.60 (0.16)
3.9 — 4.4	2.50 (0.17)
4.4 — 5.0	2.29 (0.17)
5.0 — 5.7	2.36 (0.19)
5.7 — 6.6	2.45 (0.22)
6.6 — 7.7	2.53 (0.25)
7.7 — 9.0	2.02 (0.22)
9.0 —10.8	1.72 (0.22)
10.8 —13.2	1.72 (0.29)

energy of cascading gamma rays. It was extracted from the spectrum by means of the calibration curve for centroids without
unfolding the raw spectrum. The results are given in Table 2.

The two-parametric spectrum was further used to extract neutron spectrum in coincidence with 1434 keV gamma rays. The latters refer to strong $2^+ \rightarrow 0^+$ gs transition in ^{52}Cr which cumulates almost all (n,n'γ) cross-section. Pertinent neutrons, therefore, represent practically complete (n,n') component with no admixture from (n,2n) channel. The resulting exclusive neutron spectrum is given in Fig. 3.

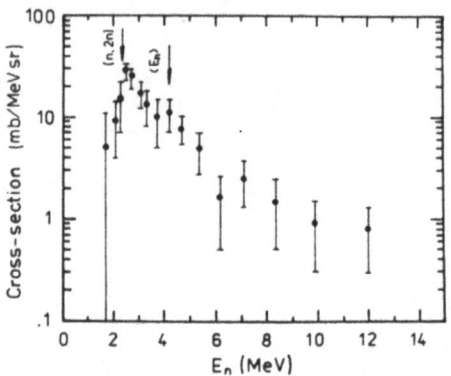

FIG. 3. Exclusive neutron spectrum. Average neutron energy indicated by the arrow is 4.19 (0.17) MeV.

4. DISCUSSION

To our best knowledge all data included into this report have been measured for the first time. Still, however, certain comparison with earlier data is possible in two instances.

Dickens et al. /4/ have reported average gamma-ray energy observed on Cr(n,xγ) reactions as a function of energy of incident neutrons. Their value, $\bar{E}_\gamma \approx 2.3-2.4$ MeV, taken from figure at 14.6 MeV, seems to be in very good accord with our results. One, of course, should keep in mind that we report practically the ^{52}Cr(n,n'γ) component only.

Our exclusive neutron spectrum at spectral energies above about 2.5 MeV can be compared with full neutron spectra as measured by Hermsdorf et al. /5/ at 14.6 MeV, 90° and by Takahashi et al. /6/ at about 14.0 MeV, 103° on natural Cr. Accord with these data seems to be fairly good, too.

Theoretical analysis was started within the statistical model of nuclear reactions using the updated version of the STAPRE code /7/. As input we used quite standard set of parameters, such as for optical model, level densities, gigantic dipole gamma-ray strength functions and realistic decay schemes of low-lying levels.

One point of interest concerned the gamma competition in the decay of unbound states of ^{52}Cr. Preliminary results of our, so far incomplete, analysis indicate that the calculated competition is lower than suggested by the experimental data.

Special attempt to analyse average gamma multiplicities following emission of neutrons of specified energy was initiated by Běták and Dobeš. Their analysis, reported elsewhere in this Conference /8/, is of interest because they consider in it the pre-equilibrium mechanism of gamma-ray emission.

REFERENCES

1. S. Hlaváč and P. Obložinský, Nucl. Instr. Meth. 206 (1983) 127.
2. Š. Gmuca and I. Ribanský, Jad. Energ. 29 (1983) 56.
3. H. Klein et al., Proc. IAEA Consultants' Meeting, Smolenice, 1983, see Report INDC (NDS)-146, Vienna, 1983, p. 191.
4. J. K. Dickens et al., Nucl. Sci. Eng. 62 (1977) 515.
5. D. Hermsdorf et al., Report ZfK-277(U), TU Dresden, 1975.
6. A. Takahashi et al., OKTAVIAN Report A-83-01, University Osaka, 1983.
7. M. Uhl and B. Strohmaier, Code STAPRE; in Nuclear Theory for Applications, ICTP, Trieste, 1980.
8. E. Běták and J. Dobeš, this volume.

TEMPERATURE SHIFT OF NEUTRON RESONANCES

V.K. Ignatovich, *A. Meister, *S. Mittag, *W. Pilz,
*D. Seeliger and *K. Seidel

Joint Institute for Nuclear Research, Dubna, USSR
*Technical University Dresden, GDR

ABSTRACT. Temperature-induced shift of neutron resonances have been experimentally observed studying low-energy resonances of ^{103}Rh, ^{109}Ag, ^{161}Dy and ^{163}Dy with a time-of-flight spectrometer at the Dubna pulsed reactor. The shifts are regarded as an analogue to the second-order Doppler effect in Mossbauer gamma-ray spectroscopy.

1. INTRODUCTION

The interaction cross-sections of slow neutrons with atomic nuclei have resonances shaped by both the properties of the nuclei and the thermal motion of the target atoms relative to the incident neutrons. The latter brings Doppler broadenings of the resonances described as soon as in 1937 by Bethe and Placzek /1/ for gaseous motion and in 1939 by Lamb /2/ for neutron capture in crystals. In 1960, a so-called Doppler effect of second order has been observed in experiments with gamma rays /3, 4/. The Mossbauer spectroscopy has shown that the gamma peak is shifted if the temperatures of absorber (T_a) and emitter (T_e) are different

$$\Delta E_\gamma = - \frac{E_\gamma}{Mc^2} \Delta \langle \varepsilon_k \rangle \ , \tag{1}$$

where $\Delta \langle \varepsilon_k \rangle = \langle \varepsilon_k \rangle_a - \langle \varepsilon_k \rangle_e$ is the difference between the mean kinetic energies of the absorber and the emitter atoms which is at high temperatures $\approx 3/2 k_B (T_a - T_e)$; M is the mass of the atoms and k_B is the Boltzmann constant.

If eq. (1) is formally transformed to the absorption of neutrons, a corresponding neutron resonance should be depending on temperature and shifted by /5/

$$\Delta E_n = - \frac{1}{A} \Delta \langle \varepsilon_k \rangle , \tag{2}$$

where A is the mass number of the absorbing atoms and 1 stands for mass number of the neutrons; $A \gg 1$. However, in contrast to Mossbauer gamma peaks, neutron resonances have widths $\Gamma \gg \Delta E_n$ and are Doppler broadened; that means, a recoilless component is not observed separately. The resonance position may be defined as

$$\overline{E}_n = \int_{E_1}^{E_2} \sigma_R(E_n) E_n \; dE_n \bigg/ \int_{E_1}^{E_2} \sigma_R(E_n) \; dE_n , \tag{3}$$

where σ_R is the resonance part of the cross-section and E_1, E_2 are chosen so that $\sigma_R(E_1) = \sigma_R(E_2) = 0$. Then, the temperature shift follows as

$$\Delta \overline{E}_T = \overline{E}_n(T_2) - \overline{E}_n(T_1) . \tag{4}$$

Theoretical considerations have resulted in /6/

$$\Delta \overline{E}_T = + \frac{1}{3A} \Delta \langle \varepsilon_k \rangle \qquad \text{for free-gas atoms} \tag{5}$$

and

$$\Delta \overline{E}_T = - \frac{1}{3A} \Delta \langle \varepsilon_k \rangle \qquad \text{for absorption in crystals,} \tag{6}$$

if $\Gamma \gg \Gamma_n$ and $\Gamma \gg h\nu$, the energy of the crystal phonons. These temperature shifts of neutron resonances are searched in the following experiment.

2. EXPERIMENT

Neutron transmission spectra were measured with metallic samples of Rh, Ag, and Dy at various temperatures in the region of low-energy resonances using a time-of-flight spectrometer

at the Dubna pulsed reactor IBR-30 operated in booster mode
with the linac LUE-40. The experimental arrangement is shown
in Figs. 1-3.

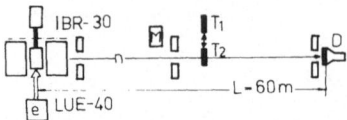

FIG. 1. Experimental arrangement. T_1, T_2 - samples at temperatures T_1 and T_2, respectively, D - ^{6}Li-glass scintillation detector, M - neutron monitor.

The samples heated to different temperatures were alternately placed into the beam for 10 min. The temperatures of the

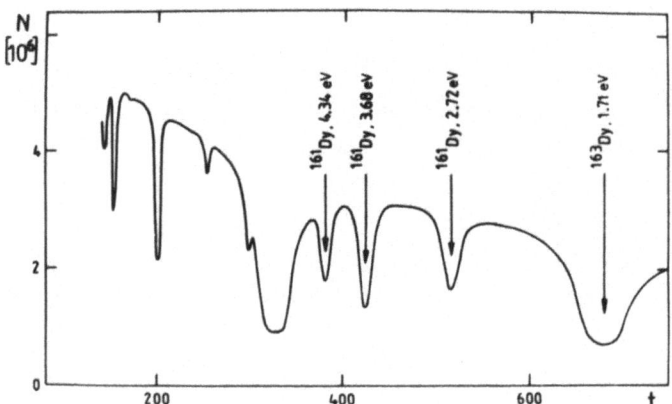

FIG. 2. Time-of-flight spectrum of the Dy sample at 303 K. The
width of the time channels is 4 μs.

samples were interchanged every 12 h. At the end of a period
of the order of one week, all spectra measured at T_1 and T_2,
respectively, were added. In supplementary runs, the spectrum
of the neutron beam without the samples was measured. The background was determined with thick resonance probes.

The temperature shifts were extracted from the experimental
data using two independent methods as follows:

a) From the transmission spectra, $\Delta\sigma_R(E_n) = \sigma_R(T_2,E_n) - \sigma_R(T_1,E_n)$ was calculated with which the shift is

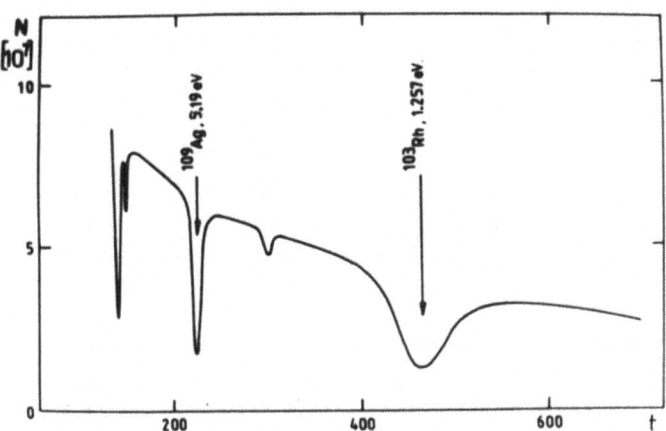

FIG. 3. Time-of-flight spectrum of the sample with Rh and Ag at 294 K. The width of the time channels is 8 μs.

$$\Delta \bar{E}_T = \int_{E_1}^{E_2} \Delta \sigma_R(E_n)E_n dE_n \bigg/ \int_{E_1}^{E_2} \sigma_R(T_1,E_n)dE_n \, , \qquad (7)$$

where

$$\int_{E_1}^{E_2} \sigma_R(T_1,E_n)dE_n = \int_{E_1}^{E_2} \sigma_R(T_2,E_n)dE_n \, .$$

b) The time-of-flight spectra, $N(T_1,t)$ and $N(T_2,t)$, were compared with the aid of chi-square fits as at the analysis of chemically induced shifts /7/. Then, the obtained difference curves were interpreted as differences of the Doppler broadenings, using the theoretical approach of Nelkin and Parks /8/ with the Einstein model for the phonon spectra, but with a shift parameter in one of the both resonance cross-sections.

3. RESULTS

The obtained resonance shifts are compiled in Fig. 4. The expected theoretical behaviour (eq. (6)) is also inserted. If $\alpha_T = \Delta \bar{E}_T A / \langle \varepsilon_k \rangle$ is calculated corresponding to eq. (6), the

307

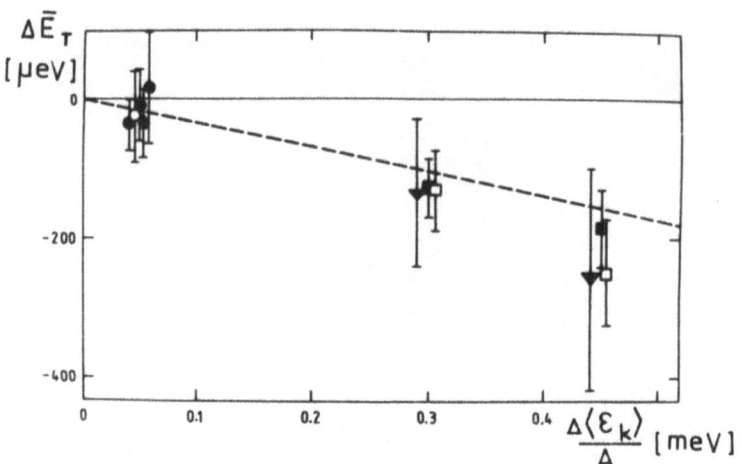

FIG. 4. Experimental shift values and theoretical prediction
(eq. (6), ---)). ■ ^{103}Rh, E_0=1.257 eV, T_1=294±2 K, T_2=667±10
K and T_1=294±2 K, T_2=538±10 K; ▼ ^{109}Ag, E_0=5.19 eV, T_1=294±2
K, T_2=667±10 K and T_1=294±2 K, T_2=532±10 K; ● ^{163}Dy, E_0=1.713
eV, ^{161}Dy, E_0=2.72 eV, 3.68 eV, 4.34 eV, T_1=303±1 K, T_2=370±3
K; open signs, values extracted by method a); full signs, values extracted by method b).

average of the data obtained using the method a) is $\overline{\alpha_T^a}$ =
= -0.49±0.13, when the method b) is used then α_T^b = -0.43±0.09,
respectively. In Fig. 5, the α_T of all resonances analysed are
plotted versus the neutron resonance energies. No systematic

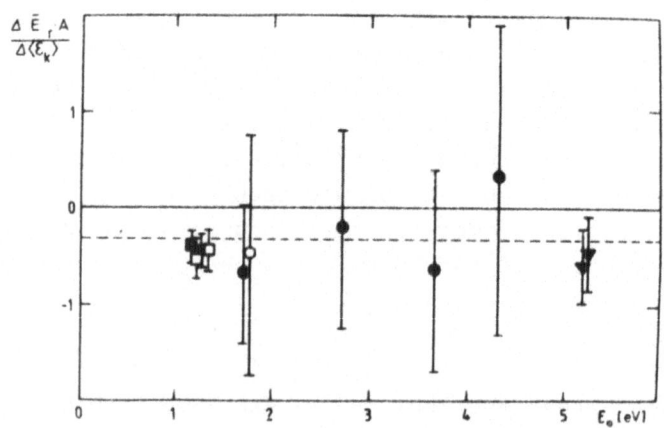

FIG. 5. Temperature shifts normalized to $\langle \varepsilon_k \rangle$/A plotted versus
the resonance energies.

308

correlations with the recoil energies, the resonance parameters, or the phonon spectra which could imitate the temperature shifts are evident in the frame of the experimental error bars.

REFERENCES

1. H. Bethe and G. Placzek, Phys. Rev. 51 (1937) 462.
2. W. E. Lamb, Phys. Rev. 55 (1939) 150.
3. R. Pound and G. A. Rebka, Phys. Rev. Lett. 4 (1960) 274.
4. B. D. Josephson, Phys. Rev. Lett. 4 (1960) 341.
5. V. K. Ignatovich et al., Report JINR P-7296, 1973.
6. K. Seidel et al., Report JINR P3-85-17, 1985.
7. A. Meister et al., this volume.
8. M. S. Nelkin and D. E. Parks, Phys. Rev. 119 (1960) 1060.

CHEMICALLY INDUCED SHIFTS OF URANIUM NEUTRON RESONANCES

A. Meister, S. Mittag, *L. B. Pikelner, W. Pilz,
D. Seeliger and K. Seidel

Technical University Dresden, GDR
*Joint Institute for Nuclear Research, Dubna, USSR

ABSTRACT. Transmission spectra were measured with a time-of-flight spectrometer at the Dubna pulsed reactor for several chemical compounds of uranium. The spectra were compared in the regions of low-energy resonances to observe chemically induced shifts which are interpreted as an effect of mean-square charge radius changes of the nuclei by neutron capture.

1. INTRODUCTION

The charge distributions of nuclei in the ground states have been investigated in many experiments /1/; data have also been obtained for low-lying excited states. But the experimental methods used — electron scattering, measurements of muonic X-rays, electronic X-ray isotope shifts, optical isotope shifts, or nuclear isomer shifts with the Mossbauer spectroscopy, etc. - cannot be applied at higher excited nuclei, hence data on their charge distributions are practically not available.
Slow neutrons captured by nuclei, excite those with their binding energy. Moreover, their energy can be measured with excellent resolution by the time-of-flight method. To obtain data on the proton distribution of the nucleus at the neutron binding energy, it is necessary to determine energy amounts of the order of the electronic hyperfine interaction between protons and electrons of an atom. To this aim experiments were carried out to measure chemically induced shifts of uranium neutron resonances /2, 3/. The neutron energies of a resonance with a

nuclide in the chemical compounds I and II differ approximately by /4/

$$\Delta E_o = \frac{e^2}{6\varepsilon_o} Z \; \Delta \rho_e \; \Delta \langle r^2 \rangle \qquad (1)$$

where $\Delta \rho_e = \rho_{eI} - \rho_{eII}$ is the difference of the electron densities in the nuclear region, and $\Delta \langle r^2 \rangle$ is the change of $\langle r^2 \rangle = \int_o^\infty \rho(r) \; r^2 \; dv$ of the proton distribution due to neutron capture, i.e. the difference between the $\langle r^2 \rangle$ of the compound nucleus state and of the ground state of the target nucleus. Z is the proton number of the nucleus.

2. EXPERIMENT AND DATA HANDLING

The experimental arrangement is shown in Fig. 1.

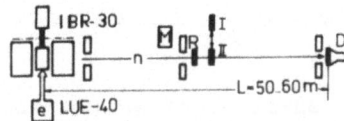

FIG. 1. Experimental arrangement. I, II — uranium samples, R — reference sample, D — ^{6}Li-glass scintillation detector, M — BF$_3$-counter as monitor.

Time-of-flight transmission spectra of the samples were measured with the pulsed reactor IBR-30 operated in booster mode with the linac LUE-40. The samples of different chemical compounds of uranium, I and II, were alternately placed into the neutron beam for about five minutes. After each short-run, the obtained spectrum was checked for any instability in the reactor power and for drifts in the electronic equipment, using the counts within a suitable channel range of the spectrum, the counts of the monitor and the number of the reactor pulses. Only such spectra were accumulated for which the experimental conditions were identical within a few per cent as compared with the preceding runs. Reference samples were permanently put in the beam. Their resonances observed in the transmission spectra

311

(obtained with the different uranium samples) were analysed in the same way as the U-resonances in order to reveal all shifts and any other differences between the measured spectra. A typical spectrum obtained with a U-sample in natural isotopic composition, is shown in Fig. 2. In other measurements, samples enriched in ^{235}U to about 90% were used; the time-of-flight spectrum is shown in Fig. 3. The resonances marked were analysed.

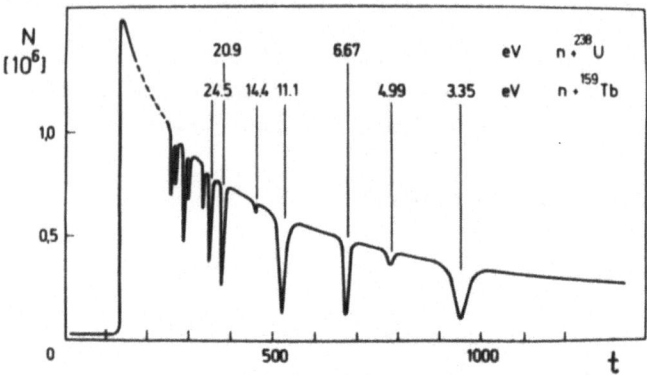

FIG. 2. Time-of-flight spectrum of an uranium sample in natural isotopic composition. The width of the time channels t is 2 μs.

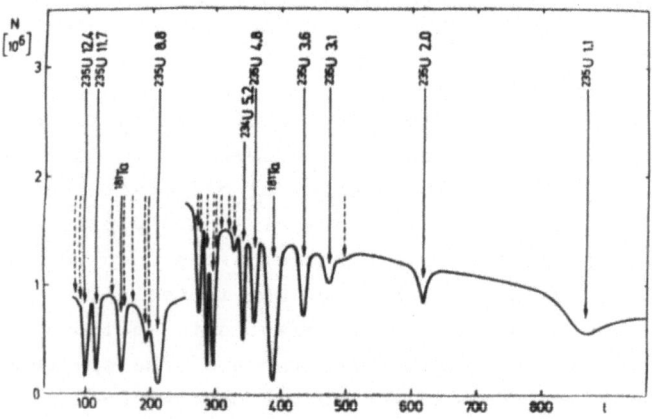

FIG. 3. Time-of-flight spectrum of an uranium sample enriched in ^{235}U to about 90%. The widths of the time channels t are 2 μs for t < 256 and 4 μs for t ≥ 256.

In order to extract the resonance shifts, the spectra of two samples, $N_I(t)$ and $N_{II}(t)$, measured in the same period were compared for given channel ranges, from t_n to t_m, located symmetrically with respect to the resonance dips, using chi-square fits with variation of the resonance positions Δt, a and N_B in N_{II} relative to N_I

$$N'_{II} = a \left[N_{II}(t + \Delta t) + N_B \right] ,$$

and minimizing

$$\chi^2 = \sum_{t=t_n}^{t_m} \frac{\left[N_I(t) - N_{II}(t) \right]^2}{N_I(t) + N_{II}(t)} .$$

The parameters, a and N_B, take into account minor differences in the transmitted beam and background intensities, respectively. The obtained Δt were (for instance, at the 6.67 eV resonance in the spectrum, shown in Fig. 2) of the order of a hundredth of a channel or smaller. The time resolution of the spectrometer was about 2 channels, and the resonance half-width caused by the Doppler broadening, was about 15 channels. Therefore, the statistical uncertainties of the spectrum points had to be very small, i.e. with detector pulse rates of some 10^5 s^{-1}. The measuring time per one sample pair amounted to almost 100 h.

The Δt include, besides the wanted shifts induced by the hyperfine interaction, a falsifying contribution arising from different Doppler broadenings in the compared polycrystalline samples. Its separation is described elsewhere /2, 5/.

3. RESULTS AND DISCUSSION

According to eq. (1), the next step is a plot of all found shifts ΔE_o versus the electron density differences $\Delta \rho_e$ for each resonance of the sample pairs used. The slope of a straight line through these points and the origin yields $\Delta \langle r^2 \rangle$. An example is shown in Fig. 4.

The $\Delta \rho_e$ were estimated using available experimental data on chemical X-ray shifts ($K_{\alpha 1}$ and $L_{\alpha 1}$) in uranium compounds and on Mössbauer isomer shifts in isovalent neptunium compounds, as well as data from relativistic Hartree–Fock–Slater calculations for free ions /2/.

The $\Delta\langle r^2\rangle$ obtained are shown in Fig. 5. For comparison with theoretical calculations on the change of the mean-square charge radius of a nucleus with excitation, $\Delta\langle r^2\rangle_n$, the experi-

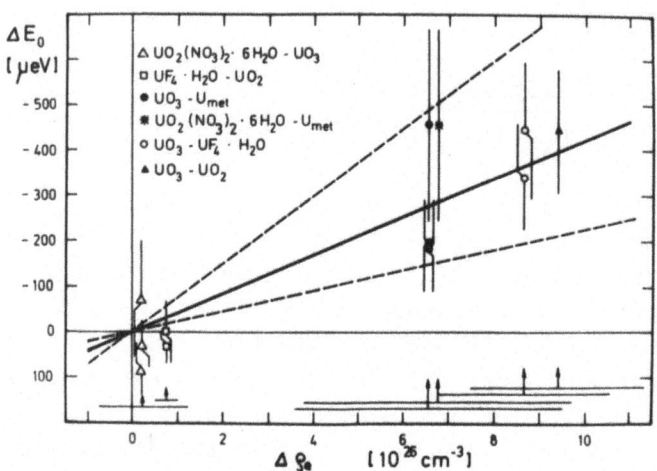

FIG. 4. Chemically induced shifts of the 6.67 eV resonance of 238U plotted versus the electron density differences of the sample pairs used.

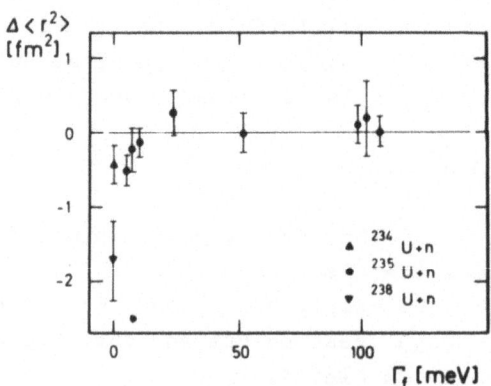

FIG. 5. Change of the mean-square charge radius of uranium nuclei by the capture of neutrons plotted versus the fission width of the resonances.

mental $\Delta \langle r^2 \rangle$ must be reduced by the difference between the mean-square charge radii of the ground states of the compound and of the target nucleus $\Delta \langle r^2 \rangle_i$; $\Delta \langle r^2 \rangle_i$ have been deduced from optical isotope shifts /6/ amount to $+0.05 \pm 0.02$ fm^2 when a neutron is added to ^{234}U or ^{238}U and $+0.10 \pm 0.03$ fm^2 for ^{235}U.

Theoretical estimation of $\Delta \langle r^2 \rangle_n$ was carried out by Bunatian /7/ who used the finite temperature Green functions. He has found that in the region of uranium at temperatures corresponding to the neutron binding energy, $\Delta \langle r^2 \rangle_n$ is of the order of -0.1 fm^2. It has been found that the reduction of $\langle r^2 \rangle_n$ through excitation is connected with the release of the pair correlation, whereas at higher nuclear temperatures, $\langle r^2 \rangle_n$ increases still more as generally expected for excited fermion systems. This value calculated using a statistical model, may be compared with the experimentally obtained data on ^{235}U where eight resonances were analysed so that an average may be estimated: $\Delta \langle r^2 \rangle_n = -0.18 \pm 0.10$ fm^2 is in agreement with the prediction.

Compared to it, the two single $\Delta \langle r^2 \rangle_n$ of even-even target nucleus resonances are smaller.

Weak correlation of the $\Delta \langle r^2 \rangle_n$ of ^{235}U resonances with the fission width seems to be apparent. For states with $\Gamma_f > \frac{1}{2} \Gamma_\gamma$, for instance, the average is $\Delta \langle r^2 \rangle_{nf>} = -0.02 \pm 0.12$ fm^2, whereas for those with $\Gamma_f < 1/2 \Gamma_\gamma$, $\Delta \langle r^2 \rangle_{nf<} = -0.38 \pm 0.14$ fm^2. (The quoted uncertainties include a systematical contribution of about 0.06 fm^2 in both cases). However, in comparison with the ground state value $\langle r^2 \rangle_{no} = 33.8$ fm^2 /1/ and with the $\langle r^2 \rangle_n$-change during the fission ($\Delta \langle r^2 \rangle_n$ is at deformations corresponding to the second well of the fission barrier 4–5 fm^2 and at the outer saddle point about 10 fm^2) the $\langle r^2 \rangle_n$ of the compound nucleus states being very close to the ground state value.

REFERENCES

1. Atomic and Nuclear Data Tables $\underline{14}$ (1974) No.5—6.
2. A. Meister et al., Nucl. Phys. $\underline{A362}$ (1981) 18.
3. S. Mittag et al., Nucl. Phys. $\underline{A435}$ (1985) 97.
4. V. K. Ignatovich et al., Report JINR P4-7296, 1973.
5. K. Seidel et al., Report JINR P3-11741, 1978.
6. K. Heilig and A. Steudel, At. Nucl. Data Tables $\underline{14}$ (1974) 613.

GDR EXCITATION IN ^{58}Ni(n,p)^{58}Co REACTION AT 17.3 MeV NEUTRON ENERGY

J. Rondio, B. Mariański, K. Czerski, A. Korman and L. Zemło

Institute for Nuclear Studies, Warsaw, Poland

ABSTRACT. The energy and angular distributions of protons emitted in the ^{58}Ni(n,p)^{58}Co reaction have been measured at 17.3 MeV neutron energy. The reaction mechanism has been analysed.

1. INTRODUCTION

The ^{58}Ni(n,p)^{58}Co reaction has been investigated at different neutron energies. This reaction offers some possibilities for studying analogue states of GDR /1–3/. The excitation of possible candidates for analogue states of GDR in the ^{58}Ni(n,p) ^{58}Co at 18.5 MeV neutron energy has been reported in our previous paper /4/. A theoretical fit to the measured angular distribution for proton energy region corresponding to these transitions has been attempted using a DWBA calculation with l = 1 transfer. An isotropic contribution had to be added to get an agreement with the experimental data. It suggests that another mechanism must play a role in the excitation of these transitions, beside the direct one. We have assumed that this mechanism is a multistep excitation with GDR as a doorway state. GDR is excited by the neutron captured in the target nucleus beside other states and then can de-excite by proton emission. The energy of such a proton should be equal to the excitation energy of GDR minus binding energy of a proton in a compound nucleus and it should be independent of incident neutron energy. The proton groups seen in proton spectra in the energy region corresponding to these transitions in reaction at 18.5 MeV neutron energy should be compared to those at a different incident energy in order to see this independence. So, the differential cross-sections for the ^{58}Ni(n,p)^{58}Co reaction at a neutron energy of 17.3 MeV have also been measured.

2. EXPERIMENTAL

The experiment was carried out using an eight-telescope set-up which made it possible to measure simultaneously eight proton energy spectra at eight different angles /5/. A slow-fast coincidence system was employed to eliminate the background and two-dimensional analysis to identify the particles.

Neutrons that had an average energy of 17.3 MeV were produced in the ^{3}T(d,n)^{4}He reaction, the deuterons being accelerated to 1.5 MeV in a Van de Graaff accelerator. The target used was ^{58}Ni isotope enriched to 99.7% which was deposited electrolytically on a 1 mm tantalum backing. The measurement of the protons recoiled from the polyethylene target provided for an absolute normalization of the ^{58}Ni(n,p)^{58}Co cross section. The energy scale of the proton spectra was corrected for energy losses in the target and proportional counters.

3. RESULTS

The proton spectra of the ^{58}Ni(n,p)^{58}Co reaction, measured at 17.3 MeV and 18.5 MeV neutron energies, are shown in Fig. 1. It is easy to see, that only in the high energy part of the spectrum a certain shift of energy scales is needed to obtain an agreement between the shapes of the spectra. The situation varies with energy. The peaks at 8.6, 9.6 and 11.8 MeV proton energies are seen in both spectra at the same energies and no shift between energy scales is necessary. This confirms the assumption that the reaction mechanism is a multistep one.

The proton spectrum at 17.3 MeV is compared (Fig. 2) with the (γ,p) excitation function /6/ which reveals very distinctly the GDR excitation with proton emission. But, a close similarity of the shapes of both the spectra is not to be expected because of the different types of excitation in both reactions,

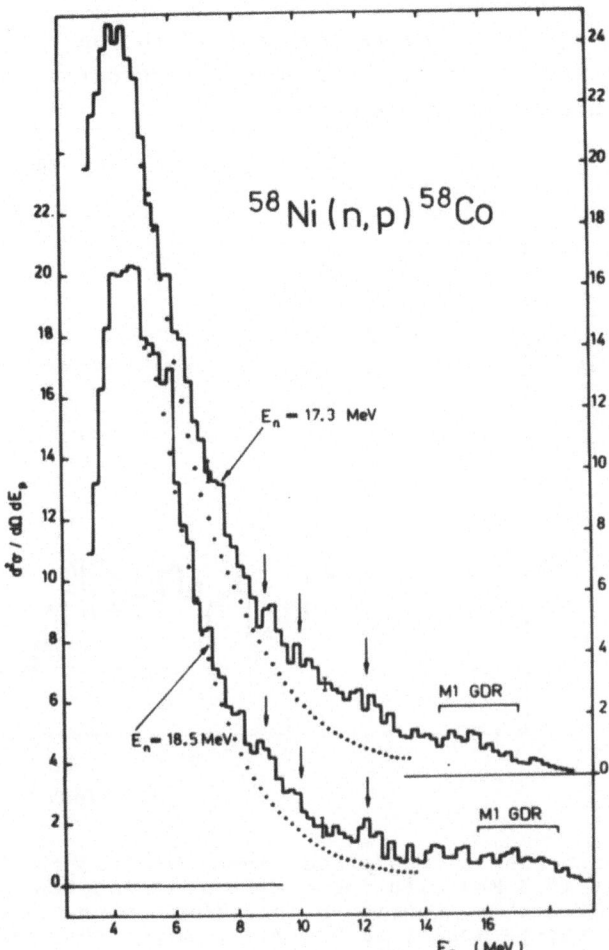

FIG. 1. The proton spectra from the ^{58}Ni(n,p) reaction at
17.3 MeV and 18.5 MeV neutron energies. The spectra are summed
over four angles from 10° to 40° to improve the statistics.
Ordinates are in arbitrary units. The dotted lines represent
the calculated continuum. The proton groups seen in the 8–14
MeV energy region are indicated by arrows.

different isotopes excited and also because the GDR states in
^{59}Ni share the excitation with another excited states. The energy scales of (γ,p) and (n,p) spectra, shown in Fig. 2, had
to be shifted by an energy that was 0.4 MeV greater than the
binding energy of a proton in ^{59}Ni, what seems to be reasonable when taking into account all the differences governing
the both reactions. Comparison of both spectra allows us to

conclude that the GDR is really excited in the (p,n) reaction
at above neutron energies. A diagram representing the proposed
multistep mechanism is shown in Fig. 3.

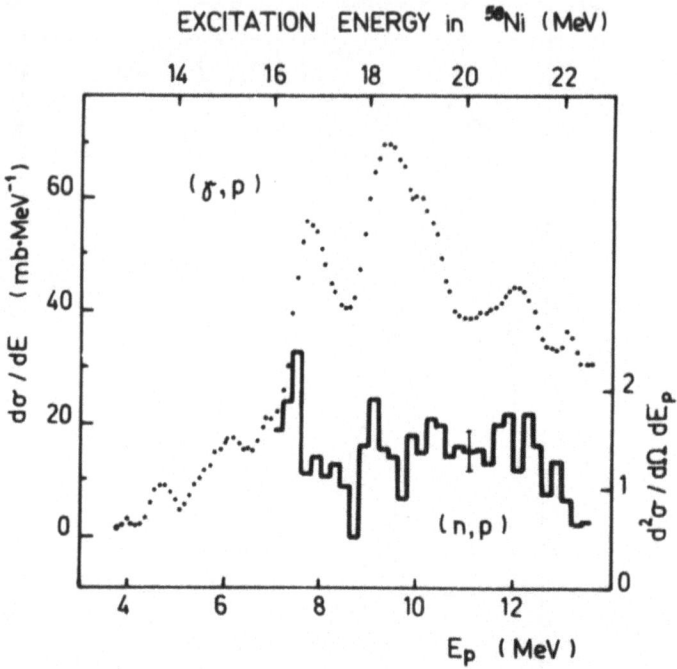

FIG. 2. Comparison of the proton spectrum from the ^{58}Ni(n,p)^{58}Co
reaction at 17.3 MeV with the excitation function of the ^{58}Ni
(γ,p) photoreaction. Continuum contribution has been subtracted.
A shift of energy scales is 0.4 MeV greater than the binding
energy of a proton in ^{59}Ni.

FIG. 3. Diagrammatic representation of the semidirect scattering process with virtual formation of a giant resonance. The
Bohr-Mottelson notation has been used.

REFERENCES

1. K. Debertin and E. Rössle, Nucl. Phys. 70 (1965) 89.
2. C. F. Clement and S. M. Perez, Nucl. Phys. A165 (1971) 569.
3. J. L. Ullmann, F. P. Brady, C. M. Castaneda, D. H. Fitzgerald, G. A. Needham, J. L. Romero and N. S. P. King, Nucl. Phys. A427 (1984) 493.
4. J. Rondio, B. Mariański and A. Korman, J. Phys. G: Nucl. Phys. 11 (1985) 549.
5. B. Mariański, J. Rondio and A. Korman, Nucl. Instr. Meth. 206 (1983) 421.
6. B. S. Ishkhanov, J. M. Kapitonov, J. M. Piskarev, V. G. Shevchenko and O. P. Shevchenko, Sov. J. Nucl. Phys. 11 (1970) 272.

MEASUREMENT AND THEORETICAL CALCULATION OF THE ^{252}Cf SPONTANEOUS-FISSION NEUTRON SPECTRUM

H. Märten and D. Seeliger

Technical University Dresden, GDR

ABSTRACT. Concerning the ^{252}Cf(sf) neutron spectrum remarkable progress in experiment and theory have been made during the last three years. Experimental techniques and analysis procedures have been improved. The precise measurement of the standard neutron spectrum from spontaneous fission of ^{252}Cf requires the optimum experimental arrangement corresponding to the energy range to be measured. Several types of data corrections have to be considered with care. The most important requirements to be met in a ^{252}Cf(sf) neutron spectrum measurement are summarized briefly. We consider the high-energy range especially.

Theoretical models for the calculation of fission neutron spectra are based on the predominant emission mechanism, i.e. the evaporation from fully accelerated fragments. It is emphasized that an exact evaporation theory of fission neutron spectra should take into account the fragment distribution in nucleonic numbers, excitation energy, kinetic energy, and nuclear spin as well as the cascade neutron emission from highly excited, neutron-enriched fragments in competition to gamma-ray emission. However, practical applications require several approximations. Some approaches which were studied in the framework of both the Weisskopf formalism and the Hauser—Feshbach theory are discussed. We point out some of the deviations in spectrum calculation if neglecting or approximating typical characteristics of fission neutron emission.

The results of new Cf spectrum calculations are compared with recent experimental data which confirm a Maxwellian spectrum at energies below $\sim$ 1 MeV for T = 1.42 MeV. Between 1.5 and 4 MeV, measured data tend to exceed this Maxwellian by about 3%. Significant deviations from a Maxwellian with T = 1.42 MeV appear between 6 and 20 MeV where a fit of experimental data yields a value of T close to 1.37 MeV.

Recent theoretical calculations of the Cf neutron spectrum agree very good with measured data. Especially, the complex cascade evaporation model permits a conformable description of recent experimental data in the whole energy range (1 keV—20 MeV) if introducing the CMS anisotropy of emission.

1. INTRODUCTION

The ^{252}Cf(sf) neutron spectrum recommended as a standard by an
IAEA Consultants´ Meeting /1/ is of high importance for practical applications as well as for theoretical studies of the
fission process and the mechanism of fission neutron emission.
It is employed as a reference in both microscopic and macro-
scopic measurements. Cf sources are widely used for instrument
calibration.

The status of measurement, evaluation and theoretical analysis of the ^{252}Cf spontaneous-fission neutron spectrum is
viewed by the IAEA (INDC) resulting in regular summaries and
recommendations /2, 3/. Blinov presented a comprehensive review
in 1980 /4/. Therefore, the present review paper is focused on
recent investigations of the Cf neutron spectrum.

As already outlined in the conclusions of the IAEA Consul-
tants´ Meeting /3/, the accuracy of experiments as well as the
theoretical description of the Cf spectrum have been improved
substantially. Experimental data cover the energy range from
1 keV to 30 MeV at present and have been deduced from measurements with different types of both fission fragment (FF) detectors and neutron (n) detectors in overlapping energy ranges.
Experimental arrangements have been developed to keep corrections as small as possible. The data analysis was refined;
corrections are considered with more care.

The spontaneous fission of ^{252}Cf has been studied in many
theoretical and experimental works concerning fragment distributions in kinetic energy, nucleon numbers, excitation energy,
nuclear spin, etc., concerning the diversity of particle
emission processes in fission also. However, several fundamental problems are still open. In particular, the mechanism of
fission neutron emission has been a subject of the research in
nuclear physics (see review in ref. /5/ and summary concerning
scission neutron emission in ref. /6/). According to our present
knowledge, at least 80% of fission neutrons are evaporated from
fully accelerated fission fragments. The nature of the remainder,
i.e. scission neutrons or central component, has not yet been
clarified in spite of many investigations. Therefore, hitherto

published fission neutron spectrum calculations had been based on evaporation models considering the complexity of fission up to a certain degree. Nevertheless, recent activities give rise to an optimistic outlook and seem to terminate the stagnation appeared since the sixties. Here, the spontaneous fission of ^{252}Cf is the preferred subject.

2. ON THE EXPERIMENTAL DETERMINATION OF THE Cf NEUTRON SPECTRUM

2.1. Status of Experiments

High effort concerning experimental technique as well as data analysis has been devoted to the precise determination of the ^{252}Cf(sf) neutron spectrum. However, the discrepancies between the data of various experiments are much higher than the estimated uncertainties.

The most frequently employed method has been the neutron time-of-flight (TOF) technique based on neutron detection with /2, 4/

(i) organic scintillators (E $\gtrsim$ 0.2 MeV),

(ii) ^{6}Li-including materials (lithium glasses, ^{6}LiI crystals, E $\lesssim$ 2 MeV,

(iii) ^{235}U fission chambers (whole energy range),

(iv) black neutron detectors /9/ (organic scintillator, E $\gtrsim$ 0.2 MeV).

Further techniques involve the registration of recoil protons as well as of reaction products from ^{3}He(n,p) and ^{6}Li(n,α).

The improved measurements presented after 1980 /7—15/ have been carried out by employing TOF spectrometers.

The experiments dated before 1979 had been reviewed by Blinov /4/. He took note of large discrepancies concerning the deduced Maxwellian temperature T (spectrum hardness parameter) as well as the spectrum shape at very low and very high energies especially. The published T values cover the energy range from 1.18 to 1.57 MeV. More recent measurements /7—10, 14/ confirm Blinov's conclusion that the ^{252}Cf(sf) neutron spectrum can approximately be described by a Maxwellian distribution with T = 1.42 MeV at low energies and in the intermediate energy range. Remarkable deviations from a Maxwellian with T = 1.42 MeV appear in the energy range from 6 to 20 MeV /11—13/ where a Maxwellian fit yields a value of T close to 1.37 MeV. Recent experimental data /7—9, 11/ exceed the Maxwellian with T = 1.42 MeV in the range from 1.5 to 4 MeV by about 3% (see section 4).

Blinov pointed out that the NBS evaluation /16/ is not suited to reproduce the experimental data below 1 MeV espacially. A new evaluation was strongly recommended at the IAEA Consultants' Meeting /3/.

2.2. Requirements with Regard to TOF Measurements. Data Correction

Fission neutron TOF spectroscopy involves the registration of both fission fragments and neutrons with a time resolution $\lesssim$ 1 ns. Chalupka et al. /17/ discussed the requirements to be met by a FF detector. The preferred types are avalanche detectors, gas scintillators, and ionization chambers which provide a sufficient discrimination against the associated alpha activity of Cf sources. To keep the corrections of spectrum contributions due to scattering effects and neutron producing reactions very small, a low mass FF detector is necessary.

The measurable ^{252}Cf(sf) neutron spectrum is influenced by the anisotropic FF detection efficiency caused by absorption in the sample plane /6, 9, 11/ due to the deposit thickness and backing roughness. The total fragment detector efficiency ε_{FF} should be close to 100%. The ratio of the measurable neutron spectrum to the total (undisturbed) one was calculated in refs. /6, 13/. A typical example is shown in Fig. 1. Similar studies have been performed by other groups /9, 15/. Corrections with regard to ε_{FF} are nearly independent of E at about 60° (angle with regard to sample plane normal) /6, 17/. The value ε_{FF} can be deduced from a FF-n-coincidence measurement as a function of β (angle of n detector position) /11/ or from a $N(E, \beta = 90^{\circ})/N(E, \beta = 0^{\circ})$ measurement /9/.

In- and out-scattering effects concern the air column between source and n detector, the collimator and shielding system, and

the construction materials of both the FF detector and the n
detector. An improved version of a collimator and shielding
system suited to keep scattering effects small was presented
by Poenitz and Tamura /9/. The most simple correction is that
one for neutron transmission on the base of total cross-sections.
The background caused by scattered neutrons, which cover the
energy range below about 1—4 MeV predominantly (depending on
geometrical conditions), should be studied by Monte Carlo calculations (MCC) /7, 9, 11/. A shadow cone experiment yields
limited information only.

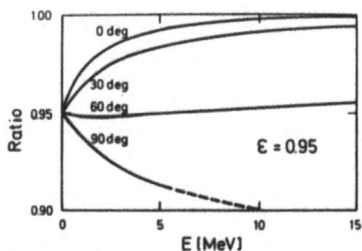

FIG. 1. Ratio of the measurable to the undisturbed Cf neutron
spectrum represented for selected angles of neutron detector
position with regard to the Cf sample plane normal. Calculation
result for ε_{FF} = 0.95 /6, 13/.

The central task of a fission neutron spectrum measurement
is the precise determination of the n detector efficiency ε_n.
In the case of n detectors based on the ^{6}Li(n,α) or the
^{235}U(n,f) reaction, ε_n is determined according to the standard
cross-sections /2/. For special corrections, we refer to original papers. The efficiency of a black neutron detector (BND)
is close to 100% due to its operation mode. Deviations from
100% can be deduced from MCC. Therefore, uncertainties in ε_n
are very low ($\approx$ 1—2%) /9/. However, the energy resolution of
a BND has to be considered carefully. The uncertainties of the
measured TOF spectrum become large at high energy especially
due to the strongly increasing spectrum gradient. Organic scintillators with pulse shape discrimination properties are very
often employed for fission neutron TOF spectroscopy at energies
above 0.5 MeV (0.2 MeV). Such n detectors are characterized by

an intermediate efficiency ($\sim$ 1–30%) to be determined in separate experiments. A rather substantial effort is necessary to obtain $\varepsilon_n(E)$ with an uncertainty below 5%. ε_n studies are commonly added by MCC to interpolate or extrapolate measured efficiency data. Here, the light output (LO) resolution has to be considered in a reliable way (for LO calibration purposes on the base of response functions especially). The gamma background of organic scintillators can considerably be suppressed by pulse shape discrimination. The dynamic range of the used discriminator limits the measurable spectrum range at low neutron energy especially. The reliability of n/γ discrimination can be enhanced if introducing the pulse shape amplitude as a further parameter in the measurement. For instance, the PTB group /11/ carried out a four-dimensional measurement of neutron TOF, scintillator LO, pulse shape amplitude, and FF detector amplitude (ΔE signal from the fission chamber). An alternative method is at least the check-up of the n/γ discriminator carried out simultaneously with the measurement (spectroscopy of the pulse shape amplitude for a selected LO window /13/). The flight paths in previous ^{252}Cf(sf) spectrum measurements are spread over the range from 20 cm (low-energy range) to 12 m (high-energy range). The flight path L is defined as the distance between the source and the average detection depth in the n detector. Consequently, it can be E-dependent in special cases. In general, the time resolution of the TOF spectrometer is also a function of E due to the detector depth /10, 13/. The time resolution and, sometimes, the TOF bin width disturb the measured spectrum if the TOF distribution increases or decreases strongly, i.e. in the n detector bias region as well as at the high-energy spectrum end. This annoying effect depends on L mainly. Corrections can be carried out using unfolding methods. Nevertheless, L should be chosen sufficiently high to keep corrections in the region of several per cent. In the case of organic-scintillator n detectors, the determined gamma-peak position should be checked by an additional measurement with and without n/γ discrimination to avoid a possible annoying influence due to the amplitude-dependent timing, which occur in spite of developed timing methods in the case of large LO ranges to be analysed. The time

scale calibration should be checked and verified (for instance,
C transmission measurement /9/).

The background due to non-correlated STOP signals (FF detector) was considered in recent experiments only. This partial
(TOF-channel dependent) background can be avoided by a pile-up
rejector /7/ tolerating an extended dead time, or it can be
deduced analytically /11, 18/ on the base of time scale characteristics and the Cf source strength.

2.3. Problems of the Measurement of the ^{252}Cf(sf) Neutron Spectrum in the High-energy Range

Large discrepancies of experimental data concerning the high-
energy end of the Cf spectrum had been outlined by Blinov /4/.
To obtain a rather good energy resolution at high E a sufficiently high flight path is required. Further, one should use an n
detector with a high ε_n at a large solid angle $\Delta\Omega$. ^{235}U fission
chambers are not suited because of their low efficiency. The
use of a BND is questionable due to the high-energy resolution
effects on the TOF spectrum at high E. Consequently, an optimum
arrangement seems to be

(i) the use of an organic scintillator n detector at a comparatively high flight path,

(ii) the application of efficient background suppression
methods (pulse shape discrimination, heavy shielding),

(iii) the application of a rather strong Cf source,

(iv) the two-dimensional (TOF, LO) spectroscopy which permits
the selection of the optimum LO bias depending on E generally,
i.e. a maximum foreground/background ratio can be achieved.

These requirements have been met in refs. /11—13/ especially,
but item (iv) was only considered in refs. /12, 13/. In these
two works, an efficient pulse shape discriminator was used to
suppress both the gamma rays and the cosmic-muon background.
The shape of the ^{252}Cf(sf) neutron spectrum at the high-energy
end /12/ is shown in Fig. 2. This result was confirmed in an
experiment finished recently /13/. The data represented in
Fig. 2 are in very good agreement with the PTB measurement /11/
(concerning the energy range below 15 MeV).

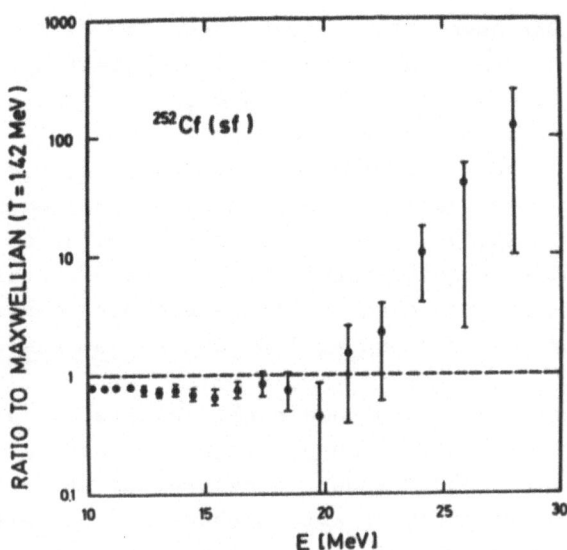

FIG. 2. The high-energy end of the Cf neutron spectrum represented with reference to a Maxwellian distribution with
T = 1.42 MeV. The figure was taken from ref. /12/.

3. THEORY OF THE ^{252}Cf(sf) NEUTRON SPECTRUM

3.1. The Fission Process and Fission Neutron Emission

It was found early that the fission neutron emission probability
is enhanced close to the fission axis /19/. Consequently, evaporation of fission neutrons from fully accelerated fragments
was considered as the predominant emission mechanism. On the
base of this assumption, fission neutron spectra have been
described quite well (refs. /20, 21/ and references therein).
More precise experimental studies had shown that a ~ 10% component of fission neutrons is emitted isotropically in the
laboratory frame (LS). Further investigations have been summarized recently /6/. Published results are contrary. Therefore,
the picture of scission neutron emission is not clear at present.
The yield of scission neutrons as well as their differential
emission probability are not known with sufficient accuracy.
The partial spectra of the different eventual kinds of scission

neutrons /23—26/ have not yet been found theoretically. In
particular, the rapid changes in nuclear potential as the fissioning nucleus moves from the saddle of the fission barrier
to the scission point and at the transition of strongly deformed
fragments into their equilibrium form may result in strong
single-particle excitations and, hence, nucleon emission from
non-equilibrium states. The reliability of hitherto existing
analysis results is questionable due to the uncertainty in the
model parameters concerning the time-dependent parametrization
of the nuclear potential as well as due to the restrictions of
the models themselves.

The fission fragments become highly excited after dissipation of their deformation energy which they have at scission.
Energy dissipation and fragment acceleration are two simultaneous processes which occur within about 10^{-20} s after scission
/27/. Therefore, a possible emission mechanism of fission neutrons could be the evaporation (or emission from non-equilibrium states?) during fragment acceleration.

Fission is partly accompanied by charged-particle emission
(ternary fission). Some of these light nuclei are instable.
In particular, 11% of the alpha particles are originally
emitted as ^{5}He nuclei which decay ($T_{1/2} \approx 8 \times 10^{-22}$ s) into an
alpha particle and a neutron /28/.

3.2. Statistical-model Approach to Fission Neutron Spectra

Hitherto, calculations of the energy spectrum of fission neutrons which include all possible emission mechanisms have been
infeasible. The bare main one, i.e. the evaporation from fully
accelerated fission fragments, is a rather complex process.
In this case, detailed calculations of emission spectra in the
framework of statistical models have to account for many
characteristics of fission and fission neutron emission:

(i) nucleon (N,Z,A = N+Z), excitation energy (E^X), kinetic
energy (E_k) and spin (I) distribution of the fission fragments
(which depends on the features of the fissioning nucleus),
i.e. $P(E^X,I,A,Z,TKE)$ (TKE - total kinetic energy of the fission
fragments; E_k is defined by A and TKE according to moment
conservation);

(ii) cascade neutron emission from highly excited, neutron-enriched fragments in competition to gamma emission.

The fragment distribution of item (i) is not derivable from fission theory completely and/or with sufficient accuracy. Therefore, one has to consider experimental data and/or special assumptions.

Further, the emission of fission neutrons is not isotropic in the centre-of-mass system (CMS) of the fragments due to the fragment spin ($\overline{I} \approx 6$—8 $\hbar$) /29, 30/. This fact is often neglected.

Synopsis 1 represents the general scheme which has to be accounted for the calculation of fission neutron spectra on the base of a statistical model. The CMS probability of neutron emission $\varphi(\varepsilon)$ is calculated in the framework of either the Weisskopf formalism /31/ (without P(I) consideration) or the Hauser—Feshbach theory /32/. In any case, the level density $\varrho(U,I)$ of the residual nuclei (excitation energy U) and the transition probabilities (inverse cross-section σ_c of compound-nucleus formation or transmission coefficients T_{lj}, respectively) have to be taken into account. The dependence $\varphi(\vartheta)$ (CMS angular distribution) has been deduced by Gavron /30/ by statistical calculations (integral in regard of ε !). He obtained a distribution which can be approximated by

$$\varphi(\vartheta) \sim (1 + \beta \cos^2 \vartheta), \qquad \beta \approx 0.1. \qquad (1)$$

The sum over i (emission step index) in the first equation of Synopsis 1 means consideration of cascade emission. The transformation of the CMS distribution $\varphi(\varepsilon,\vartheta:A,Z,TKE)$ into the LS and the following weighted concentration taking into account the fragment occurrence probability P(A,Z,TKE) yields N(E). However, the whole scheme of Synopsis 1 has not yet been observed. Hitherto published treatments take not into account the model dependence on Z. The calculations are carried out taking the average $\overline{Z}$ for a given A. The dependence $\varphi(\vartheta)$ was only considered roughly without correlation to ε, E^x, I, A, Z on the basis of eq. (1) /33, 34/. The initial distribution in fragment spin /29/ can only be introduced in Hauser—Feshbach calculations /35—37/. Using the Weisskopf ansatz one can approximately consider the influence of the spin distribution on spectrum

SYNOPSIS 1

Statistical-model approach to the ^{252}Cf(sf) neutron spectrum

CMS $\varphi(\varepsilon,\vartheta:A,Z,TKE) = \sum_i \sum_I \int dE^x\, P_i(E^x,I:A,Z,TKE)\varphi_i(\varepsilon,\vartheta:E_y^x I,A,Z)$

⇓

Transformation into LS using $E_f = E_k/A = TKE(1/A - 1/252)$

⇓

LS $N(E,\theta:A,Z,TKE)$

⇓

$N(E) = \sum_{A,Z} \int dTKE \int d\Omega\, P(A,Z,TKE)N(E,\theta:A,Z,TKE)$

---

shape if assuming $\varrho(U,I) = \varrho(U,I = 0)$ /38/. Most of the hitherto published theoretical studies of fission neutron spectra have been characterized in ref. /6/. Here, we focus on recent improved treatments.

3.3. Recent Developments of Models

The model proposed by Madland and Nix /39,40/ was based on rough approximations concerning the description of level density and excitation energy distributions, but it is easily applicable to any fission reactions. It was successfully used to describe neutron spectra from induced fission reactions as well as from spontaneous fission of ^{252}Cf. A rather good agreement with experimental data can be achieved if adjusting the level density parameter

$a = A/C.$ (2)

The neglect of the diversity of scission configurations is

332

compensated by a higher value of C than physically reasonable. However, C is the only one free parameter of the model. Further input data are known well. Synopsis 2 includes the basic formulae of two versions of the Madland—Nix model (MNM). We know that the CMS spectra depend on A actually. This fact is neglected in the MNM but the consideration of σ_C for the two fragment groups in version II.

SYNOPSIS 2

Madland—Nix model

(D. G. Madland and J. R. Nix, LASL)

+++   Weisskopf formalism

+++   Dependence of CMS spectra on A, Z and TKE neglected

+++   CMS anisotropy neglected

+++   Constant-temperature description of $\varphi(U)$

+++   Optical-model calculation of σ_C

+++   Triangular-shape distribution in nuclear temperature of residual fragments (maximum value T_m), i.e. rough implicite consideration of $P_o(E^x)$ and cascade emission

VERSION I

+++ $N(E) = N(E, \overline{E}_f, \sigma_C^{\overline{A}})$

VERSION II

+++ $N(E) = \frac{1}{2} \left[N(E, E_f^L, \sigma_C^L) + N(E, E_f^H, \sigma_C^H) \right]$

   Recent studies /42/ have been carried out in the framework of a generalized Madland—Nix model (GMNM, Synopsis 3).
   Version III characterized in Synopsis 3 considers the different maximum values T_m of the rest-temperature distribution as well as the different weights of the light and heavy fragment group (L, H). The full dependence of the input parameter on A

SYNOPSIS 3

Generalized Madland—Nix model

(H. Märten and D. Seeliger, TUD)

+++   AS MNM, but consideration of

     I. $T_m(A)$

    II. $E_f(A)$

   III. $\widetilde{\sigma}_C(A)$

    IV. Weight $P(A)\bar{\nu}(A)$

          (fragment yield and neutron yield, respectively)

VERSION III

+++ $N(E) = \dfrac{1}{\bar{\nu}_{tot}} \left[\bar{\nu}^L N(E, E_f^L, \widetilde{\sigma}_C^L, T_m^L) + \bar{\nu}^H N(E, E_f^H, \widetilde{\sigma}_C^H, T_m^H) \right]$

VERSION IV

+++ $N(E) = \dfrac{1}{\bar{\nu}_{tot}} \sum_A \bar{\nu}(A) P(A) N(E, E_f(A), \widetilde{\sigma}_C(A), T_m(A))$

---

is taken into account in version IV /42/.

The different model versions of MNM/GMNM have been compared for a fixed C = 8.0 MeV /42/. Figure 3 shows how the spectrum is changed if using a more complex model for the calculation of the Cf neutron spectrum. The spectrum at high and low energy increases and at intermediate energy decreases as the complexity of the model increases. The deviations are large at high energy especially. They can be compensated by adjusting C. However, the spectrum shape is changed if requiring an equal average LS emission energy (cf. section 4).

An improved version of the cascade evaporation model (CEM, Synopsis 4) was used to calculate the ^{252}Cf(sf) neutron spectrum in the energy range from 1 keV to 40 MeV /34/. Previous studies /33/ were concentrated on the high-energy range only. The new calculation is based on a complex consideration of

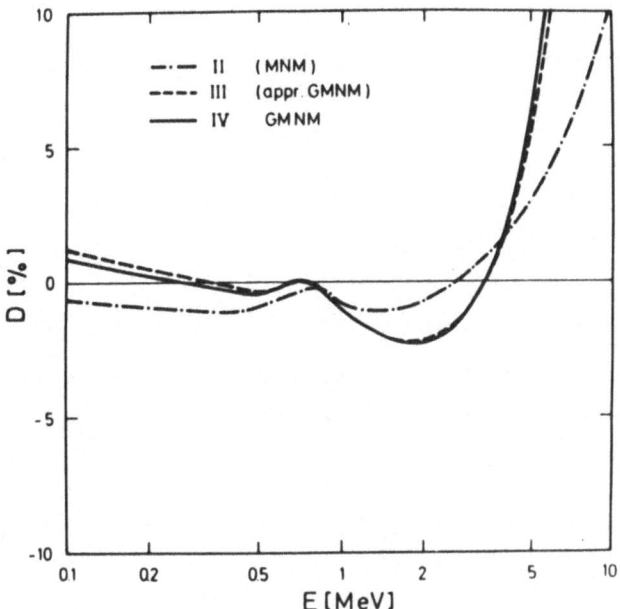

FIG. 3. The percentage deviation D of the Cf neutron spectra
calculated in the framework of the versions II, III and IV
from the version I distribution /42/. The calculations have
been carried out for a fixed parameter C (8.0 MeV).

scission configurations defined by A and TKE, i.e. asymmetry
and elongation of the fissioning nucleus at scission, respectively. Neglecting the TKE dependence one obtains a changed
spectrum description at high energy especially (Fig. 4).

The consideration of the CMS anisotropy (eq. (1)) yields
an enhancement of the low-energy spectrum part in particular
(cf. section 4). The CEM was also used to study the influence
of the input data on the spectrum /6/.

Applying the Hauser–Feshbach theory for fission neutron spectrum calculations one is able to consider the competition of
neutron emission to gamma-ray emission as well as the initial
distribution in fragment spin which is known roughly /29/.
The first study was presented by Browne and Dietrich /35/.
Recent calculations have been carried out by Gerasimenko et
al. /36, 37/ and Rubchenya /26/ (Synopsis 5). They neglected
the probability distribution in TKE for a given A as well as

Complex cascade evaporation model

(H. Märten and D. Seeliger, TUD)

+++ Weisskopf formalism, cascade emission

+++ Dependence of CMS spectra on Z neglected

+++ Rough consideration of CMS anisotropy

+++ Semiempirical description of $\varsigma(U, I = 0)$ based on the Fermi-gas formula with entropy

$$S = 2\sqrt{\tilde{a}(U + \delta U' f(U) + \delta P' h(U))}$$

($\delta U'$, $\delta P'$ – shell and pairing correction, respectively)

+++ Optical model calculation of σ_C

+++ $P_o(E^x : A, TKE)$ Gaussian shape
(deduced from experimental $\bar{\nu}, \sigma_\nu^2, \bar{E}_\gamma$ data)

+++ $N(E) = \sum\limits_{A} \int dTKE \, P(A, TKE) N(E : A, TKE)$

+++ Study of multiple-differential emission probabilities
$N(E, \theta : A, TKE)$

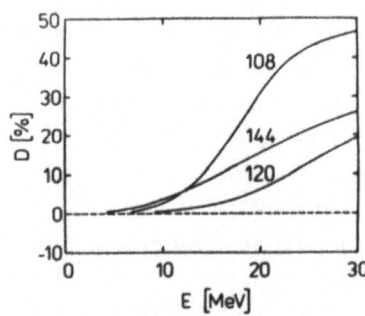

FIG. 4. The effect of the approximative spectrum calculation without consideration of the correlation between the CMS spectra and TKE (for fixed A) with reference to the "exact" one. Percentage deviations are represented in the case of typical fragment mass numbers /6, 33/.

SYNOPSIS 5

Hauser—Feshbach calculation

(B. F. Gerasimenko and V. A. Rubchenya, RIL)

+++ Hauser—Feshbach formalism, cascade emission

+++ Dependence of CMS spectra on Z and TKE neglected

+++ CMS anisotropy neglected

+++ Semiempirical description of $\varphi(U,I)$ based on the Fermi-gas formula with the entropy

$$S = 2\sqrt{\tilde{a}(1 + f(U)\delta W/U)(U - \delta)}$$

(δW, δ - shell and pairing correction, respectively)

+++ Optical model calculation of T_{lj}

+++ $P_o(E^x:A)$ Gaussian shape
(deduced from experimental $\bar{\nu}$, σ_ν^2 data)

+++ $P_o(I:A)$ considered, $\bar{I} \approx 6-8\ \hbar$

+++ $N(E) = \sum_A P(A)N(E:A)$

+++ Study of multiple-differential emission probabilities
$N(E,\theta:A,TKE)$

---

the CMS anisotropy.

Synopses 2—5 indicate that CEM and HFC involve the realistic consideration of the initial distributions in E^x shown in Fig. 5 for typical fragment mass numbers, i.e. $P_o(E^x:A)$ which is integral in regard of TKE. Both models are based on improved methods of level density description which take into account shell and pairing corrections (CEM /43/, HFC /44/).

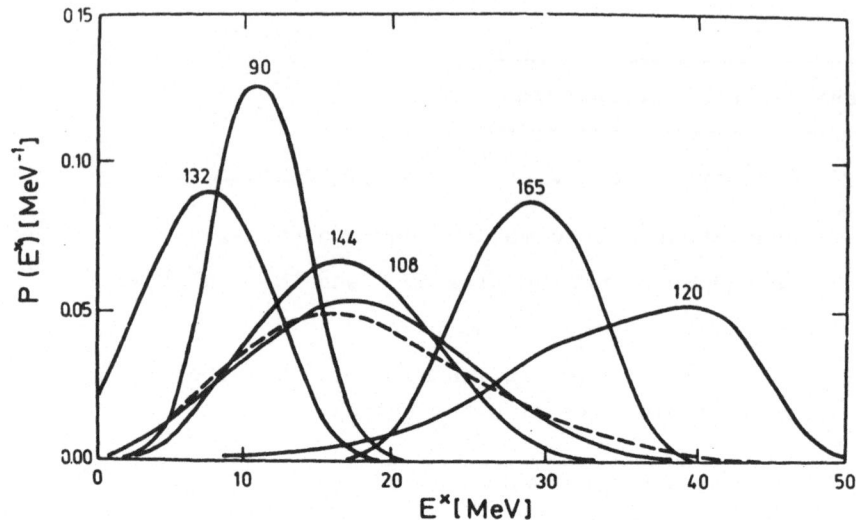

FIG. 5. Initial distribution in excitation energy for some
typical mass numbers of fragments from spontaneous fission of
252Cf. The dashed line represents the weighted average. Figure
taken from ref. /6/.

4. COMPARISON BETWEEN RECENT EXPERIMENTAL DATA
AND THEORETICAL CALCULATIONS

Figures 6 and 7 represent recent experimental data on the Cf
neutron spectrum in comparison with the results of different
calculations in the energy range between 10 keV and 20 MeV.
Concerning the experimental results one can conclude:

(i) Recent data are close to a Maxwellian distribution with
$T = 1.42$ MeV for $E < 1$ MeV. The ANL data tend to lower emission
probabilities (about -8% deviation at 0.2 MeV).

(ii) Recent experimental data are in very good agreement
(with the exception of ref. /11/ (?)) in the energy range
between 1 and 5 MeV. They exceed the Maxwellian with $T = 1.42$
MeV by about 3%.

(iii) In the 5—20 MeV region, the data tend to lower emission
probabilities obviously with reference to the $T = 1.42$ MeV
Maxwellian. Typical deviations are 0% at 5 MeV, -20% at 10 MeV,
-25% at 16 MeV.

(iv) Excess neutrons have been found above 20 MeV /12, 13/.

Figure 6 shows the results of different calculations in the framework of both the MNM (version II) and the GMNM (version III and version IV). Here, C of eq. (2) Was adjusted so that a fit of the calculated spectrum to a Maxwellian yields T = 1.42 MeV in each case /42/. Similar results are obtained in the 0.1—6 MeV energy region. Further, the result of a calculation.

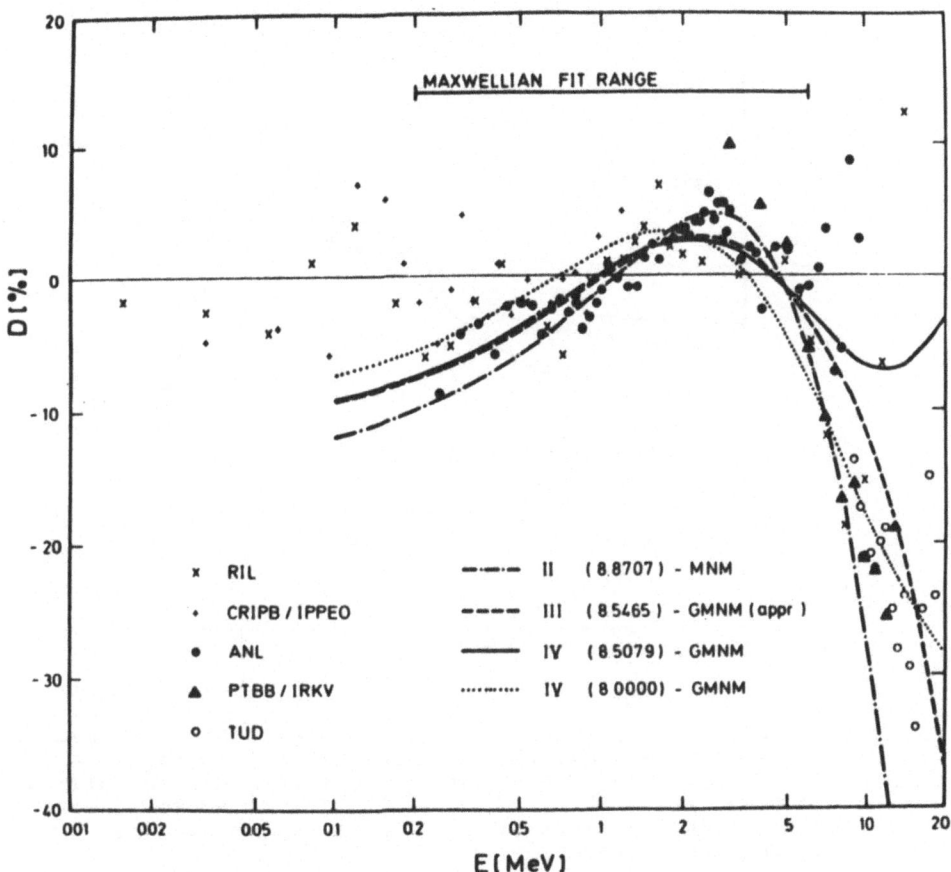

FIG. 6. Percentage deviations of recent experimental data (RIL /7/, CRIPB/IPPEO /8/, ANL /9/, PTBB/IRKV /11/, TUD /12/) on the Cf neutron spectrum from a Maxwellian distribution with T = 1.42 MeV. With the exception of calculation IV (8.0000), the C values (included in parenthesis) have been adjusted to obtain T = 1.42 MeV if fitting the calculated spectra to a Maxwellian in the shown energy range. The experimental errors are not represented for clearness. We refer to the original papers. Figure taken from ref. /42/.

using C = 8.0 MeV (the most probable value with reference to
level density systematics) is represented. In this case, the
high-energy data of recent experiments are reproduced. Above
5 MeV, version II and version III yield a spectrum slope which
is contrary to experimental data. This is a consequence of
rough approximations (see Synopses 1—3).

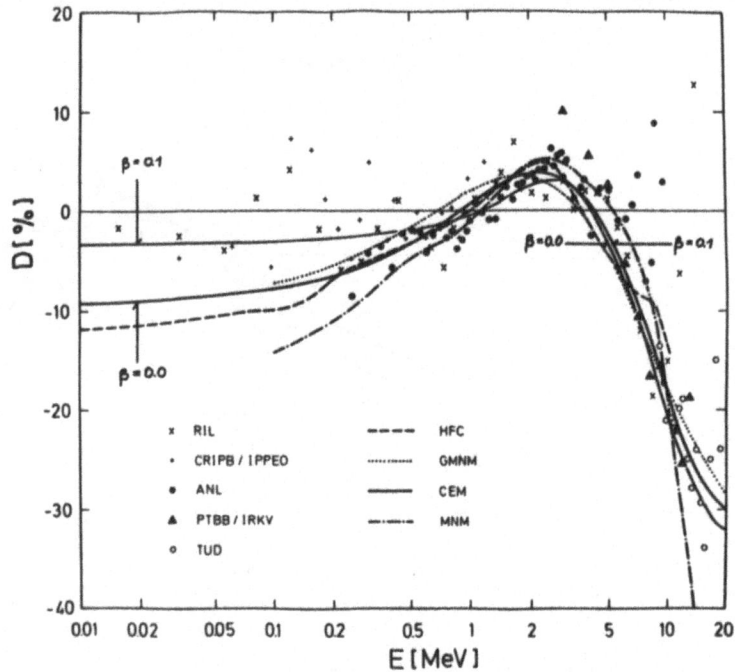

FIG. 7. As Fig. 6 concerning the experimental data. The results
of the MNM and the HFC calculation were taken from refs. /41/
and /10/, respectively. The shown GMNM spectrum was calculated
for C = 8.0 MeV /ref. /42/, cf. Fig. 6). The Cf energy spectrum
calculated in the framework of the CEM is represented for two
CMS anisotropy parameter β.

In Fig. 6, we compare recent Cf spectrum calculations /34, 37, 41, 42/ with experimental data. Obviously, the MNM under-estimates the experimental data at low energy (E < 0.5 MeV) as well as in the high-energy region (E > 10 MeV). The results of the GMNM, HFC and CEM (β = 0.0) calculations are in very good agreement, but they tend to underestimate the low-energy data (-10% deviation). Considering the CMS anisotropy roughly (β = 0.1, eq. (1)), the spectrum is enhanced by about 5.5% at 10 keV. The high-energy tail is also enhanced by about 2%. Between 0.8 and 3 MeV, the calculation with β = 0.1 yields a somewhat lower emission probability (cf. study in ref. /6/, first consideration in ref. /21/). The CEM calculated spectrum with β = 0.1 agrees very good with experimental data in the whole energy range (1 keV—20 MeV). However, this version takes into account the CMS anisotropy approximately, because eq. (1) means that $f(\nu)$ is considered without correlation to ε. It is emphasized that the CEM calculation was <u>not</u> based on arbitrary normalizations or parameter adjustment. The measured spectrum above 20 MeV (Fig. 2) cannot be described by an evaporation model. The found neutron excess should be attributed to non-equilibrium emission /33/.

5. CONCLUSIONS

The presented review paper refers to recent experimental data which have been deduced from improved measurements. Discrepan-cies between these recently measured Cf neutron spectra are much less than known from previous experiments (cf. Blinov's review).

The progress made concerning Cf spectrum calculations is emphasized. All statistical-model versions considered (MNM, GMNM, CEM, HFC) yield similar results in the energy range from 0.5 to 10 MeV. The MNM underestimates experimental data at both spectrum ends (E < 0.5 MeV, E > 10 MeV). However, it can easily be applied for practical purposes. In particular, neutron spectra from induced fission reactions can be described with an sufficient accuracy in the most important energy region of 0.5—10 MeV. The complexity of CEM and HFC calculations makes it

necessary to consider a multitude of input data. According to
our experience, precise consideration of nuclear structure is
as important as the consideration of the fission process complexity including probability distributions, the exact CMS-LS transformation of the spectra (i.e. introduction of P(TKE) correlated
with CMS spectrum), etc. However, such detailed studies are only
applicable to well-investigated fission reactions (at present!).

The reviewed calculation methods don't take into account
other possible emission mechanisms which might influence the
total energy spectrum in certain energy ranges. Differential
emission probabilities of neutrons which are evaporated during
fragment acceleration as well as of neutrons originating from
the ^{5}He decay have been estimated recently /45/. It was found
that these processes are less important.

It was already pointed out /6, 46/ that further detailed
investigations of multiple-differential emission probabilities
of fission neutrons are necessary to clarify the open problems
of fission neutron emission.

REFERENCES

1. Prompt Fission Neutron Spectra, Proc. Consultants' Meeting,
   Vienna 1971, IAEA, Vienna, 1972.
2. Nuclear Data Standards for Nuclear Measurements, Technical
   Reports Series No. 227, IAEA, Vienna 1983.
3. IAEA Consultants' Meeting on the ^{235}U Fast-neutron Fission
   Cross-section and the ^{252}Cf Fission Neutron Spectrum, Proc.
   INDC(NDS)-146/L, 1983.
4. M. V. Blinov, Proc. IAEA Consultants' Meeting on Neutron
   Source Properties, Debrecen, 1980, INCD(NDS)-114/GT, 1980.
5. J. Terrell, Proc. IAEA Symp. on Physics and Chemistry of
   Fission, Salzburg, 1965, IAEA, Vienna, Vol. 2, p. 3.
6. H. Märten et al., see ref. /3/, p. 199.
7. M. V. Blinov et al., Proc. Int. Conf. on Nuclear Data for
   Science and Technology, Antwerp, 1983, ed. by K. H. Böckhoff,
   D. Reidel, Dordrecht, 1983, p. 479.
8. A. Lajtai et al., see ref. /3/, p. 177.
9. W. P. Poenitz and T. Tamura, see ref. /7/, p. 465 and
   ref. /3/, p. 175.
10. M. V. Blinov et al., preprint, 1984.
11. R. Böttger et al., see ref. /7/, p. 484.
12. H. Märten et al., see ref. /7/, p. 488.
13. H. Märten et al., this volume.
14. J. Boldeman, Trans. Am. Nucl. Soc. 32 (1979) 733 and
    information about new measurements received via NDS, H. D.
    Lemmel, Note 84/5/11.

15. Mon Jiangshen et al., Chin. J. Nucl. Phys. 3 (1981) 163.
16. J. Grundl and C. Eisenhauer, Nat. Bur. Stand., Spec. Publ. NBS-493, 1977.
17. A. Chalupka et al., see ref. /3/, p. 187.
18. H. Klein et al., see ref. /3/, p. 191.
19. J. S. Fraser, Phys. Rev. 88 (1952) 536.
20. B. E. Watt, Phys. Rev. 87 (1952) 1037.
21. J. Terrell, Phys. Rev. 113 (1959) 527.
22. H. R. Bowman et al., Phys. Rev. 126 (1962) 2120.
23. V. S. Stavinski, Sov. Phys.-JETP 36 (1959) 629.
24. R. W. Fuller, Phys. Rev. 126 (1962) 684.
25. Y. Boneh and Z. Fraenkel, Phys. Rev. C10 (1974) 893.
26. V. A. Rubchenya, Leningrad report RI-28, 1974.
27. B. C. Samanta et al., Phys. Lett. B108 (1982).
28. E. Cheifetz et al., Phys. Rev. Lett. 29 (1972) 805.
29. J. B. Wilhelmy et al., Phys. Rev. C5 (1972) 2041.
30. A. Gavron, Phys. Rev. C13 (1976) 2561.
31. V. F. Weisskopf, Phys. Rev. 52 (1937) 295.
    J. M. Blatt and V. F. Weisskopf, "Theoretical Nuclear Physics", J. Wiley a. Sons, New York, 1952.
32. W. Hauser and W. Feshbach, Phys. Rev. 87 (1952) 366.
33. H. Märten and D. Seeliger, J. Phys. G10 (1984) 349.
34. H. Märten and D. Seeliger, this volume.
35. I. C. Browne and F. S. Dietrich, Phys. Rev. C10 (1974) 2445.
36. B. F. Gerasimenko et al., Proc. All-Union Conf. Neutron Physics, Kiev, 1980, Moscow, 1980, Vol. 3, p. 137.
37. B. F. Gerasimenko et al., Proc. All-Union Conf. Neutron Physics, Kiev, 1983, Moscow, 1984, Vol. 1, p. 349.
38. E. Nardi et al., Phys. Lett. 43B (1973) 259.
39. D. G. Madland and J. R. Nix, Nucl. Sci. Eng. 81 (1982) 213.
40. D. G. Madland and J. R. Nix, see ref. /7/, p. 473.
41. D. G. Madland and R. J. LaBauve, preprint LA-UR-84-129, 1984.
42. H. Märten and D. Seeliger, INDC(GDR)-30/L.
43. K. H. Schmidt et al., Z. Phys. A308 (1982) 215.
44. A. V. Ignatyuk et al., Yad. Fiz. 21 (1975) 485.
45. H. Märten et al., see ref. /37/ and INDC(GDR), 1984, in press.
46. M. V. Blinov et al., see ref. /3/, p. 161.

(n,p), (n,α) AND NEUTRON INDUCED TERNARY FISSION REACTIONS

C. Wagemans*

Nuclear Physics Laboratory, Gent and
SCK-CEN, Mol, Belgium

ABSTRACT. A survey is given of the most recent measurements of (n,p) and (n,α) reactions with thermal and resonance neutrons. Attention is given to the basic physical aspects of these reactions and to applications in the field of astrophysics and nuclear technology. Also recent progress in the field of thermal neutron induced ternary fission is discussed.

1. INTRODUCTION

In the present paper, a survey is given of the latest developments in the field of thermal and resonance neutron induced (n,p) and (n,α) reactions. Measurements with MeV neutrons and the so-called standard reactions ^{6}Li(n,α)t and ^{10}B(n,α)^{7}Li are not considered here. Furthermore, some recent thermal neutron induced ternary fission experiments are reported, since also in this reaction alphas and other light charged particles are produced. Attention is given to the basic physical aspects of these reactions, and to applications in the field of astrophysics and nuclear technology.

2. THERMAL NEUTRON INDUCED (n,p) AND (n,α) REACTIONS

The status of the thermal neutron induced (n,p) and (n,α) reactions has been reviewed by Emsallem /1/ in 1979 and updated by Popov /2/ in 1982. Since then, several new results became available.

---

*NFWO.

At the Antwerp Conference in 1982, Gledenov et al. /3/ presented measurements on the ^{22}Na(n,p)^{22}Ne reaction in the energy region from 0.01 to 1000 eV and D'hondt et al. /4/ discussed (n_{th},p) and (n_{th},α) reactions as a possible source of gas production in nuclear reactors. At the Münster Meeting in 1983, new results on ^{32}S and ^{33}S (n_{th},α) were presented by Wagemans et al. /5/ and on ^{40}Ca, ^{41}Ca and ^{43}Ca by D'hondt et al. /6/. At the conference in Florence in 1983, results on ^{22}Na($n_{th},\gamma p$) ^{22}Ne were reported by Asghar et al. /7/ and on ^{64}Zn, ^{65}Zn and ^{67}Zn(n_{th},α) by Emsallem et al. /8/. In the meantime, the latter results have been published in detail.

In the following alineas we will discuss these data together with results on ^{26}Al(n_{th},p) /9/, on ^{40}K(n_{th},p), (n_{th},α) and ($n_{th},\gamma\alpha$) /10/ and with unpublished results obtained at the ILL (Grenoble) /11/. All these new data (with the exception of ref. /3/) have been obtained at the end of a curved neutron guide at the ILL using silicon surface barrier detectors. The neutron beam was intense (6×10^8 neutron/cm^2 s), very thermalized ($\Phi_{th}/_{fast}\Phi\approx10^6$) and very clean, since the gamma background was reduced by a factor 10^6.

^{22}Na(n_{th},p)^{22}Ne and ^{22}Na($n_{th},\gamma p$)^{22}Ne

In a previous paper, Kvitek et al. /12/ determined a ^{22}Na(n_{th},p_1) reaction cross-section value of $(30.6\pm2.6)\times10^3$ b. This means that despite the very high cross-section values involved, the ^{22}Na thermal neutron induced cross-section data are inconsistent. If we consider, e.g. the most recent values reported in CINDA /12/, the thermal absorption cross-section $(51.1\pm3.1)\times10^3$ b is much smaller than the sum of the capture cross-section $(35.9\pm1.2)\times10^3$ b and the (n_{th},p) cross-section mentioned above. Kvitek et al. /12/ also observed a (n_{th},p_0) transition which was much weaker, $\Gamma_{p_0}/\Gamma_{p_1}$ being equal to $(7.4\pm0.2)\times10^{-3}$. These results could be explained after a measurement in the resonance region /3/, in which the existence of a strong (n,p_1) resonance at 145 eV neutron energy and with parameters $\Gamma_n^0 = 1.6\pm0.2$ eV, $\Gamma_{p_1} = 114\pm20$ eV and $\Gamma_{p_0}/\Gamma_{p_1} = (6\pm2)\times10^{-3}$ was demonstrated. So far, no experimental evidence for the existence of the

($n_{th}, \gamma p$) reaction has been found, although several nuclei show the presence of the ($n_{th}, \gamma \alpha$) reaction. In view of its large Q_p-value and its relatively small (n_{th}, p_0) cross-section, ^{22}Na is expected to be an ideal case for the study of the ($n_{th}, \gamma p$) reaction. Although some doubts exist due to the uncertainty on the background correction, the results of Asghar et al. /7/ indicate that the $^{22}Na(n_{th}, \gamma p)$ reaction might be present with a cross-section of 1.28 ± 0.32 b.

$^{26}Al(n_{th}, p)^{26}Mg$

As will be explained in section 3, this reaction is of great interest in astrophysics. It was recently studied at the Grenoble High Flux Reactor by Trautvetter et al. /9/ using a neutron beam with an average energy of 40 meV. They observed the p_0-transition at 4.78 MeV with a cross-section of 26 ± 10 mb and a strong p_1-transition at 2.97 MeV with a cross-section of 1.85 ± 0.15 b. The energetically possible $^{26}Al(n_{th}, \alpha)$ reaction ($Q_\alpha = 2.967$ MeV) has not been observed, which is maybe due to the background conditions. Anyhow, this transition is not favoured since it requires $l_\alpha = 4$.

$^{32}S(n_{th}, \alpha)^{29}Si$ and $^{33}S(n_{th}, \alpha)^{30}Si$

For the $^{33}S(n_{th}, \alpha)$ reaction ($Q_\alpha = 3.493$ MeV), cross-section values varying between 52 and 180 mb have been reported up to now (Table 1). These strong differences are probably caused by the fact that sulphur sublimates under vacuum, even at room temperature. Since also the $^{33}S(n, \alpha)$ reaction is of interest

TABLE 1. The ^{32}S and $^{33}S(n_{th}, \alpha)$ cross-section (mb)

Reference	$^{32}S(n_{th}, \alpha)$	$^{33}S(n_{th}, \alpha)$
Munnich /14/	6.8 ± 0.11	80 ± 80
Benisz /15/	3.9 ± 0.51	51 ± 22
Harris /16/		52 ± 1
This work	≤ 0.5	115 ± 10

346

for astrophysics, a series of measurements was performed at the Grenoble High Flux Reactor, using sealed sulphur layers. So an accurate value of 115±10 mb was obtained /5/ for the $^{33}S(n_{th},\alpha)$ reaction cross-section. Also the $^{33}S(n_{th},\gamma\alpha)$ reaction seems to occur, for which cross-section an upper-limit of 1 mb has been determined (Fig. 1).

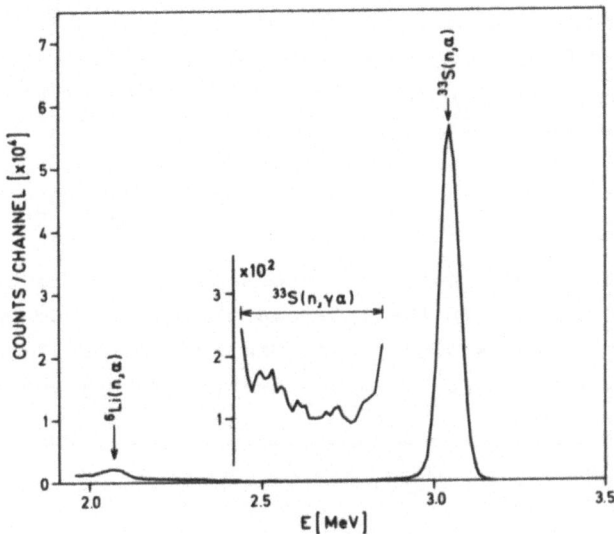

FIG. 1. $^{33}S(n_{th},\alpha)$ energy distribution.

For the $^{32}S(n_{th},\alpha)$ reaction, suprisingly large cross-section values have been reported taking into account its relatively low Q_{α}-value (1.526 MeV). We did not observe this (n_{th},α) transition /11/. Anyhow, from a detailed examination of the results of Munnich /14/ and Benisz et al. /15/ one can conclude that they probably detected background $^{10}B(n,\alpha_1)$ particles $(E_{\alpha_1} = 1.48$ MeV).

$^{39}K(n_{th},\alpha)^{36}Cl$, $^{40}K(n_{th},p)^{40}Ar$ and $^{40}K(n_{th},\alpha)^{37}Cl$

A detailed study of the $^{40}K(n_{th},p)$ and $^{40}K(n_{th},\alpha)$ reaction has been reported by Emsallem et al. /10/. The most striking result of this work was the observation of an important $^{40}K(n_{th},\gamma\alpha)$ contribution with a cross-section of 26±4 mb. Also the multipolarity and the energy of the primary low-energy gamma rays have been determined. Krusche et al. /17/ searched for these gamma transitions in their $^{40}K(n_{th},\gamma)$ measurements, but they could identify only one such transition.

For the $^{39}K(n_{th},\alpha)$ reaction, a cross-section of 4.3 ± 0.5 mb has been reported in the literature /15/. We did not observe this transition in our experiments /11/ ($\tilde{\sigma}_\alpha \leq 0.2$ mb). Similar to the ^{32}S case discussed before, we believe that background ^{7}Li (E = 1.01 MeV) or α_1 (E = 1.48 MeV) particles emitted in the $^{10}B(n_{th},\alpha)$ reaction have not been fully discriminated from the $^{39}K(n_{th},\alpha)$ particles (E = 1.23 MeV).

$$^{40}Ca(n_{th},\alpha)^{37}Ar, \quad ^{41}Ca(n_{th},\alpha)^{38}Ar \text{ and } ^{43}Ca(n_{th},\alpha)^{40}Ar$$

For the $^{40}Ca(n_{th},\alpha)$ reaction, a cross-section value of 2.4 ± 1.1 nb has been reported in the literature /14/. We did not observe this transition ($\tilde{\sigma}_\alpha \leq 30\,\mu b$) /6/. Similar to the ^{32}S and $^{39}K(n_{th},\alpha)$ measurements discussed before, we believe that

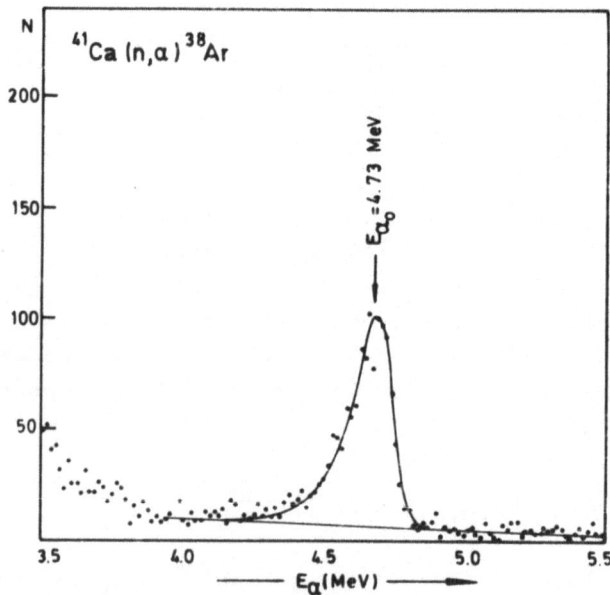

FIG. 2. $^{41}Ca(n_{th},\alpha)$ energy distribution.

background ^{10}B(n,α_1) particles (E = 1.48 MeV) have not been
discriminated from the ^{40}Ca(n_{th},α) particles (E = 1.58 MeV) in
these nuclear emulsion measurements.

For ^{41}Ca(n_{th},α) (Q_α = 5.223 MeV, Q_p = 1.203 MeV) and
^{43}Ca(n_{th},α) (Q_α = 2.277 MeV) no data are available in the literature. For the latter reaction, an upper limit $\sigma_\alpha \leq 47\,\mu$b has
been determined /6/. In Fig. 2 the ^{41}Ca(n_{th},α_0) transition is
shown /6/, for which a preliminary $\sigma(n_{th},\alpha)$ value of 23 mb has
been determined. The poor sample enrichment (1.33% ^{41}Ca) did
not permit the identification of (n_{th},p) transitions. Anyhow,
a new experiment using a 81.69% enriched ^{41}Ca sample is being
prepared.

^{64}Zn(n_{th},α), ^{65}Zn(n_{th},α) and ^{67}Zn(n_{th},α)

Results for these zinc isotopes (and also for ^{77}Se) have been
reported by Emsallem et al. /8/. For the stable ^{64}Zn and ^{67}Zn
only the (n_{th},α_1) transitions appear to be present at 3.566
and 3.330 Mev, respectively, with $\sigma(n_{th},\alpha_1)$ values of 11 and
159 μb, respectively. Especially the absence of the ^{67}Zn(n_{th},α_0)
transition ($\sigma_\alpha \leq 10\,\mu$b) seems surprising, since Gledenov et al.

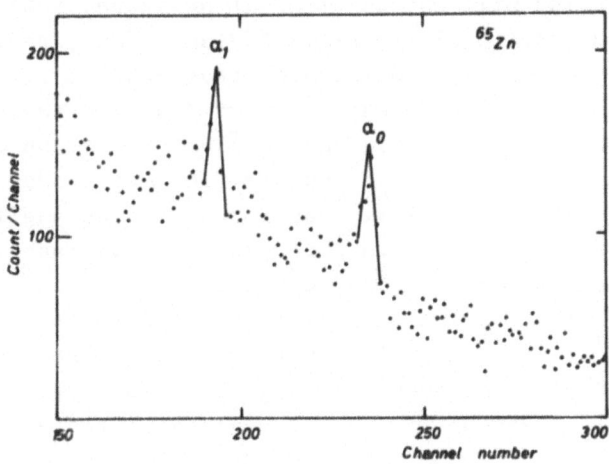

FIG. 3. Pulse-height spectrum for the ^{65}Zn(n_{th},α) reaction.
The α_0 and α_1 transitions occur at 6.089 and 4.987 MeV, respectively.

/18/ recently observed significant (n, α_o) transitions with
resonance neutrons. A Breit—Wigner extrapolation of these
resonance data yields a $\sigma(n_{th}, \alpha_o)$ value of 370 μb, in contradiction with the experimentally determined upper level of
10 μb. This is an indication for the existence of destructive
interference between the resonances. The most striking results,
however, are the equally strong (n_{th}, α_o) and (n_{th}, α_1) transitions $(\sigma_{\alpha_o} = \sigma_{\alpha_1} = 1\pm0.1$ b) induced in the radioactive ^{65}Zn
(Fig. 3). These high cross-sections will have implications for
reactor materials containing ^{64}Zn, as will be discussed below.

Gas Production in Thermal Fission Reactors

Although the fission process itself is the major source of gas
production in reactors, also (n_{th}, α) reactions in structural
materials, fission products and actinides can contribute to it.
Especially gas production in structural and canning materials
can create mechanical problems.

In Table 2 a survey of recent (n_{th}, α) measurements on structural materials and fission products is given. Also the
^{50}V(n_{th}, p) reaction has been observed /4/ with $\sigma(n_{th}, p)$
= 0.40 ± 0.02 mb. Obviously, the (n_{th}, α) reactions on the fission
products studied will not be of great nuisance. This is not the
case for the structural materials ^{50}V and ^{64}Zn (leading to ^{65}Zn
after neutron capture), in which serious amounts of hydrogen
resp. helium can be produced. ^{64}Zn and ^{58}Ni are the only cases
known so far in which significant helium production occurs
after double thermal neutron capture. The gas production via
the (n_{th}, α) reaction on actinides appears to be negligible.
This is shown in Table 3, in which the results reported by
D'hondt et al. /4/ are updated with our most recent data /11/.
For nuclei with very low thermal fission cross-sections as
^{232}Th and ^{238}U, the (n_{th}, α) detection limit is better than 1 μb.
The other nuclei mentioned have high σ_f values, so the (n_{th}, α)
particles are mixed up with ternary alpha particles.

TABLE 2. (n_{th}, α) cross-sections for selected structural materials and fission products

Nucleus	E_α (MeV)	σ_{exp} (μb)	References
^{57}Fe	2.23	≤ 0.1	/11/
^{65}Zn	6.03	2×10^6 a	/9/
^{73}Ge	3.70	1^b	/11/
^{91}Zr	5.42	5^b	/11/
^{95}Mo	6.14	30 ± 4	/4/
^{97}Mo	5.15	0.44 ± 0.18	/4/
^{101}Ru	5.57	≤ 0.15	/4/
^{105}Pd	6.09	0.54 ± 0.18	/4/
^{113}Cd	4.77	≤ 1	/4/
^{115}Sn	5.99	58 ± 15	/4/
^{125}Te	6.34	2^b	/11/

a $- \sigma_{\alpha_0} + \sigma_{\alpha_1}$, b - preliminary value.

TABLE 3. (n_{th}, α) cross-sections in the actinide region

Nucleus	^{232}Th	^{233}U	^{235}U	^{238}U	^{239}Pu	^{241}Pu	^{241}Am
σ_α (μb)	≤ 1	≤ 200	≤ 100	1.5	≤ 300	≤ 210	≤ 210
Ref.	/4/	/11/	/4/	/4/	/11/	/11/	/11/

3. RESONANCE NEUTRON INDUCED (n,p) and (n,α) REACTIONS

At the Capture Gamma Conference in Grenoble, 1981, the resonance neutron induced (n,α) reactions have been reviewed by
Popov /19/. At the same meeting, a few other more detailed
papers have also been presented. Also the review article of
Pikelner et al. /20/ partially deals with the same subject.

Besides the previously mentioned results of Gledenov et al.
/3, 18/ on ^{22}Na(n,p) and ^{67}Zn(n,α), three (n,p) or (n,α) measurements in the resonance region have been reported since i.e.
on ^{40}K /21/, on ^{143}Nd and ^{147}Sm /22/, and on ^{26}Al /9/. In the
following alineas these results will be discussed together
with recent data on ^{33}S and ^{41}Ca obtained at GELINA /23/.

The (n,p) and (n,α) reactions on ^{26}Al, ^{33}S, ^{40}K and ^{41}Ca have a common field of interest namely astrophysics. Especially the understanding of the reactions leading to ^{26}Al production and destruction ($^{26}Al(n,p)$) in the interstellar medium of our galaxy may help to clarify details of the explosive nucleosynthesis and may give information about the formation of our solar system. Also the three other nuclei have a direct impact on the explosive nucleosynthesis calculations. To have a viable theory for the origin of the natural elements, it is indeed imperative to account for the existence of all the naturally occurring isotopes, no matter how small their abundance. Starting from the conversion of ^{32}S and ^{36}Ar, initially present in the star during explosive carbon burning, calculations lead to an overproduction of the neutron-rich isotope ^{36}S. $^{33}S(n,\alpha)$ being the major destructive reaction in the reaction sequences considered, the lack of accurate data on this cross-section is a handicap for the astrophysicists. The same kind of calculations lead to a serious overproduction of ^{46}Ca. Also here reliable experimental data on the main destructive reactions, i.e. ^{40}K and $^{41}Ca(n,\alpha)$ are strongly requested.

$^{26}Al(n,p)^{26}Mg$

Trautvetter et al. /9/ studied the $^{26}Al/n,p_0$) and (n,p_1) reactions in Maxwell—Boltzmann like neutron spectra with kT = 31 and 71 keV, respectively, and in a nearly homogeneous spectrum with E_n = 310±40 keV, all produced at the Karlsruhe Van de Graaff accelerator. These neutron energies correspond to stellar temperatures of the order of 10^9 K. At 31 keV they determined a $\sigma(n,p_0)$ of 14±8 mb and a $\sigma(n,p_1)$ of 145±26 mb. The corresponding values were 16±13 mb and 84±14 mb at 71 keV, and 21±8 mb and 72±17 mb at 310 keV.

These results are the only ones available so far. They do, however, not fully satisfy the astrophysicists' requests, since no data are given below 30 keV, i.e. for temperatures which are typical of the red giant phase in stars. Furthermore, in this energy region resonances are likely to occur with an estimated spacing D(l=0) ~ 30 keV, so a simple 1/v extrapolation of the cross-section data available is certainly to rough an approach.

^{33}S(n,α)^{30}Si

Auchampaugh et al. /24/ performed a measurement of the ^{33}S(n,α) cross-section at the LASL Van de Graaff accelerator. Since the resolution of these measurements was too poor to resolve individual neutron resonances, a higher resolution measurement was performed at GELINA (Geel Linear Accelerator). At the 1981 Grenoble Capture Gamma Conference preliminary results were reported /25/ which, however, are too low by about a factor of three. This was due to sublimation effects in the open sulphur sample used. In the meantime, a new series of measurements has been performed using various sealed sulphur samples. Figure 4 shows the ^{33}S(n,α) cross-section from 10 to 1000 keV deduced from one of these measurements /23/. These data are being

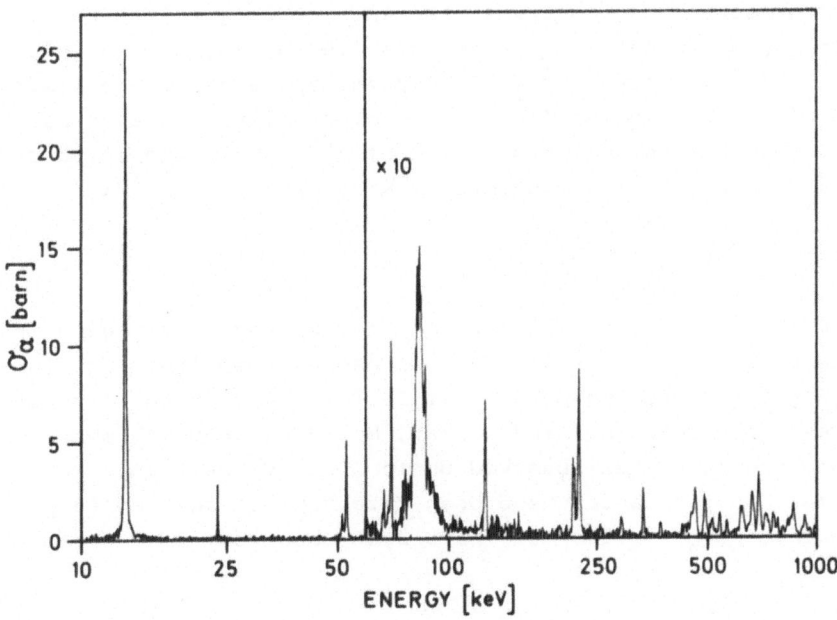

FIG. 4. The ^{33}S(n,α) cross-section in the energy region from 10 to 1000 keV.

combined with the results of a recent ^{33}S total cross-section measurement performed at ORELA /26/. Although some problems persist, a consistent set of resonance parameters is likely to become available soon for most of the resonances.

^{40}K(n,p)^{40}Ar and ^{40}K(n,α)^{37}Cl

To the best of our knowledge, ^{40}K is the only case to date where significant (n,p) and (n,α) transitions have been simultaneously observed in the neutron resonance region. Measurements by Weigmann et al. /21/ at GELINA revealed a pronounced resonance structure above 1 keV with peak cross-sections up to 5 b. Moreover, the relative proton and alpha yield strongly fluctuates from resonance to resonance. These fluctuations could be explained in terms of the involved angular momenta. None of the observed resonances shows properties required to explain the observed /10/ thermal (n,p) cross-section of 4.4$\pm$0.9 b, which is thus interpreted as being due to a 7/2$^-$ bound state. From a comparison of the observed resonance spacing to the density of the spin 1/2 compound states observed by Keyworth et al. /27/ in the inverse ^{40}Ar(p,n)^{40}K reaction, a direct determination of the spin cut-off parameter of the level density was obtained. This rather large value of σ^2 = 16 is supported by recent ^{40}K(n$_{th}$,γ) measurements of Krusche et al. /17/.

^{41}Ca(n,α)^{38}Ar

A study of the ^{41}Ca(n,α) reaction in the resonance region is being performed at GELINA /23/. As can be seen from the time-of-flight spectrum shown in Fig. 5, four intense (n,α) resonances occur between 3 and 20 keV. Besides, about 40 smaller resonances have been observed up to 500 keV. Between 3 keV and thermal, no resonances have been detected. So far, no (n,p) resonances (Q_p = 1.203 MeV) have been observed.

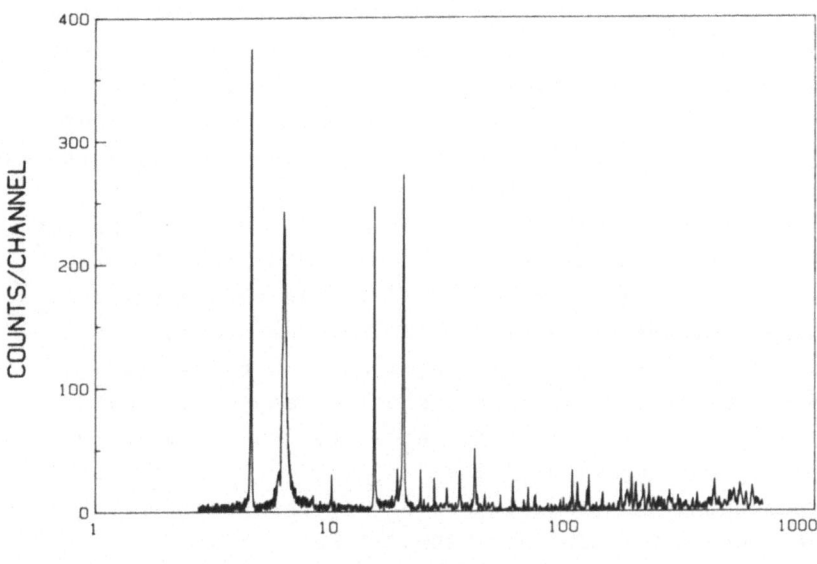

FIG. 5. The ^{41}Ca(n,α) time-of-flight spectrum from 3 to 500 keV.

^{143}Nd(n,α)^{140}Ce and ^{147}Sm(n,α)^{144}Nd

These reactions were studied by Antonov et al. /22/ at the Fakel
linear electron accelerator at Kurchatov with the aim to investigate the neutron energy dependence of the alpha widths. In
the case of ^{147}Sm(n,α), these authors observed strong fluctuations of Γ_α from resonance to resonance (between 0.14 and 34.2
μeV). Although these Γ_α values are not in contradiction with
expectations based on the statistical theory, Antonov et al.
/22/ do not exclude the presence of some non-statistical process
in the ^{147}Sm(n,α) decay channel.

4. DETERMINATION OF REACTION Q-VALUES VIA
   (n,p) AND (n,α) REACTIONS

Although the (n,p) and (n,α) reactions, with thermal as well as
with resonance neutrons, are generally not performed with the
aim of determining reaction Q-values, this quantity is obtained
as kind of a by-product. Often a decent accuracy can be reached,

355

which, in selected cases becomes even better than the accuracy obtained with devices dedicated to this purpose. In cases where mass-spectrometers fail (e.g. for short-living isotopes), these reactions can also help to determine atomic masses /28/. In the following alineas we will shortly discuss a few selected examples.

The (n_{th},p) reactions provide an easy tool for the determination of the mass difference of mass doublets. For the $^{40}K(n_{th},p)^{40}Ar$ reaction, e.g. a ^{40}K-^{40}Ar mass difference of 1506.3$\pm$3 keV was calculated /10/ from the experimental Q_p-value of 2288.7$\pm$3 keV, assuming a neutron-proton mass difference of 782.39 keV. The corresponding mass-spectrometric value is 1505$\pm$3 keV. An even better accuracy can be obtained when combining the energies of the $^{40}K(n,p)^{40}Ar$ resonances with the corresponding energies of the inverse reaction $^{40}Ar(p,n)^{40}K$. In this way a Q_p-value of 2286.7$\pm$1 keV has been obtained yielding a ^{40}K-^{40}Ar mass difference of 1504.3$\pm$ 1 keV /21/.

The (n_{th},α) reaction on the radioactive ^{153}Gd (241.6 d) is a good example of the usefulness of this type of reactions. From the energy position of the $^{153}Gd(n_{th},\alpha_0)$ transition, a Q_α-value of 9.79$\pm$0.03 MeV was determined /28/, which was 0.21 MeV higher than the value calculated from the 1977 atomic mass tables of Wapstra and Boss. In the 1985 edition, Wapstra and Audi /29/ stated that the 1977 data were wrong, since they were based on electron capture decay energy measurements, which turned out to be in error.

As a last example we consider the $^{33}S(n,\alpha)^{30}Si$ reaction in the resonance region. In Table 4, the neutron resonance energy positions are compared with a few corresponding alpha energies from the inverse $^{30}Si(\alpha,n)^{33}S$ reaction (Wiechers et al. /30/). By combining both results, a Q-value of 3497.7$\pm$5 keV is obtained for the $^{33}S(n_{th},\alpha)$ transition, in agreement with the value of 3493.1$\pm$0.6 keV calculated from the mass tables of Wapstra and Audi /29/. It should be noticed that the accuracy on E_α determines that of the Q-values, the error on the E_n-measurement being very small.

TABLE 4. Comparison of resonance energies in $^{30}Si(\alpha,n)^{33}S$ and $^{33}S(n,\alpha)^{30}Si$

E_α (keV) /30/	E_n (keV) /23/	$Q(n,\alpha)$ (keV)
3979±5	13.45±0.07	3497.87
3990	23.95±0.09	3497.41
4042	70.9	3497.86
4104	127.7	3497.60
4215	228.7	3497.80

5. THERMAL NEUTRON INDUCED TERNARY FISSION REACTIONS

The thermal neutron induced ternary fission process is an important source of alpha particles, production cross-sections of the order of several barns being reached. About 90% of the emitted ternary particles are alphas.

The remaining fraction mainly consists of tritons, although also protons, deuterons, ^6He and even heavier particles are emitted. Obviously, the ternary fission process constitutes an important source of gas production in a fission reactor. From a fundamental physics point of view, the ternary fission characteristics are expected to yield information on the dynamics of the fission process itself. Hence this process has been intensively studied. It is far beyond the scope of the present paper to give a complete review of the ternary fission studies available to date. An elaborated review of the situation up to 1971 has been reported by Halpern /31/. Since then, a lot of mostly experimental results have been reported. A very detailed study of the energy and angular distributions of the ternary alpha particles as a function of the mass and kinetic energy of the fission fragments has been performed for the thermal neutron induced fission of ^{235}U by Guet et al. /32/, by Choudhury et al. /33/ and very recently by Pannicke et al. /34/.

Although these results certainly improved our understanding of the ternary fission process, a comprehensive theoretical approach is still lacking. In the scope of the present paper, the suggestion of Cârjan /35/ provides an interesting working hypothesis. This author describes the ternary alpha emission (and also that of the other charged light particles) as an alpha decay in the last phases of the scission process.

In the following alineas we will try to interpret recent results on the characteristics of the tritons and the alpha particles emitted in the thermal neutron induced fission of various actinides in terms of Cârjan's theory, but also in a more classical way, i.e. based on the liquid drop model. These results are part of a systematic study being performed at the Grenoble High Flux Reactor.

The energy distribution and the emission probabilities of the tritons and the alpha particles emitted in the thermal

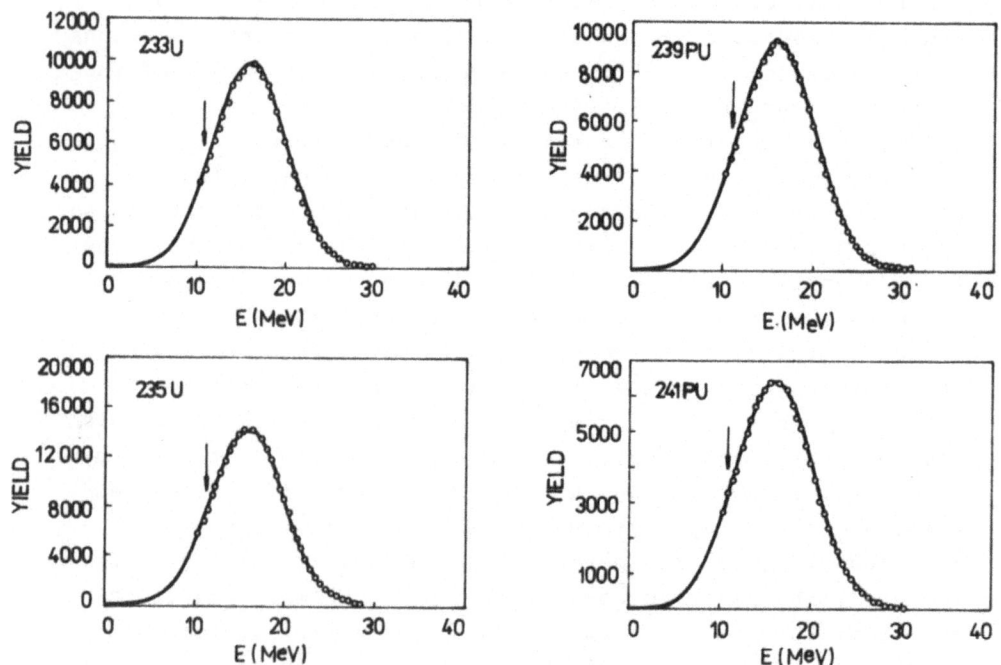

FIG. 6. Ternary α energy distributions. The arrows indicate the calculated position of the (n_{th}, α_0) transitions.

neutron induced ternary fission of ^{233}U, ^{235}U, ^{239}Pu and ^{241}Pu /36/ have been determined using surface barrier ΔE-E detectors and the particle identification technique. The energy distributions of the ternary alpha particles are shown in Fig. 6. Similar spectra have been obtained for the tritons, with, however, a smaller FWHM and a lower E. These results can be completed with preliminary values obtained for ^{241}Am(n_{th},f) /11/ and with results on the ternary alpha emission for ^{237}Np(n_{th},f) /37/ obtained under the same circumstances.

Figure 7 gives a global view of the main results of these measurements, i.e. the absolute ternary alpha emission yields LRA/B, the absolute triton emission yields t/B and the FWHM of the triton and alpha-energy distributions (E_α and E_t being constant). The data are plotted as a function of Z^2/A and of $-\log \lambda$ (λ being the radioactive alpha constant) of the fissioning system. Figure 7 shows that the general trend for all these observables is an increase with increasing Z^2/A and a decrease with increasing values of $-\log \lambda$.

The quantity Z^2/A appears in the liquid drop model as the ratio of the electrostatic to the surface energy of the drop. So Z^2/A is a measure of the fissility of the drop. Moreover, liquid drop model calculations show an increase in the deformation at scission with increasing Z^2/A. Since there is experimental evidence that the fragment deformation has a strong impact on the emission of the ternary particles, a correlation of the ternary particle emission characteristics with Z^2/A is very plausible.

Finally, we want to discuss the apparent correlation between the ternary fission characteristics and $\log \lambda$. In the framework of the alpha-decay mechanism suggested by Cârjan /35/, the ternary alpha emission is expected to be correlated with the Coulomb barrier penetrability P and with the reduced alpha emission width δ^2 (i.e. the probability of having an alpha particle inside the nucleus. Since $\lambda = \delta^2/\hbar$ ($\hbar$ being Planck's constant)), a correlation between λ and the ternary alpha emission is likely to occur in the framework of Cârjan's model, although the λ-values used in Fig. 7 are for ground-state transitions and the (n_{th},f) reactions considered are leading

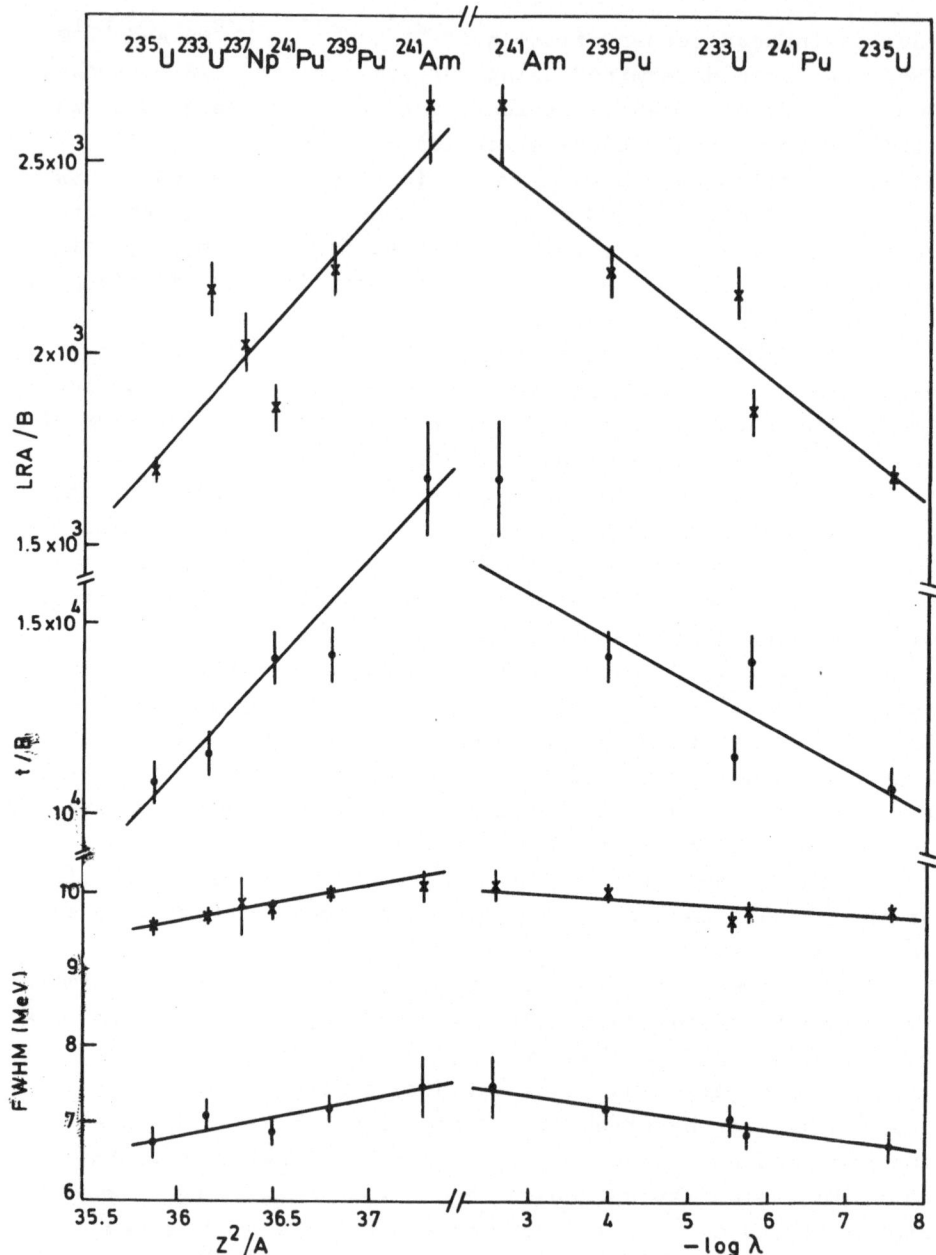

FIG. 7. Absolute ternary α and triton emission yields and FWHM of the α(x) resp. triton (●) energy distribution as a function of Z²/A and of −log λ (λ being the ground-state radioactive alpha decay constant) of the fissioning system. Here λ is given in year⁻¹.

to a fissioning system in a very deformed and excited state.

However, before drawing far-going conclusions, one should state that the parameters Z^2/A and λ are not uncorrelated. From a least-squares adjustment in the region $84 \leqslant Z \leqslant 98$, Viola and Seaborg /38/ determined the following relation between λ and Z

$$\log 0.693 - \log \lambda = (2.11Z - 48.99)Q_\alpha^{-1/2} - (0.39Z + 16.95),$$

where λ being in s^{-1} and Q_α being in MeV. So clearly Z is the determining parameter in Z^2/A as well as in $-\log \lambda$.

Dr H. Weigmann is acknowledged for reading the manuscript and for useful suggestions. Dr H. Trautvetter et al. are acknowledged for making their results available before publication.

REFERENCES

1. A. Emsallem, DrSc Thesis, Lyon University, France, 1979.
2. Y. Popov, Proc. 3rd Int. Symp. on Neutron Induced Reactions, Smolenice, 1982, p. 121.
3. Y. Gledenov, J. Kvitek, S. Marinova, Y. Popov, J. Rigol, and V. Salatsky, Proc. Int. Conf. on Nuclear Data for Science and Technology, Antwerp, 1982, p. 150.
4. P. D'hondt, C. Wagemans, E. Allaert, A. De Clercq, A. Emsallem and R. Brissot, see ref. /3/, p. 147.
5. C. Wagemans, P. D'hondt, R. Brissot and A. Emsallem, Verh. DPG 18 (1983) 1111.
6. P. D'hondt, C. Wagemans, A. Emsallem and R. Brissot, Verh. DPG 18 (1983) 1110.
7. M. Asghar, A. Emsallem, P. D'hondt, C. Wagemans and R. Brissot, Proc. Int. Conf. on Nuclear Physics, Florence, 1983, p. 454.
8. A. Emsallem, M. Asghar, P. D'hondt and C. Wagemans, see ref. /7/, p. 445; Z. Phys. A315 (1984) 201.
9. H. Trautvetter, H. Becker, U. Heinemann, C. Rolfs, F. Käppeler, M. Baumann, H. Freiesleben, P. Geltenbort and F. Gönnenwein, Nucl. Phys., submitted for publication.
10. A. Emsellem, M. Asghar, C. Wagemans and H. Weigmann, Nucl Phys. A368 (1981) 108.
11. C. Wagemans et al., unpublished results.
12. J. Kvitek, V. Hnatowicz, J. Cervena, J. Vacik and V. Gledenov, Z. Phys. A299 (1981) 187.
13. CINDA, IAEA, Vienna, 1984.
14. F. Munnich, Z. Phys. 153 (1958) 106.
15. J. Benisz, A. Jasielska and T. Panek, Acta Phys. Pol. 6 (1965) 763.
16. N. Harris, Diss. Abstr. B27 (1966) 2077.
17. B. Krusche, K. Lieb, L. Ziegler, H. Daniel, T. von Egidy, R. Rascher, G. Barreau, H. Börner and D. Warner, Nucl. Phys. A417 (1984) 231.

18. Y. Gledenov et al., JINR Dubna preprint P3-84-370, 1984.
19. Y. Popov, Proc 4th Int. (n,γ) Symposium, Grenoble, 1981, p. 439.
20. L. Pikelner, Y. Popov and E. Sharapov, Sov. Phys. Usp. 25 (1982) 298.
21. H. Weigmann, C. Wagemans, A. Emsallem and M. Asghar, Nucl. Phys. A368 (1981) 117.
22. A. Antonov, Y. Gledenov, S. Marinova, Y. Popov and H. Rigol, Sov. J. Nucl. Phys. 39 (1984) 501.
23. C. Wagemans and H. Weigmann, unpublished results.
24. G. Auchampaugh, J. Halperin, R. Macklin and W. Howard, Phys. Rev. C12 (1975) 1126.
25. C. Wagemans and H. Weigmann, see ref. /19/, p. 462.
26. G. Coddens, M. Salah, J. Harvey and N. Hill, Conf. on Neutron-Nucleus Collisions. Am. Inst. of Phys., New York, 1985, p. 302.
27. G. Keyworth, G. Kyker, E. Bilpuck and H. Newson, Nucl. Phys. 89 (1966) 590.
28. C. Wagemans, E. Allaert, G. Barreau, A. Emsallem and P. D'hondt, Nucl. Instr. Meth. 190 (1981) 167.
29. A. Wapstra and G. Audi, Nucl. Phys. A432 (1985) 1.
30. G. Wiechers, W. McMurray and I. Van Heerden, Nucl. Phys. A92 (1967) 175.
31. I. Halpern, Ann. Rev. Nucl. Sci. 21 (1971) 245.
32. C. Guet, C. Signarbieux, P. Perrin, H. Nifenecker, M. Asghar, F. Caïtucoli and B. Leroux, Nucl. Phys. A314 (1979) 1.
33. R. Choudhury, S. Kapoor, D. Nadkarni and P. Rama Rao, Nucl. Phys. A346 (1980) 473.
34. J. Pannicke, M. Mutterer, J. Theobald, P. Heeg, K. Wein-gärtner, G. Barreau, B. Leroux and F. Gönnenwein, Proc. Int. Conf. on Nuclear Data for Basic and Applied Science, Santa Fé, 1985.
35. N. Cârjan, J. Phys. 37 (1976) 1279.
36. C. Wagemans, P. D'hondt, P. Schillebeeckx and R. Brissot, Nucl. Phys., submitted for publication.
37. C. Wagemans, E. Allaert, F. Caïtucoli, P. D'hondt, G. Barreau and P. Perrin, Nucl. Phys. A369 (1981) 1.
38. V. Viola and G. Seaborg, J. Inorg. Nucl. Chem. 28 (1966) 741.

PECULIARITIES OF MEAN KINETIC ENERGY DEPENDENCE OF FRAGMENTS AT NUCLEI FISSION BY NEUTRONS

A. A. Goverdovsky, B. D. Kuzminov, V. F. Mitrofanov
and A. I. Sergachev

Institute of Physics and Power Engineering, Obninsk, USSR

ABSTRACT. Mean kinetic energy $\overline{E}_k$ of fission fragments from the fission of ^{234}U by 0.8 MeV neutrons has been measured. The results suggest a decrease of $\overline{E}_k$ if the fission goes through a vibrational resonance.

Mean kinetic energy of fission fragments $\overline{E}_k$ is basically determined by the Coulomb repulsive energy in the instant of their formation. However, the $\overline{E}_k$ effects due to nuclei dynamics, i.e. the phenomena occurring at an earlier stage of fission than a discontinuity point, can be influenced somehow.

Here, it is interesting to investigate the vibrational state effects in the second potential well, which represent collective oscillations of fissile degree of freedom upon various fission process responses. The reflection of these resonances in nuclear fissility was found both in fission reactions by neutrons /3/ and in direct reactions /4/.

The present work involves the $\overline{E}_k$ measurements at ^{234}U fission by neutrons at energies in the vibrational resonance region E_n = 0.8 MeV. The measurements employed a ^{234}U onesided target. The working layer thickness was $\sim$ 60 $\mu g/cm^2$. The target was 2 cm in diameter and contained 6% of ^{235}U adopted for energy scale calibration. The mean kinetic energy of fragments at ^{235}U fission by thermal neutrons was assumed to be 172.25 MeV. The $\overline{E}_k$ measurements were carried out in "tight" geometry using a surface-barrier silicon detector with a working surface — 2 cm in diameter and located at a 2 cm distance from the

[234]U target. The measurements were performed for two positions of the detector and the target about the neutron beam direction: in the first position, the recorded fragments moved in the direction of neutron motion, in the second one – against it. These allowed the contribution to $\overline{E}_k$ of fissile system mass centre motion to be determined.

In order to increase reliability of results, the [234]U fission cross-sections (about [235]U σ_f) and angular anisotropy in the same neutron energy range were measured.

The results obtained are given in Fig. 1. In the vibrational resonance region in a fission cross-section at E_n = 0.8 MeV the kinetic energy of fragments decreased by 0.6 MeV. Here an essential growth of angular anisotropy A, is observed. Hence, the correlation of σ_f, A and $\overline{E}_k$ takes place that gives grounds for assuming on the connection of $\overline{E}_k$ measurements with fission through vibrational resonance. The angular anisotropy value and

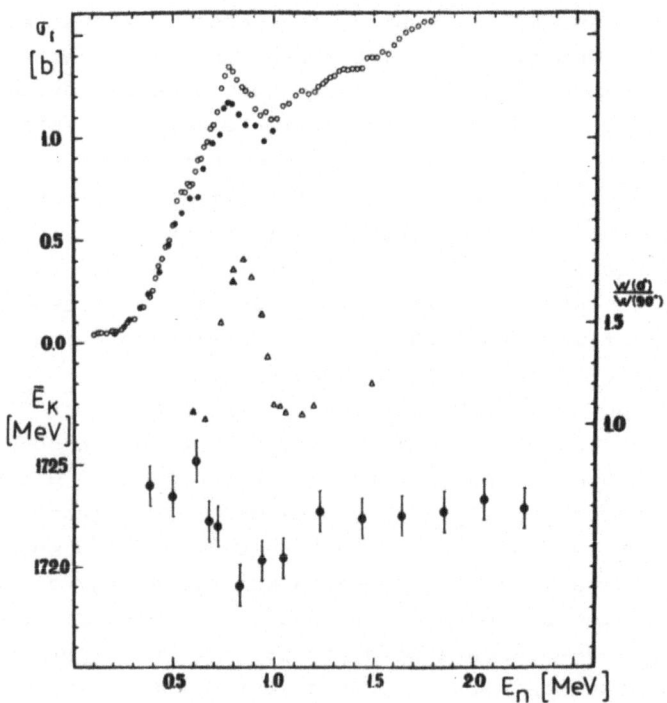

FIG. 1. Experimental results: σ_f, A and $\overline{E}_k$ •, ▲ – present work, ○ – /1/, △ – /2/.

the form of fission cross-section energy dependence indicate
possibility of non-resonance fission impurity in the energy
range in question.

Under these conditions, the kinetic energies of fragments
measured at 0° and 90° to the direction of the neutron motion
must differ according to the difference of contributions of resonance and non-resonance fissions in these directions. The resonance fission contribution increases at 0° and decreases at
90°. These investigations were performed. Angular resolution
was $\pm 10^o$. The measured fragment kinetic energy values $\overline{E}_k$ moving at 0° and 90° to the direction of the neutron motion are
given in Table 1.

TABLE 1

E_n (MeV)	$\overline{E}_k$ (0°) (MeV)	$\overline{E}_k$ (90°) (MeV)	A
0.6	172.43±0.38	171.90±0.50	1.07±0.03
0.8	170.98±0.35	172.18±0.35	1.70±0.05

At the same time the angular anisotropy of fission fragments
was measured. The measured results, A, obtained in the present work (see Table 1) and the result of ref. /2/ agree within the limit of errors. This is to say, $\overline{E}_k$ measured in this
work do correspond to the fragments moving at 0° and 90° to the
direction of the neutron motion. The results of these measurements indicate that it is just the fission through vibrational
resonance that the lower kinetic energy of fragments corresponds to.

The reduction of $\overline{E}_k$ at a neutron energy corresponding to
the vibrational resonance excitation (E_n = 0.3 MeV) (Fig. 2),
was obtained when the energy dependence of mean kinetic energy
of ^{236}U nucleus target fragments was investigated in ref. /3/.

The results of ref. /4/ (see Fig. 3), obtained in the investigation of the ^{233}U(p,pf) reaction also point to the $\overline{E}_k$ reduction at the excitation energy corresponding to the vibrational resonance.

Thus, one may note the common phenomena of $\overline{E}_k$ reduction at fission through the vibrational resonances for uranium isotopes. More elongated forms of a nucleus in the moment of rupture is likely to correspond to fission through the vibrational resonance.

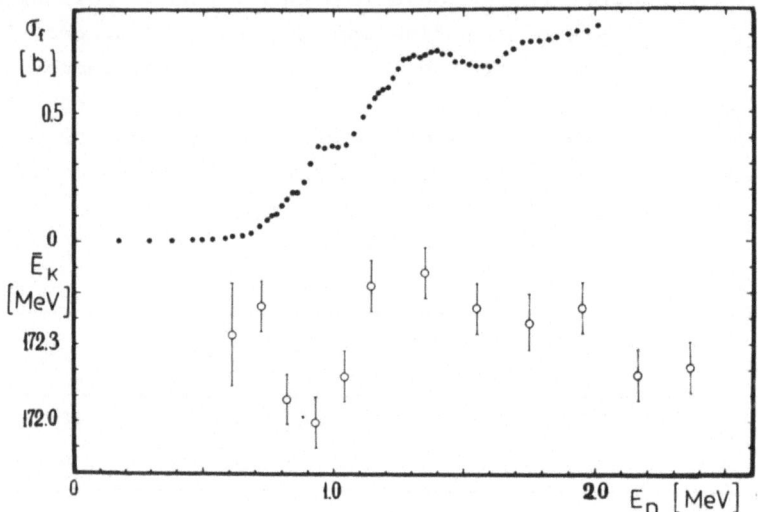

FIG. 2. ^{236}U fission cross-section /1/; mean kinetic energy of ^{237}U fission fragments in ^{236}U(n,f) reaction /3/.

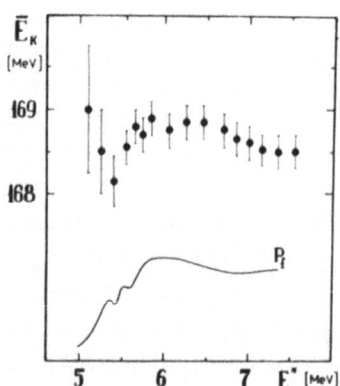

FIG. 3. Fission probability and $\overline{E}_k$ of ^{234}U fission fragments in ^{233}U(d,pf) reaction /4/.

To refine the nature of this phenomenon, in-depth investigations of the fragment kinetic energy at nuclei fission of various elements through the vibrational resonances are necessary.

REFERENCES

1. J. W. Behrens and G. W. Carlson, Nucl. Sci. Eng. $\underline{63}$ (1977) 250.
2. R. W. Lamphere, Proc. Int. Conference on Physics and Chemistry of Fission, Salzburg, 1965, Vol. 1, p. 63.
3. N. P. Dyachenko, B. D. Kuzminov et al., Yad. Fiz. $\underline{30}$ (1979) 904.
4. Y. Patin et al., Nucl. Instr. Meth. $\underline{160}$ (1979) 471.

DIFFERENTIAL STUDY OF ^{252}Cf(sf) NEUTRON EMISSION

H. Märten, D. Richter, D. Seeliger, *W. D. Fromm
and *D. Neubert

Technical University Dresden, GDR
*Central Institute for Nuclear Research, Rossendorf, GDR

ABSTRACT. The double-differential probability $N(E,\theta)$ of
^{252}Cf(sf) neutron emission (E - 1s neutron energy, θ - 1s
angle of neutron emission with reference to light-fragment
direction) was measured by the use of two parallel plate avalanche counters (one position-sensitive) for fragment detection (time of flight, direction) as well as two NE213 scintillators for neutron detection (time-of-flight, scintillator
response).
   The angular resolution amounts to 3°. The $N(E,\theta)$ data cover
the full angle range from 0 to 180° by steps of 1.5°. Preliminary results are compared with calculations performed in the
framework of the complex cascade evaporation model, i.e. evaporation from fully accelerated fragments is considered as
alone emission mechanism. A $N(E,\theta)$ valley appearing at $\theta \rightarrow 0^{\circ}$
and close to E_f (fragment kinetic energy per nucleon) is ex-
plained as a kinematic effect which was reproduced in the calculations.
   Specifically measured distributions are much more anisotropic than previous data at high energy.

1. INTRODUCTION

For the study of the mechanism of fission neutron emission,
precise measurements of multiple-differential emission probabilities and their comparison with calculations in the framework of statistical models which include the diversity of
fragment configurations as well as the characteristics of fission neutron emission (cascade emission) are required.

Only one rather complete measurement of $N(E,\theta)$ for ^{252}Cf(sf)
is known from literature (16 angle points) /1/. Further measurements of differential emission probabilities of fission neutrons had exhibited contradictions concerning scission neutrons
/1/ especially (cf. summary in ref. /2/). Here, we describe

a new experiment aimed at the measurement of N(E,θ) for the
whole angle range and the comparison with complex statistical
model approaches /3, 4/ which refer to the most probable emission mechanism, i.e. the evaporation from fully accelerated
fragments.

2. EXPERIMENTAL SET-UP

The neutron fragment correlation experiment illustrated in
Fig. 1 schematically relies on the use of a single and a position-sensitive parallel plate avalanche counter /5/ for the
measurement of fragment time of flight (TOF) and direction as
well as two NE213 scintillators (N1,N2) for neutron detection
(TOF, scintillator light output LO). Pulse shape discrimination
is applied for background suppression (gamma rays, cosmic muons).
The distinction between the light and heavy fragment group is
possible by fragment TOF spectroscopy. The multidimensional
measurement of at least neutron TOF (for two detectors), fragment TOF, and fragment direction (position) is carried out

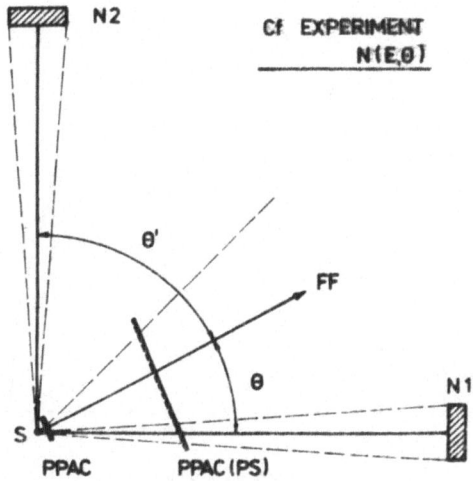

FIG. 1. Schematic representation of the experimental arrangement. Fission fragments (FF) are detected using two PPAC (one
position-sensitive) inside a chamber with low-pressure heptane.
N1 and N2 are neutron detectors at fixed positions.

applying a special data acquisition system (minicomputer-micro-
computer arrangement with magnetic disc unit).

Typical single spectra are shown in Figs. 2 and 3.

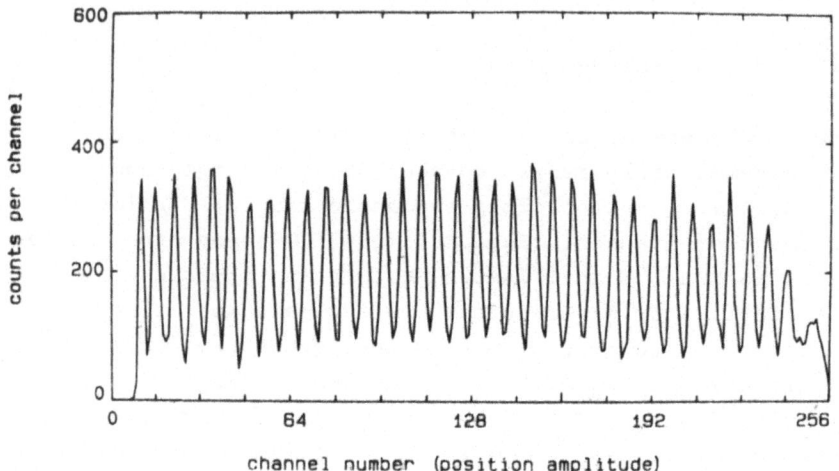

FIG. 2. Typical position amplitude spectrum.

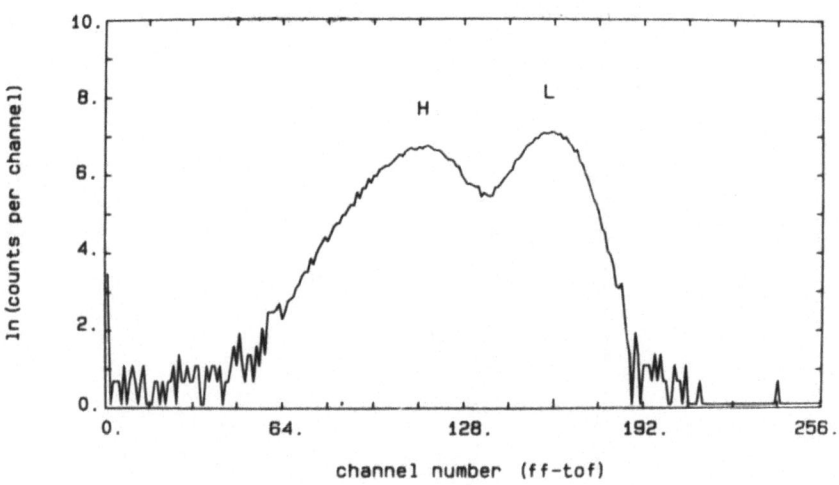

FIG. 3. Typical fragment TOF spectrum for a selected fragment
direction (L, H - light and heavy fragment group, respectively).

3. N(E,θ) RESULTS FOR ^{252}Cf(sf)

The first measurement of N(E,θ) has been carried out according
to Fig. 1 for a fixed LO bias (120 angle points, 3.1° angular
resolution). Non-concentrated data are shown in Figs. 4 and 5

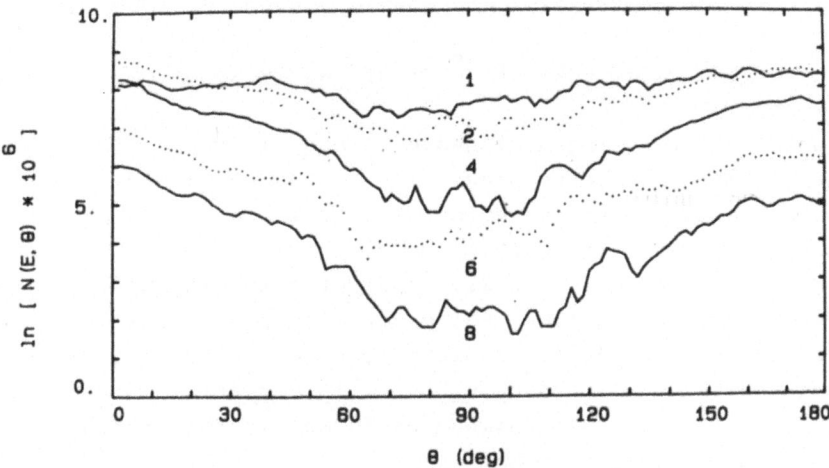

FIG. 4. Measured angular distributions of ^{252}Cf(sf) neutrons
shown for selected E (parameter in MeV).

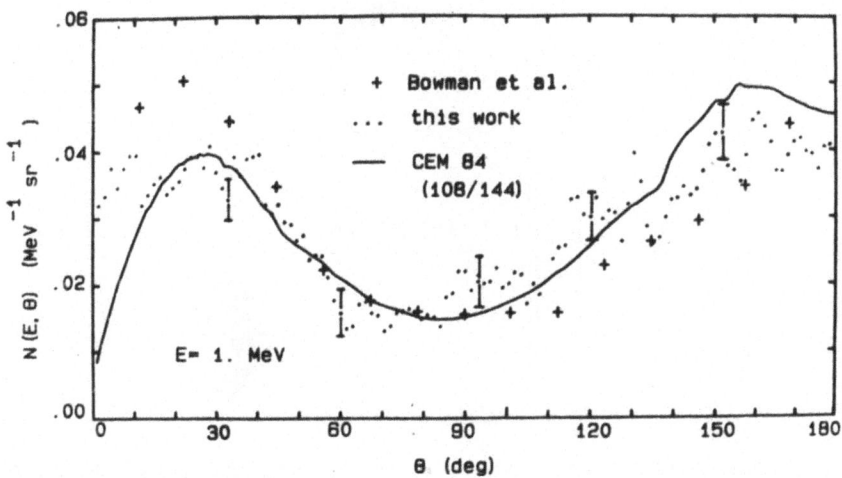

FIG. 5. Measured angular distribution at 1 MeV in comparison
with Bowman s data as well as a CEM calculation concerning
the most probable fragment mass split.

for selected energies. A N(E,θ) valley is indicated at θ→0°
and E close to E_f (fragment kinetic energy per nucleon). This
effect can be explained as a kinematic one (singular point
of the CMS-LS transformation formula /2/). It was reproduced
in the framework of the complex cascade evaporation model /3/
(CEM, see Fig. 5 concerning the most probable mass split 108/144)
as well as the generalized Madland–Nix model /4/ (GMNM).

4. THE 0°/90° ANISOTROPY OF ^{252}Cf(sf) NEUTRONS

Compared to the arrangement shown in Fig. 1, the position-sensitive PPAC was located perpendicular to the horizontal plane
between the source and detector N1 to measure the 90° spectrum
with better statistics (detector N2).

The experimental set-up is represented in Fig. 6.

The corresponding (TOF, LO) spectra have been analysed as
described in ref. /6/ (in addition correction for angular resolution). Results are shown in Fig. 7 in comparison with other
data /1, 7, 8/ and calculations performed in the framework of
the CEM as well as the GMNM.

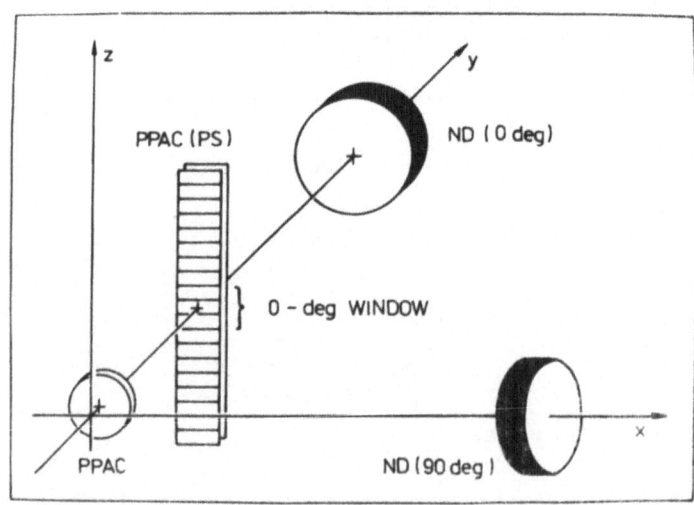

FIG. 6. Schematic representation of the experimental arrangement for the 0°/90° anisotropy measurement. The Cf source is
located in the zero point of the coordinate system.

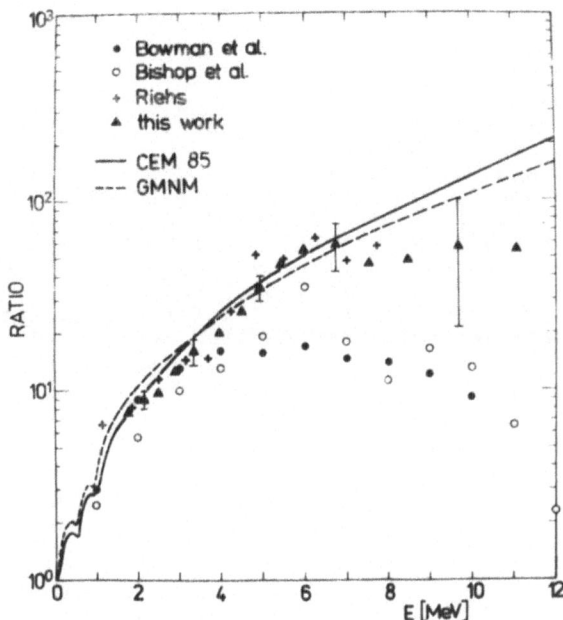

FIG. 7. 0°/90° anisotropy ratio of ^{252}Cf(sf) neutron emission versus E. Experimental data /1, 7, 8/ and calculated curves (CEM /3/, GMNM /4/) are represented for comparison (data deduced without distinction of light and heavy fragment group).

5. CONCLUSIONS

The described multiparameter experiment enables the effective measurement of $N(E,\theta)$ and $N(E,\theta:E_f)$. Systematic errors are avoided widely. The measured distributions are much more anisotropic at high E than previous data /1, 7/ (cf. Fig. 7). This trend has been confirmed by the CBNM group recently /9/. The energy dependence of the present 0°/90° intensity ratio is conformable to the assumption that all prompt fission neutrons are evaporated from the fully accelerated fragments. Specifically the existence of a hard scission neutron component which was concluded from an analysis /2/ of the Bowman data /1/ must be cancelled (cf. ref. /9/).

Further measurements and comprehensive comparisons between experiment and theory are in progress.

REFERENCES

1. H. R. Bowman et al., Phys. Rev. <u>126</u> (1962) 2120.
2. H. Märten et al.,Proc. IAEA Consultants' Meeting, Report INDC(NDS)-146/L, 1983, p. 199.
3. H. Märten and D. Seeliger, J. Phys. <u>G10</u> (1984) 349.
4. H. Märten and D. Seeliger, Report INDC(GDR)-30/L, 1984.
5. W. Neubert et al., Nucl. Instr. Meth. <u>204</u> (1983) 453.
6. H. Märten et al., Report INDC(GDR)-28/L, 1984.
7. C. J. Bishop et al., Nucl. Phys. <u>A198</u> (1972) 161.
8. P. Riehs, Acta Phys. Austr. <u>53</u> (1981) 271.
9. C. Budtz-Jorgensen and H.-H. Knitter, Proc. Int. Conf. on Nuclear Data for Basic and Applied Science, Santa Fé, 1985, in press.

SHELL EFFECTS IN THE SPONTANEOUS FISSION AND THE THERMAL NEUTRON INDUCED FISSION OF SEVERAL Pu-ISOTOPES

P. Schillebeeckx, C. Wagemans, *A. J. Deruytter
and *R. Barthélémy

Nuclear Physics Laboratory, Gent and SCK-CEN, Mol, Belgium
*CEC-JRC, CBNM, Geel Establishment, Geel, Belgium

ABSTRACT. The fission fragments energy and mass characteristics and their correlations have been studied for the spontaneous fission of ^{238}Pu, ^{240}Pu and ^{242}Pu and the thermal neutron induced fission of ^{239}Pu and ^{241}Pu. The results are interpreted in terms of the static scission point model.

1. INTRODUCTION

In the frame of a systematic study of the fission fragment mass and energy characteristics, the spontaneous fission of ^{238}Pu, ^{240}Pu, ^{242}Pu and the thermal neutron induced fission of ^{239}Pu and ^{241}Pu have been studied. The measurements were performed at the BR1-reactor of the Nuclear Energy Centre at Mol. The energy and mass characteristics of the fission fragments were determined with the so-called double-energy method /1/. In this method the fission fragments are detected by two colinear surface barrier detectors. The analysis was based on the mass- and momentum conservation relations and the Schmitt—Neiler calibration procedure, in which the detector constants can be determined from the pulse-height spectra of the ^{239}Pu or ^{241}Pu(n_{th},f) reaction. For the measurements, a homogeneous mixture of a spontaneously fissioning Pu-isotope and ^{239}Pu was used. The BR1-reactor was operated during the day and shut down at night and during the weekends. Such an operation allows one to measure a sequence of separate (sf)-(n_{th},f) runs, which can be analysed individually. In this way a very careful detector calibration with the ^{239}Pu(n_{th},f) reaction can be obtained.

2. RESULTS AND DISCUSSION

The results can be interpreted in terms of the static scission
point model of Wilkins /2/ and analogous calculations performed
by Moreau and Heyde /3/. In these calculations the probability
for the formation of a complementary pair of fission fragments
is determined by the potential energy of the system at the
scission point. This potential energy is a sum of liquid drop
terms and shell correction terms. It appears that strong negative neutron shell corrections as shown in Fig. 1 play an important role in the formation of the fragments. Taking into account
that the fission fragments conserve the charge/mass ratio of
the compound nucleus the most probable neutron number for a
given mass can be calculated.

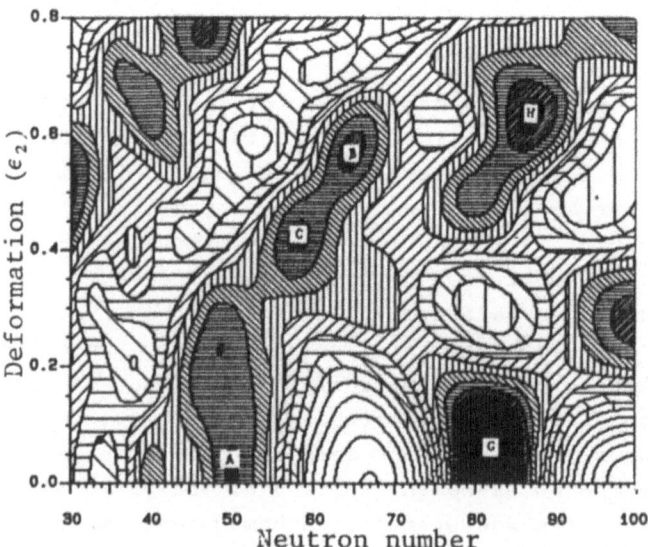

FIG. 1. Neutron shell corrections as a function of the
deformation and the neutron number /3/.

2.1. Comparison of the Mass Distributions for $238,240,242$Pu(sf)

Figure 2 shows the almost constant position of the heavy fragment mass peak, so that the increase of the mass of the compound

TABLE 1. Main energy and mass characteristics of the spontaneously fissioning plutonium isotopes, all calibrated relative to ^{239}Pu(n_{th},f)

	^{238}Pu(sf)	^{240}Pu(sf) /1/	^{242}Pu(sf)	^{239}Pu(n_{th},f)
$\bar{E}_k^\times$ (MeV)	176.5	179.1	180.4	177.7
$E_k^\times$	11.9	12.5	12.1	12.5
$E_l^\times$ (MeV)	103.1	103.4	102.9	103.3
$E_h^\times$ (MeV)	73.4	75.7	77.5	74.4
$n_h^\times$ (u)	139.2	138.7	138.2	139.7
$m_l^\times$ (u)	98.8	101.3	103.8	100.3
$6m_l^\times = 6m_h^\times$	6.0	5.7	5.4	6.0
N	1972	14842	11020	1.9×10^6

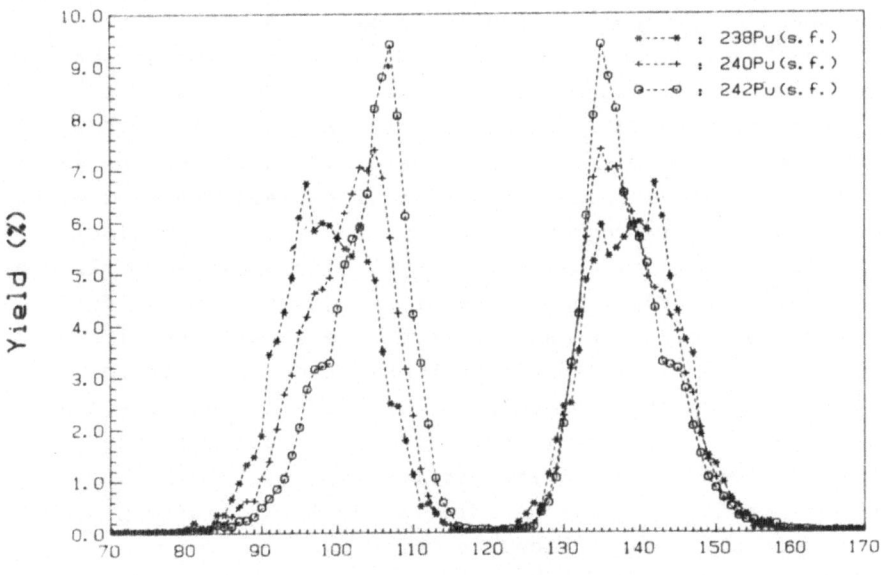

FIG. 2. Comparison of the mass yield distributions for ^{238}Pu, ^{240}Pu and ^{242}Pu(sf).

nucleus results in a shift of the light fragment peak. This is demonstrated numerically by the $\prec m_l^\times \succ$ and $\prec m_h^\times \succ$ values given in Table 1. Figure 2 also shows that many structures appear in

377

the mass distributions. These observations can be explained by
the shell effects mentioned above. In the symmetric mass region
there are no shell corrections available which can influence the
formation of the fragments. For this region one observes indeed
an almost zero yield in the mass distributions. For more asymmetric fission, the formation of the fragments will be mainly
influenced by the spherical neutron shell G and the deformed
neutron shell H in the heavy fragment. These shells cause the
constant position of the heavy fragment peak. Additional effects
may result from neutron shell corrections in the regions C-B in
the light fragment. The neutron shell corrections play an important role when the formation of both fragments is favoured by
shell effects. For ^{240}Pu(sf) and ^{242}Pu(sf) this leads to a large
peak in the mass region $m_h^x \sim 135$, due to the combined influence
of the shell G in the heavy and the shell C in the light fragment. For ^{238}Pu(sf) on the other hand, only the neutron shell G
plays a role in this mass region, which explains the smaller
peak yield observed in our measurements. However, for ^{238}Pu(sf)
we notice a high peak yield around mass 142, which can be explained by the preferential formation of a heavy fragment with
$N_h \sim 86$ and a light fragment with $N_l \sim 58$, due to the simultaneous influence of the shells H resp. B. The enhanced influence
of the shell B in the light fragment is less pronounced for
^{240}Pu(sf), but still the mass distribution shows a shoulder
around mass 142. For ^{242}Pu(sf) both shells do not coincide,
but each shell on its own causes a structure in the mass distribution: shell H in the mass region 140 and shell B in the mass
region 144. The strong decrease of the mass yields in the extreme asymmetric mass region reflects the lack of strong negative neutron shell corrections in this region.

2.2. Kinetic Energy Distributions for ^{238,240,242}Pu(sf)

A comparison of the total kinetic energy as a function of the
heavy fragment mass /Fig. 3/ shows the typical maximum in the
mass region 130—135. This maximum can be explained by the formation of spherical fragments due to the influence of the sphe-

rical neutron shell N $\sim$ 82, which is enhanced by the spherical proton shell Z $\sim$ 50.

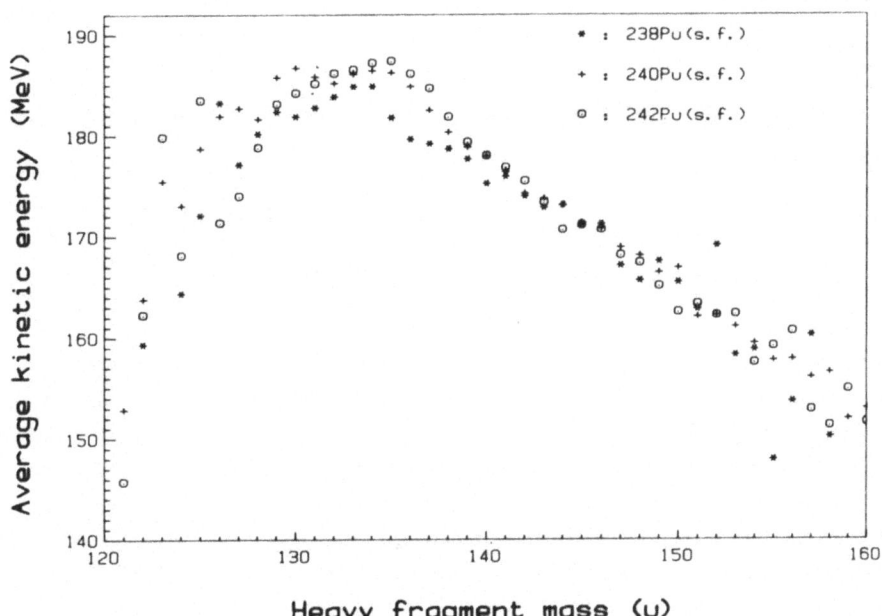

FIG. 3. Total fission fragment kinetic energy for ^{238}Pu, ^{240}Pu and ^{242}Pu(sf) as a function of the heavy fragment mass.

The coefficient of dissymetry (c.d.) or skewness of $< E_k^x >$ (m_h^x) is given in Fig. 4. One observes the highest dissymetry in the mass region 130—135, which can be explained by the presence of more than one configuration with different deformations for a given mass split m_h^x/m_l^x. This is illustrated in Fig. 5, showing the total kinetic energy distribution for mass split 135/105 for ^{240}Pu(sf) which has been decomposed in a sum of two distributions. Indeed, for a given mass split two configurations with different total deformation can appear: for mass $m_h^x \sim$ 135 spherical fragments can be formed due to the influence of the spherical neutron shell G, but also deformed fragments can be formed under the influence of an offshoot of region H. The formation of the light fragment can only be influenced by the presence of region C, which means that the light fragment has an equal deformation for both configurations of the heavy fragment.

379

Since the total kinetic energy is mainly due to the Coulomb repulsion of the two fragments, the total kinetic energy can be calculated as

$$E_k^{\times} = \frac{Z_1^{\times} Z_h^{\times} e^2 F}{D} \qquad (a)$$

with $Z_1^{\times}$ and $Z_h^{\times}$ the proton number, D the distance between the charge centres and F a shape factor. As in ref. /3/, D is calculated as

$$D = d + r_o m_1^{\times 1/3} \left[\frac{1 + 1/3\varepsilon_2^l}{1 - 2/3\varepsilon_2^l} \right]^{2/3} + r_o m_h^{\times 1/3} \left[\frac{1 + 1/3\varepsilon_2^h}{1 - 2/3\varepsilon_2^h} \right]^{2/3} \qquad (b)$$

with d a constant equal to 1.4 fm and ε_2^l and ε_2^h deformation parameters. From (a) and (b) the total kinetic energy for a given configuration $(m_1^{\times}, \varepsilon_2^l)$ and $(m_h^{\times}, \varepsilon_2^h)$ can be deduced taking into account the deformations calculated by Moreau and Heyde (Fig. 1). For a mass split 135/105 for ^{240}Pu(sf), e.g. one obtains in this way a total kinetic energy $E_k^{\times} \sim 193$ MeV for the configuration dominated by the neutron shell G, and $E_k^{\times} \sim 173$ MeV

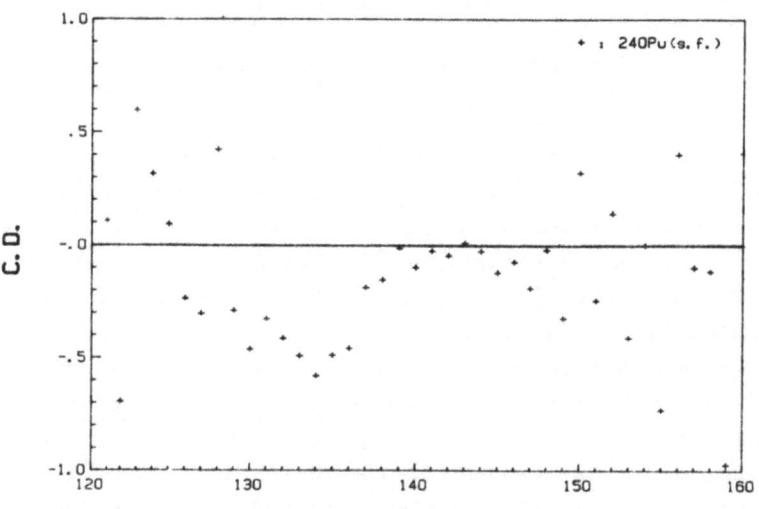

Heavy fragment mass (u)

FIG. 4. Coefficient of dissymmetry of $\prec E_k^{\times} \succ (m_h^{\times})$ for ^{240}Pu(sf) as a function of the heavy fragment mass.

for the configuration influenced by the region H. These calculated values are in good agreement with the average values from
the Gaussian fits (Fig. 5). Similar conclusions can be drawn
for the mass split 135/107 for ^{242}Pu(sf).

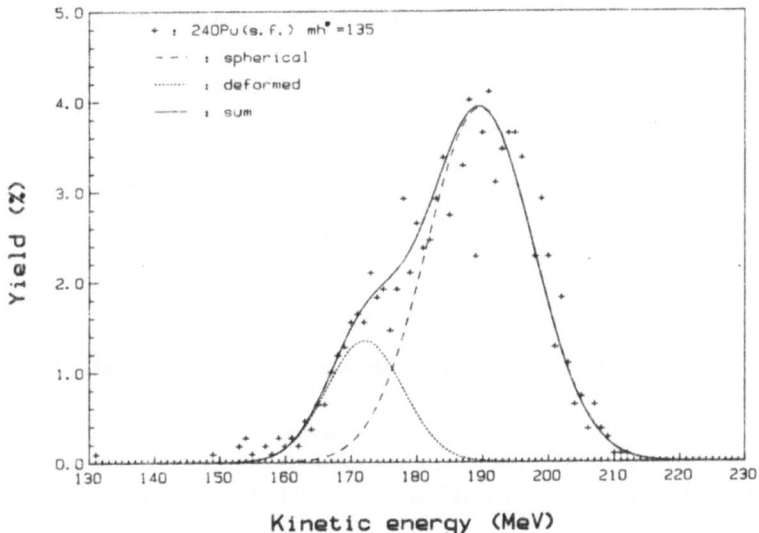

FIG. 5. Total kinetic energy distribution for mass split
135/105 for ^{240}Pu(sf).

2.3. Influence of the Excitation Energy on the Shell Effects

A comparison of the mass distribution of ^{240}Pu(sf) with
^{239}Pu(n_{th},f) (Fig. 6) shows a narrower mass distribution,
a much higher peak yield and more pronounced fine structures
for the spontaneous fission as compared to the neutron induced
fission, which reveals a decrease of the influence of the shell
corrections with increasing excitation energy. For the above-
mentioned systems, the difference of more than 6 MeV excitation
energy does not show up, the kinetic energy being even higher
for the spontaneous fission (cf. Fig. 7 and Table 1). This
points to the formation of more compact configurations for the
spontaneous as compared to the neutron induced fission and can
be attributed to a decrease of the shell corrections with increasing excitation energy of the compound nucleus. The same

381

conclusion can be drawn from a comparison of ^{242}Pu(sf) and ^{241}Pu(n_{th},f) /4/.

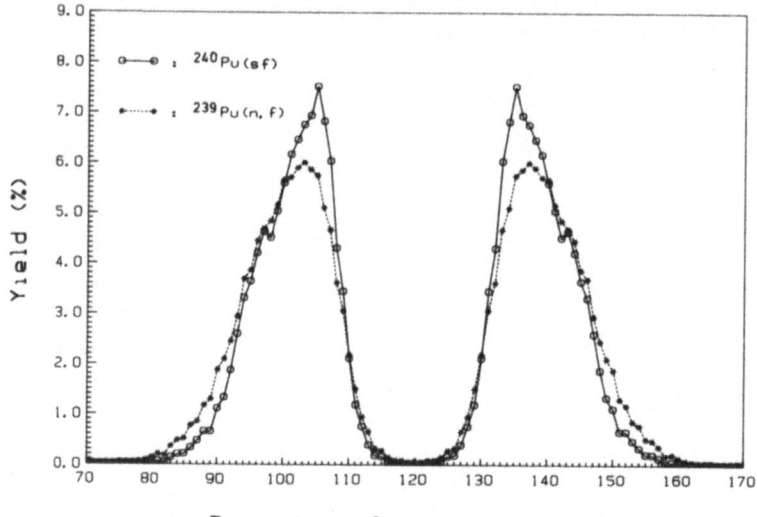

FIG. 6. Comparison of the mass yield distribution for ^{240}Pu(sf) and ^{239}Pu(n_{th},f).

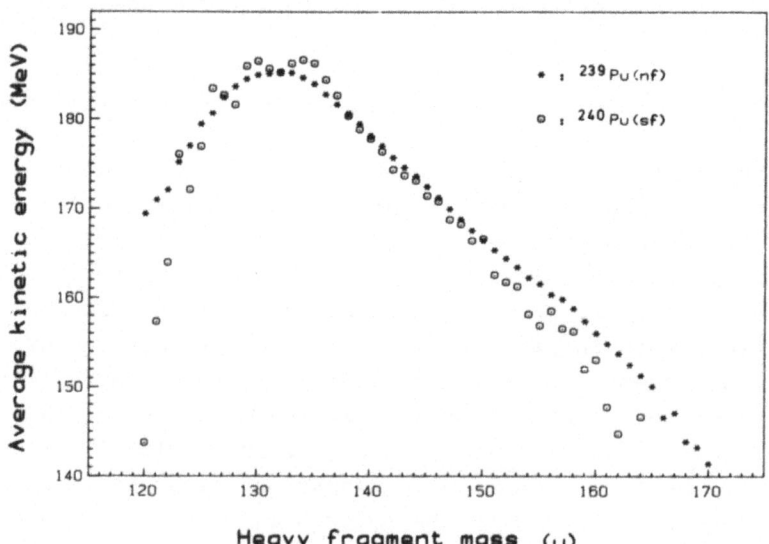

FIG. 7. Total fission fragment kinetic energy for ^{240}Pu(sf) and ^{239}Pu(n_{th},f) as a function of the heavy fragment mass.

3. CONCLUSIONS

The present results clearly illustrate that the formation of the fission fragments is very much influenced by the shell effects, especially by the strong spherical neutron shell N ∼ 82 but also by the deformed shell N ∼ 86. The influence is enhanced when <u>both</u> fragments have a stability caused by a neutron shell. It decreases, however, with increasing excitation energy of the fissioning system.

REFERENCES

1. C. Wagemans, E. Allaert, A. Deruytter, R. Barthélémy and P. Schillebeeckx, Phys. Rev. C30 (1984) 218.
2. B. Wilkins, E. Steinberg and R. Chasman, Phys. Rev. C14 (1976) 1832.
3. J. Moreau and K. Heyde, Phys. Mag. 5 (1983) 91.
4. E. Allaert, C. Wagemans, G. Wegner-Penning, A. Deruytter and R. Barthélémy, Nucl. Phys. A380 (1982) 61.

CROSS-SECTIONS OF ^{236}Np ISOMERS' FISSION BY THERMAL NEUTRONS

Yu. A. Selitsky, V. B. Funstein and V. A. Yakovlev

V. G. Khlopin Radium Institute, Leningrad, USSR

ABSTRACT. The cross-section of fission by thermal neutrons and the resonance integral of fission of the long-lived isomer ^{236}Np(l) are measured. The obtained value of the fission cross-section is compared with an analogous cross-section for the short-lived isomer ^{236}Np(s) measured by us. The dependence of the cross-section of fission by thermal neutrons on the spin of the target-nucleus is discussed.

1. INTRODUCTION

Recently we may witness an increasing interest in the study of the fission of odd-odd nuclei. This is stimulated by the fact that the phenomenon of isomerism is spread with the odd-odd nuclei, which enables us to study the fission of nuclei with identical nucleonic composition, differing only by a limited number of nuclear-physical characteristics. Besides, the odd-odd nuclides have bigger cross-sections of fission by thermal neutrons in comparison with the even-odd target-nuclei. Thus, for instance, the ^{242m}Am fission cross-section reaches 6950 ± 250 b /1/.

In work /2/ an analysis was carried out of the dependence of the cross-sections of 12 odd-odd nuclei fission by thermal neutrons on their spin and the density of compound nuclei levels. It was found that bigger cross-sections corresponded to bigger spins but no connection was discovered between the cross-section value and the compound nucleus levels density.

In our earlier work /3/ we measured the cross-section of the short-lived isomer ^{236}Np fission by thermal neutrons /3/. The fission of ^{236}Np(s) was safely identified by the drop of

fission activity in accordance with the known half-life of
^{236}Np(s) ($T_{1/2}$ = 22.5±0.4 h /4/). The obtained fission cross-
section value $\sigma_f^{Np(s)}$ = 2740±140 b appeared close to the "known-
from-literature value" of the long-lived isomer fission cross-
section $\sigma_f^{Np(1)}$ = 2500±150 b /5/, which doesn't conform with
the affirmation of ref. /2/, since ^{236}Np(s) and ^{236}Np(1) have
considerably different spins, i.e. 1$^-$ and 6$^-$, respectively /6/.
To be able to check more reliably the correlation found in /2/
we repeated the measurement of the "known-from-literature cross-
section" of the long-lived isomer ^{236}Np(1) fission, using similar methods of chemical purification of the substance and of
fragment registration as in the case of measurement of the
short-lived isomer ^{236}Np(s) fission cross-section.

2. EXPERIMENT

The isotope ^{236}Np(1) was obtained by the reaction of ^{238}U(p,3n)
where it was formed with the maximum yield /7—9/. The irradiation was carried out with protons with the energy 30 MeV and
the average current 8 μA for 60 h on the U-150 cyclotron at the
Institute of Nuclear Physics, Alma-Ata.

After cooling for 6 months ^{236}Np(1) was chemically extracted from the uranium target. To purify ^{236}Np(1) from uranium, fission products and ^{236}Pu the method of extraction chro-
matography was applied /3/. The purification coefficients were
~ 3 x 10^7, ~ 10^5 and ~ 10^6, respectively. The purified neptunium was separated by the electrolytic method on a stainless
steel backing.

The calculation of the mass of the substance on the basis
of the mass-spectrometric analysis (MSA) determining the isotopic ratio of ^{236}Np(1)/^{237}Np and the measurement of alpha
activity of ^{237}Np in the target was the first step in all studies of ^{236}Np(1). The data on absolute intensities of gamma
radiation of ^{236}Np(1) with the energies 158.3 and 160.3 keV /6/
made it possible to measure the mass of ^{236}Np(1) immediately
using its gamma radiation. As a result of gamma measurements by
means of a Ge(Li) detector the value of ^{236}Np(1) mass was
obtained equal to 37.9±1.5 ng.

In order to increase the reliability of the ^{236}Np(1) mass determination the standard method based on the measurement of the isotopic ratio of ^{236}Np(1)/^{237}Np and the alpha-activity of ^{237}Np was performed. The analysis of the isotopic composition of nanogramme quantities of a substance had been described earlier in ref. /10/. The relative content of neptunium isotopes in the target was ^{235}Np:^{236}Np(1):^{237}Np = (0.053±0.004):(0.1385 ±0.0014):1 at the time of irradiation in the reactor. The mass of ^{237}Np in the target, measured by the alpha-radiation intensity, was 254±8 ng and, therefore, of ^{236}Np(1) - 35.2±1.2 ng. The averaged mass of ^{236}Np(1), determined on the basis of the two methods, was 36.2±1.3 ng.

An insignificant admixture of ^{235}U, ^{238}U and ^{236}Pu nuclides was present in the target together with neptunium isotopes. The correction for the fission of ^{235}U was 1.6% from the fission of ^{236}Np(1). The contribution from the fission of the admixture of other nuclides did not exceed 0.01%.

The measurement of the ^{236}Np(1) fission cross-section was done relative to ^{235}U(n,f) in an outward horizontal neutron beam with a Maxwellian energy spectrum (T = 330 K) and a flux of 2×10^8 cm^{-2}s^{-1} (on the WWR-M reactor of the Leningrad Institute of Nuclear Physics). The neptunium and uranium (525 ng of ^{235}U) targets were placed in a thin walled duralumin container, with active layers close to each other, mica detectors being put between them. The cadmium ration by ^{235}U(n,f) was R = 42.7±2.4, cadmium thickness being 1 mm. Because of the presence of the overthermal component in the neutron spectrum, an additional irradiation was carried out with a cadmium protection.

3. RESULTS

The cross-section of ^{236}Np(1) fission by thermal neutrons was determined relative to the cross-section of ^{235}U fission by Maxwellian spectrum neutrons, $\sigma_f^U = \sigma_o g_T$, where σ_o was the cross-section of ^{235}U fission by neutrons with the energy 0.0253 eV, taken to equal 582.6±1.1 b /11/; g_T - Weskott factor, equal to 0.9688 for ^{235}U at T = 330 K /12/. The correction accounting for the contribution from neptunium and uranium fission by over-cadmium neutrons was 0.4%. The value of $\sigma_f^{Np(l)}$ appeared to be equal to 2760±170 b. The main contribution to the mean-square error of the cross-section (6%) was introduced by the casual component of the error in the ratio of the number of fissions in the neptunium and the uranium targets (4%) and the error in the number of nuclei of ^{236}Np(l) (3.5%) and ^{235}U (3%). The obtained value of $\sigma_f^{Np(l)}$ was in satisfactory agreement with the cross-section of 2500±150 b known from ref. /5/ and with the other less precise measurements /13—16/.

In the experiments with the cadmium shield the value of the over-cadmium ($E_n > 0.6$ eV) resonance integral of ^{236}Np(l) fission was determined to be equal to $I_f^{Np(l)} = 1030±100$ b. In calculating the $I_f^{Np(l)}$ the resonance integral of ^{235}U fission was taken to equal 275±5 b /17/; the correction for the fission of ^{237}Np and ^{235}U admixtures in the neptunium target was 4%.

4. DISCUSSION

As already mentioned in the introduction, the cross-sections of fission by thermal neutrons of odd-odd nuclei are bigger than those of even-odd nuclei. The "known-from-literature va-lues" of the cross-sections of odd-odd and even-odd target-nu-clei fission by thermal neutrons are presented in Fig. 1, in-cluding our data on the long-living and short-living isomers of ^{236}Np. As one can see from the figure, the nuclides are di-vided into two groups by the straight line, above which there appear the fission cross-sections of odd-odd nuclei. Different suppositions have been expressed in the literature about the causes of such high values of the cross-sections of odd-odd nuclei fission by thermal neutrons. Those suppositions are mainly based on the notions of high density of the levels of the compound fissile nuclei in these cases /2, 18, 19/. The in-creased density of the levels of the compound nucleus can lead to a big cross-section of fission by thermal neutrons because of a number of factors. The cross-section value is influenced by closely situated resonances in the energy dependence of the cross-section. At a high density of the levels of a compound

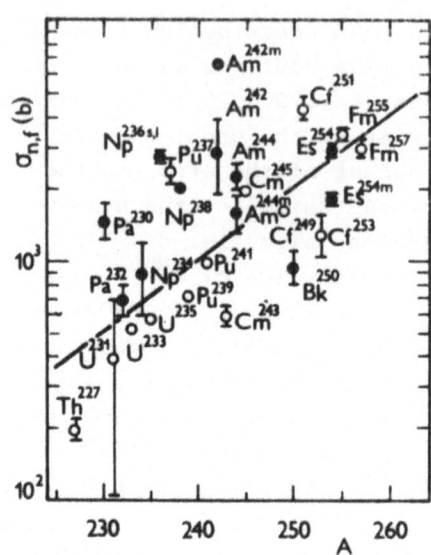

FIG. 1. Cross-sections of fission by thermal neutrons of even-odd (o) and odd-odd (•) nuclei /14, 17/. The cross-section of ^{236}Np(s) was measured by us in ref. /3/ and that of ^{236}Np(l) - in the present work.

nucleus the probability of such resonances in the region of small neutron energies increases /2, 19/. Besides, the odd-even nuclei have a higher density of transition states on the fission barrier than the even-even nuclei (see ref. /20/) and, therefore, the number of open fission channels increases.

In ref. /2/ no correlation was discovered between the cross-sections of fission by thermal neutrons and the density of compound nuclei levels in the analysis of the full totality of odd-odd nuclei. Such an analysis becomes more precise if we consider the pairs of odd-odd nuclear isomers which have the same nucleonic composition. In this case the relative value of the density of the levels is only influenced by the difference in the spins: $\varrho(I) \sim (2I+1)\exp[-(I+1/2)^2/2\mathfrak{G}^2]$. At the energy of the compound nucleus excitation equal to the energy of neutron binding, $\mathfrak{G} = 5-6$ /20/ and, therefore, the density of the levels appears to be relatively higher for the nucleus with a

bigger spin. The cross-sections of the isomers of ^{242}Am, ^{244}Am
and ^{254}Es are correlated with the spin dependence of the levels
density, but the cross-sections of ^{236}Np(s) and ^{236}Np(l),
measured by us, don't confirm it. It seems that this deviation
is stipulated by such a casual element as is the degree of
resonance proximity to the thermal energy of neutrons.

REFERENCES

1. J. W. T. Dabbs, C. E. Bemis, S. Raman, Jr. et al., Nucl.
   Sci. Eng. 84 (1983) 1.
2. H. Diamond, J. J. Hines, R. K. Sjoblom et al., J. Inorg.
   Nucl. Chem. 30 (1968) 2553.
3. E. A. Gromova, S. S. Kovalenko, Yu. A. Nemilov, Yu. A.
   Selitsky, A. V. Stepanov, A. M. Fridkin, V. B. Funshtein,
   V. A. Yakovlev, G. V. Valsky and G. A. Petrov, At. Energ.
   56 (1984) 212.
4. M. Schmorak, Nucl. Data Sheets 36 (1982) 367.
5. A. Jaffey, Report ANL-6600, 1961, 125.
6. M. Lindner et al., J. Inorg. Nucl. Chem. 43 (1981) 3071.
7. G. McCormick and B. Cohen, Phys. Rev. 96 (1954) 722.
8. M. Lefort, C.R. Acad. Sci., Ser. B 253 (1961) 2221.
9. I. Jenkine and A. Wain, Report AERE-R5790, 1968; Int. J.
   Appl. Radiat. Isotopes 22 (1977) 429.
10. B. N. Belyaev, T. P. Makarova, A. M. Fridkin et al.,
    Radiochimia 24 (1982) 185.
11. M. Divadeenam and J. R. Stehn, Ann. Nucl. Energy 11 (1984)
    375.
12. E. M. Gryntakis and J. I. Kim, Radiochim. Acta 22 (1975)
    128.
13. M. H. Studier, J. E. Gindler and C. M. Stevens, Phys. Rev.
    97 (1955) 88.
14. D. J. Hughes and R. B. Schwartz, Neutron Cross Sections,
    BNL-325, 1958.
    D. J. Hughes, B. A. Magurno and M. K. Brussel, Neutron
    Cross Sections, BNL-325, Suppl. 1, 1960.
15. C. M. Lederer and V. Shirley, "Table of Isotopes". J. Wiley
    and Sons, New York, 1978.
16. J. Gindler, L. Glendenin, E. Krapp et al., J. Inorg. Nucl.
    Chem. 43 (1981) 445.
17. S. F. Mughabghab and D. I. Garber, Neutron Cross Sections,
    Vol. 1, Resonance Parameters, BNL-325.
18. C. D. Bowman, G. F. Auchampaugh, S. C. Fultz and R. W.
    Hoff, Phys. Rev. 166 (1968) 1219.
19. J. C. Browne, R. M. White, R. E. Howe et al., Phys. Rev.
    C29 (1984) 2188.
20. S. Bjornholm and J. E. Lynn, Rev. Mod. Phys. 52 (1980) 725.

^{235}U(n,f) FRAGMENT MASS, KINETIC ENERGY AND ANGULAR DISTRIBUTION

Ch. Straede, C. Budtz-Jørgensen and H.-H. Knitter

Commission of the European Communities, CBNM Geel, Belgium

ABSTRACT. Fission fragment energies and the emission angles with respect to the incident neutron beam were measured for ^{235}U(n,f) using a twin ionization chamber. A neutron energy range from thermal to 6 MeV in steps of 0.5 MeV was covered. The preneutron emission mass distributions were evaluated and described by a five-Gaussian representation. The parametrization made visible a sudden change in the incident neutron energy dependence for all but the symmetric fission parameters between 1.5 and 2.0 MeV, which might be due to pair breaking. From the total kinetic energy as function of mass split the sum of the deformations of the two fragments was evaluated in the frame of Wilkins static scission point model. Here clear proton pairing effects became visible which seem to make the even-proton fragments less deformed than neighbouring odd-proton fragments. The present data show also that the observed drop in TKE averaged over all fragments above 4 MeV is due to changes in TKE for different mass splits with excitation energy. Angular distributions for the different fragment masses showed no mass dependence.

1. INTRODUCTION

For the thermal neutron induced fission of ^{235}U several measurements of fragment mass and kinetic energy distributions exist. For the understanding especially of the fission dynamics the changes of these distributions with the excitation energy of the compound nucleus ^{236}U are of interest. They have not yet been measured systematically. Therefore, the mass, kinetic energy and angular distributions of fission fragments from the ^{235}U(n,f) with incident neutron energies from thermal to 6 MeV were measured in steps of 0.5 MeV using the Van de Graaff accelerator of CBNM as a neutron source. About 1.5×10^7 fission

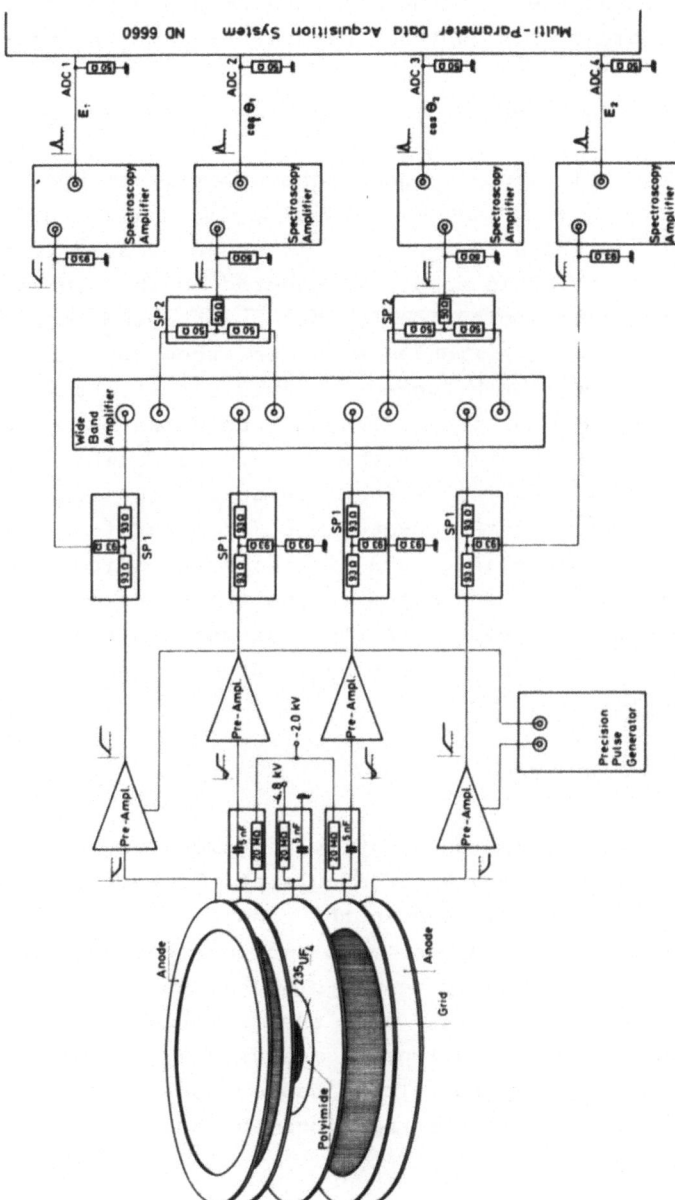

FIG. 1. Experimental setup.

events were recorded for thermal fission and $\sim 10^5$ events for each measurement with neutron energies higher than thermal.

2. EXPERIMENTAL SETUP

The experimental setup is shown schematically in Fig. 1. The ^{235}UF$_4$-layer is positioned on a thin polyimide backing which was covered by about 20 μg/cm^2 Au in order to make it electrically conducting. This sample is mounted in the centre of the cathode of a twin ionization chamber as proposed in /1/. The two fragments of each fission event are then detected simultaneously. The inserted Frisch-grids ensure that the anode signals are proportional to the fragment energies E_1 and E_2. The grid signal is a function of the E_i and $\cos \vartheta_i$ where ϑ_i is the angle between the normal of the cathode and the ion trace made by the fragment i in the counter gas /2/. Therefore the detector gives four signals for one fission event which determine the two fission fragment energies and the emission angles of the fragments with respect to the incident neutron beam. The acquired data were saved event by event on magnetic tape for further evaluation.

3. MASS DISTRIBUTIONS

From the measured fragment kinetic energies the preneutron emission mass distributions were calculated. Corrections were applied for the energy losses in the sample, for the neutron emission as function of fragment mass and incident neutron energy, and for the laboratory to centre-of-mass system transformation at incident energies larger than thermal. Figure 2 shows as examples the mass distributions as measured at thermal and 6 MeV incident neutron energies. The measured mass distributions are fitted with five Gauss distributions as indicated by the dashed curves in Fig. 2

$$Y(M) = \sum_{i=1}^{5} W_i/((2\pi)^{1/2}\sigma_i) \exp(-(M - M_{0i})/(2\sigma_i^2)) ,$$

where M is the fragment mass, M_{0i} is the centre of Gauss function i, W_i gives the weight of Gauss function i, σ_i^2 is the
variance of Gauss function i. (For all i = 1, 2, 3, 4, 5.)

It is assumed that M_{03} = 118, M_{05} = 236 - M_{01}, M_{04} = 236
- M_{02}, W_5 = W_1, W_4 = W_2, $\sum_{i=1}^{5} W_i$ = 200, $\sigma_5 = \sigma_1$ and $\sigma_2 = \sigma_4$.

The full line in Fig. 2 represents the result of a least
squares fit. The seven parameters $W_{1,2,3}$, $M_{01,2}$, $\sigma_{1,2}$ are
shown as function of the incident neutron energy in Fig. 3.

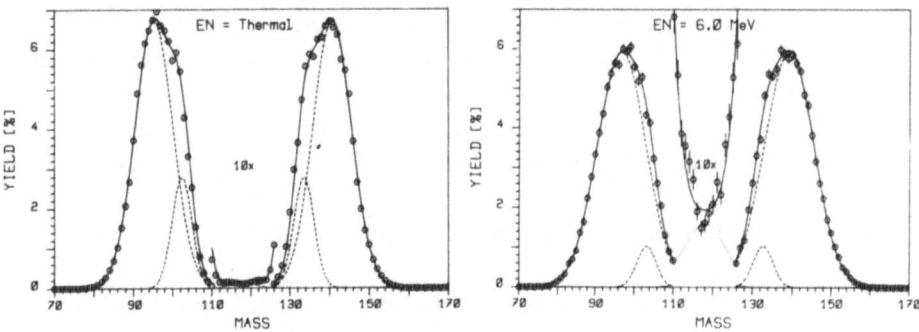

FIG. 2. Fragment mass distributions.

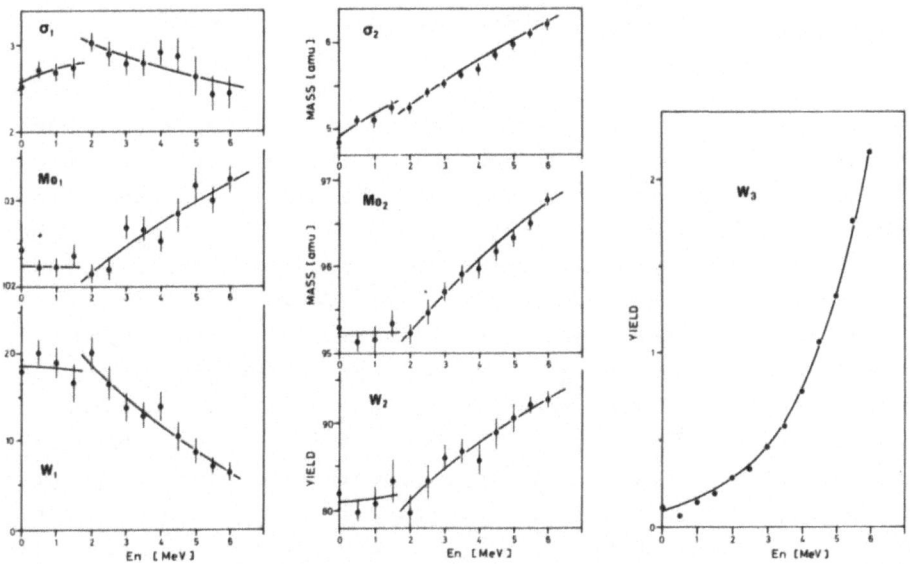

FIG. 3. Mass distribution fit parameters.

The increase of the symmetric yield is exponential. The other parameters are expected to vary linearly with the nuclear temperature, θ , at the saddle point /4/ and according to /4/ θ is at the saddle proportional to the square root of the excitation energy above the barrier $\theta \propto (E_n+B_n-E_f)^{1/2} = (E_n[\text{MeV}] + 1)^{1/2}$. Since all the parameters, as shown in Fig. 3, show a discontinuity at a neutron energy between 1.5 and 2.0 MeV, these data were fitted separately above and below this boundary with the expression $W_i M_i \bar{\sigma}_i = a_i + b_i (E_n [\text{MeV}] + 1)^{1/2}$. The results are shown as full lines in Fig. 3. The most striking behaviour of these parameters is the almost constant value for $E_n \leq 1.5$ MeV and much stronger dependence on excitation energy for $E_n \geq 2$ MeV. This sudden change might be connected with the onset of pair breaking. No previous mention of this sudden change in the mass distribution was found in the literature.

4. TOTAL KINETIC ENERGY AND FRAGMENT DEFORMATION

The total fragment kinetic energy is closely connected to the fragment deformations β_1 and β_2, since the deformations determine the average distance between the charge of the fragments at the moment of scission and therefore the Coulomb repulsion energy. The total kinetic energy is thus given by

$$TKE = Z_1 Z_2 / D(\beta_1, \beta_2) ,$$

where Z_1 and Z_2 are the proton numbers of the fragments 1 and 2, respectively and D is the average distance between the charge of the two fragments at scission. In the static scission point model /5/ D is given by $a_1(\beta_1) + a_2(\beta_2) + d$. a_1 and a_2 are the major axis of the prolate spheroids representing the deformed fragments. d is chosen as in /5/ to be 1.4 fm.

The total kinetic energy and its variations with incident neutron energy can then give information on changes in the deformation of the fragments at the scission point, and hence give a measure of the potential energy surface as function of mass split and excitation energy. Assuming that both fragments are equally deformed at the scission point, it is then possible to find the total deformation $\beta_{tot} = 2\beta$ for a given mass split by solving the above equation for TKE with respect to β .

Figure 4 shows the sum of the quadrupole deformations of the two fragments as function of mass split for thermal neutron induced fission of ^{235}U. The deformations are calculated with the above formula from the average TKE-values measured as function of mass split. The full triangles are the values from /5/ who calculated β_1 and β_2 for the mass splits 118/118, 102/134 and 96/140. The hatched areas and the vertical lines indicate masses with high predominance of fragments with even proton numbers. Figure 4 shows clearly the local dips in the total deformation caused by the proton pairing effect. The deformation as function of mass split can be understood from the proton and neutron shell corrections as given in /5/. The different shell, pairing and liquid drop terms interplay of course to give the prefixed deformation also as function of the incident neutron energy. However, these details cannot be discussed here /2/.

The total kinetic energy averaged over all fragments compared to the thermal value of (170.6±1.0) MeV, as found in the present experiment, is plotted as function of the incident neutron energy in Fig. 5 as full line. The open circles indicate the $\Delta \overline{\text{TKE}}$-values which are calculated if the change were caused

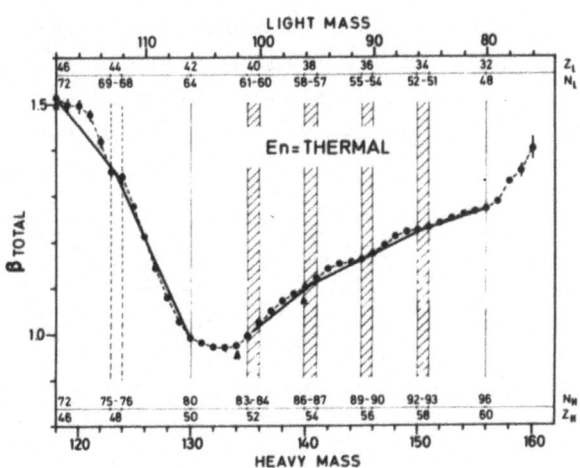

FIG. 4. Sum of quadrupole deformations of both fragments as function of mass split for ^{235}U(n,f). Region of more than 75% yield with even proton number, position of fragments with even proton number /4/.

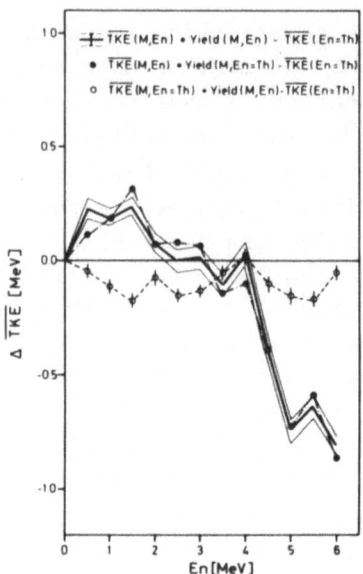

FIG. 5. Change of $\overline{TKE}$ with neutron energy.

by a change in the mass distribution only. The sudden drop at
4 to 4.5 MeV in the experimental TKE which was already observed
earlier /6/, is clearly due to changes of TKE for different
mass splits. This can be seen from Fig. 6, which gives an example for $[TKE(E_n) - TKE(thermal)]$ as function of mass split.
In this comparison to thermal energy a clear drop is seen in

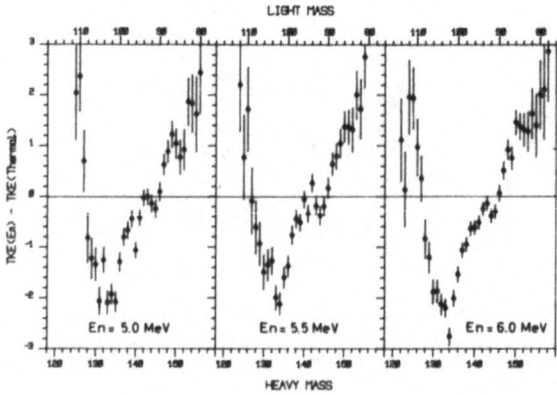

FIG. 6. Examples of $[TKE(E_n) - TKE(therm.)]$ versus mass split.

the region of the mass splits around 102/134 and some smaller
structures at other mass splits and changes in the wings of
the mass distribution.

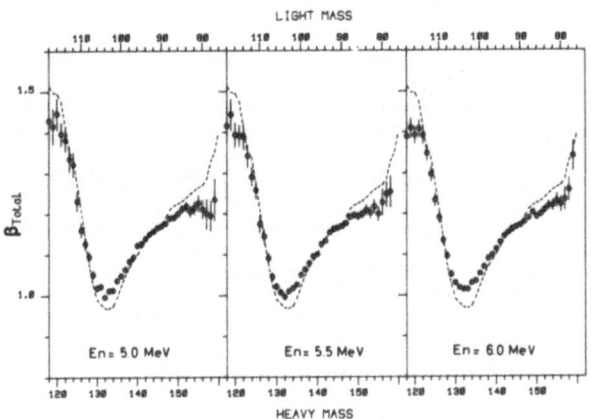

FIG. 7. Examples of total deformation versus mass split.
Dashed line E_n = therm.

Figure 7 shows some examples of the changes of the total
deformation of the fragments as function of energy. The comparison is made relative to the thermal incident neutron energy.
For the symmetric mass split a large decrease of fragment deformation is observed. This observation is in agreement with
the static scission point model /5/ which predicts an approach
to the smaller deformation one calculates when shell and
pairing corrections are set to zero. For the mass splits around
102/134 an increase in deformation at 6 MeV is observed compared to the thermal value. Also this is qualitatively in agreement with the static scission point model /5/.

5. ANGULAR DISTRIBUTIONS

Angular distributions were investigated in several previous
measurements. The measurements are mostly averaged over all
fragments. The present experiment gives the angular distributions also as function of mass splits. However, the present

angular distributions of fission fragments showed no fragment
mass dependence.

REFERENCES

1. H.-H. Knitter and C. Budtz-Jørgensen, Proc. Knoxville Conf.,
   NBS Spec. Publ. 947 (1980) 594.
2. Chr. A. Straede, Neutron Induced Fission of ^{235}U, Thesis,
   Central Bureau for Nuclear Measurements, Geel, Belgium,
   1985.
3. J. L. Cook, E. K. Rose and G. P. Trimble, Aust. J. Phys.
   29 (1976) 125.
4. A. Gilbert and A. G. W. Cameron, Can. J. Phys. 43 (1965)
   1446.
5. B. D. Wilkins, E. P. Steinberg and R. R. Chasman, Phys. Rev.
   C14 (1976) 1832.
6. J. W. Meadows and C. Budtz-Jørgensen, Nuclear Data and
   Measurement Series, ANL/NDM-64, 1982.

PROGRESS IN THE DETECTION OF LOW-ENERGY NEUTRONS

O. A. Wasson

National Bureau of Standards, Gaithersburg, MD, USA

ABSTRACT. The development of improved neutron detection for energies less than 20 MeV is required for advances in the study of neutron-induced reactions. These neutron detectors must have an accurately known response function as well as fast-timing capabilities for use at pulsed neutron sources. The accuracy in the determination of the response of most neutron detectors is ultimately limited by the accuracy of the neutron standard reaction cross sections. The exceptions to this rule include thick total absorption detectors and associated particle methods. Several examples of recent progress in neutron detection are presented.

1. INTRODUCTION

The ability to measure neutron reaction cross sections is ultimately dependent upon the measurement of neutron fluence. The accuracy of most of the various neutron detectors is limited by the knowledge of a few basic reaction cross sections which have been defined /1/ as the neutron standard reaction cross sections in the energy region below 20 MeV. These standards include various neutron reactions on H, ^{10}B, ^{6}Li, ^{12}C, ^{197}Au, and ^{235}U. The present uncertainty in these cross sections /2/ is approximately 1% (1 standard deviation) below 100 keV and greater for higher energies. The exceptions are the hydrogen scattering cross section which is known within an uncertainty of 1% (1 SD) throughout this entire energy region and the ^{235}U(n,f) cross section which is also known to within 1% at thermal and 14 MeV. This implies that the uncertainty in the detection efficiency of thin detectors is greater than 1%. The methods for overcoming this limitation utilize thick total absorption and associated-particle techniques.

399

Examples of the neutron detectors possessing fast timing
properties which have been utilized at the National Bureau of
Standards (NBS) for neutron fluence measurements in the Neutron
Measurements and Research Program are given in the remainder of
this paper.

2. THE DUAL THIN SCINTILLATOR FOR 1—15 MeV NEUTRONS

The first neutron detector utilizes the recoil protons produced
by the neutron scattering from the hydrogen contained in a thin
organic scintillator. This detector /3/ which has fast timing
was designed to operate in a collimated 1—15 MeV energy neutron
beam. Thin scintillators were chosen in order to keep multiple
neutron scattering low. The escape of protons from the scintillator typically produces a distortion in the measured proton
recoil spectrum. This distortion is eliminated in the present
detector.

The detector which is shown in Fig. 1 consists of two thin
plastic scintillators optically separated from each other and
independently coupled to phototubes. The protons which escape
from the first scintillator are detected by the second scintillator which is placed behind the first. Because of the low mul-

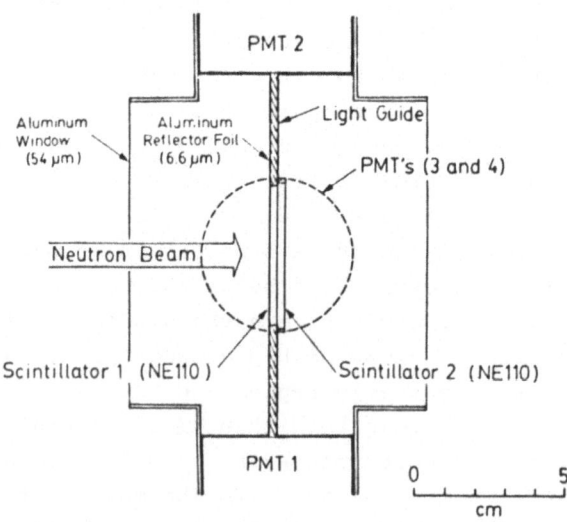

FIG. 1. Geometry of the dual-thin scintillator.

tiple scattering and the spectrum discrimination, there is relatively little dependence of the detector efficiency on the
carbon cross section and angular distributions. Therefore the
detector efficiency is essentially dependent on the hydrogen
cross section, on light tables, and on hydrogen areal density.
These parameters are known well enough to determine the detector efficiency within an uncertainty of 1%.

The proton recoil spectra and absolute detector efficiency
were measured at 2.45 and 14.0 MeV neutron energy using the associated-particle technique with the NBS Positive-Ion Van de
Graaff neutron source /4/. The measured spectra for 14 MeV neutrons, as shown in Fig. 2, are in excellent agreement with those

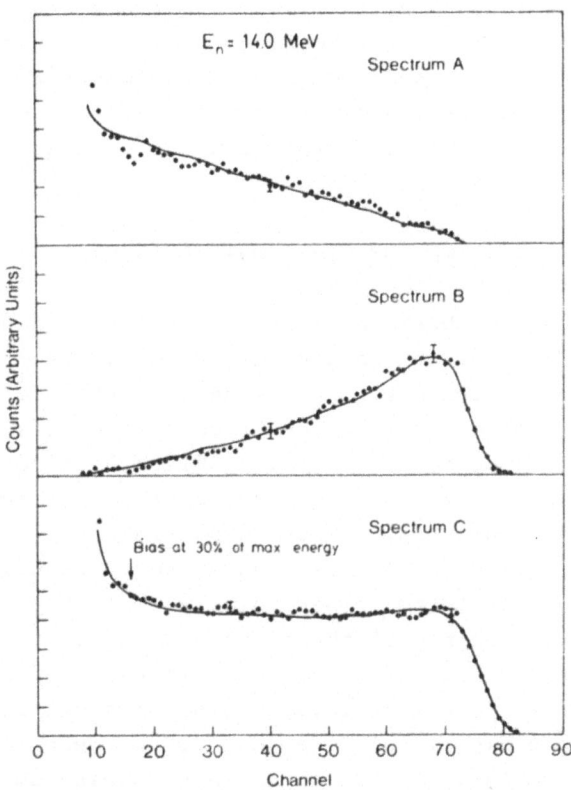

FIG. 2. Experimental proton recoil spectra obtained in an associated-particle experiment at 14.0 MeV. Spectrum A is from the
first scintillator. Spectrum B is the sum coincidence of both
scintillators. Spectrum C which is the sum of spectra A and B
approximates the ideal thin scintillator response. The solid
lines are Monte Carlo calculations.

calculated using a Monte Carlo technique. This program includes
several NBS modifications /5/ to the original program written
by Poenitz /6/ for a thick detector.

The absolute efficiency of the detector calculated for the
1—15 MeV energy region varies from approximately 0.015 to
0.007. Since the 0.7% uncertainty in measurement at 14.0 MeV is
less than the 1.7% uncertainty in the calculation, the calculation is normalized to the experiment at this energy. This increases the calculated efficiency by 1.3% at all neutron energies. The renormalized calculation is then used for the detector efficiency throughout the 1—15 MeV energy region. The re-
sultant uncertainty in the efficiency is thus 0.7% at 14.0 MeV
and increases to 1.5% at the lower neutron energy of 2.45 MeV.

This detector was later employed on the 200 m flight path of
the NBS linac pulsed neutron source to measure the ^{235}U(n,f)
cross section in the 1—6 MeV energy range. The results were in
good agreement with those obtained with the Black Detector fluence monitor /7/ in the 1—3 MeV region.

3. THE BLACK DETECTOR FOR 0.3—3 MeV NEUTRONS

Another method to obtain a neutron detector with well-characte-
rized efficiency is to use a thick, totally absorptive device
such that the efficiency is no longer directly dependent upon
the nuclear sections of the constituent material. Such a totally absorbing detector was called a "Black Detector" by Poenitz
/6/. A plastic scintillator was designed at NBS to emphasize
the 1 MeV neutron energy region. Over the years, the characterization of this detector has been extended to higher neutron
energies where it is no longer totally absorbing.

A schematic diagram of the neutron detector shows Fig. 3.
The detector consists of a 12.5 cm diam. by 19 cm long plastic
scintillator coupled to a 12.5 cm diam. photomultiplier tube. A
reentrant hole 5 cm in diameter and 2.5 cm deep increases the
neutron absorption. The dimensions were chosen so that the incident neutron would lose most of its energy in multiple collisions with the carbon and hydrogen of the scintillator. This
simplifies the calculation of the detector efficiency and redu-

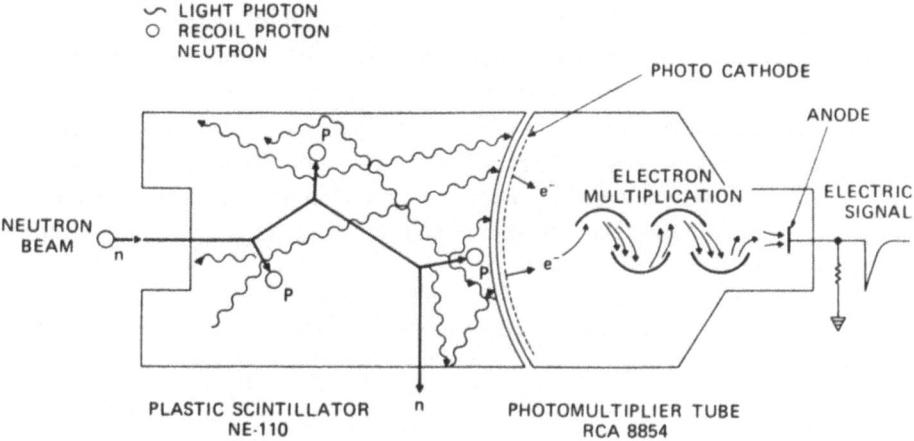

LIGHT PHOTON
RECOIL PROTON
NEUTRON

FIG. 3. Black neutron detector operation.

ces the uncertainty due to the uncertainty in the hydrogen and
carbon scattering cross sections.

This detector was utilized at the NBS 3 MV positive ion Van
de Graaff accelerator /7/. For most measurements, neutrons with
a typical energy spread of 40 keV were produced by bombarding
metallic lithium targets with pulses of protons 5 to 10 ns wide
at a 1 MHz repetition rate. The proton energy varied between
1.8 and 3.0 MeV to produce neutrons in the 0.1 to 1.3 MeV energy region. The neutron source was surrounded by a massive shield
to reduce the room background. Machined inserts collimated the
neutron beam in the forward direction to a cone with a half angle of 4.5°. The neutron detector was located in a shielded en-
closure behind a collimator positioned 5.9 m from the target.
Time-of-flight techniques were used to separate neutrons from
the gamma rays in the detector. The neutron fluence at the
1.3 m sample position was deduced from the neutrons incident on
the "Black Detector".

The quantity of interest from the detector is the efficiency
ε, which is defined as the detector yield per neutron incident
on the re-entrant hole. The efficiency is calculated from a
Monte Carlo based program originally developed by Poenitz /6/ and
modified by Meier /5/ to include the Poisson statistics for a

403

small number of photoelectrons as suggested by Lamaze et al. /8/. The shape of the detector response for monoenergetic incident neutrons is determined mainly by the nonlinearity in the scintillator light output for proton and carbon recoils and the statistics of the photoelectron emission process in the photomultiplier tube. The measured and calculated response of the detector to 540 keV neutrons is shown in Fig. 4. The agreement is excellent.

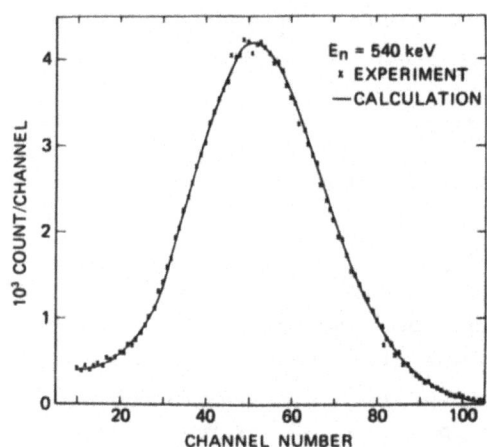

FIG. 4. Black detector response for 540 keV neutrons.

The efficiency was calculated throughout the energy region. For typical electronic biases of 10% of the peak position, the efficiency varied between 99 and 91% in the 0.2 and 1.2 MeV energy region. The efficiency calculations were experimentally verified in the 500 to 900 keV energy region by Meier /9/ who used the associated particle method to measure the number of incident neutrons.

Recently the characterization of this detector has been extended /10, 11/ to 3 MeV where it is no longer totally absorbing and the efficiency has decreased to 80%.

4. PROTON-RECOIL TELESCOPE FOR 1—20 MeV NEUTRONS

Another utilization of neutron scattering from hydrogen is the

404

annular proton telescope which was developed by Carlson and
Patrick /12, 13/ for neutron fluence measurements in the 1—20
MeV range. The detector was designed to operate at the 60 m
station of the NBS linac neutron time-of-flight facility. This
detector, shown in Fig. 5, is contained in a large vacuum chamber 45 cm diam.by 183 cm length. The proton recoils from neu-

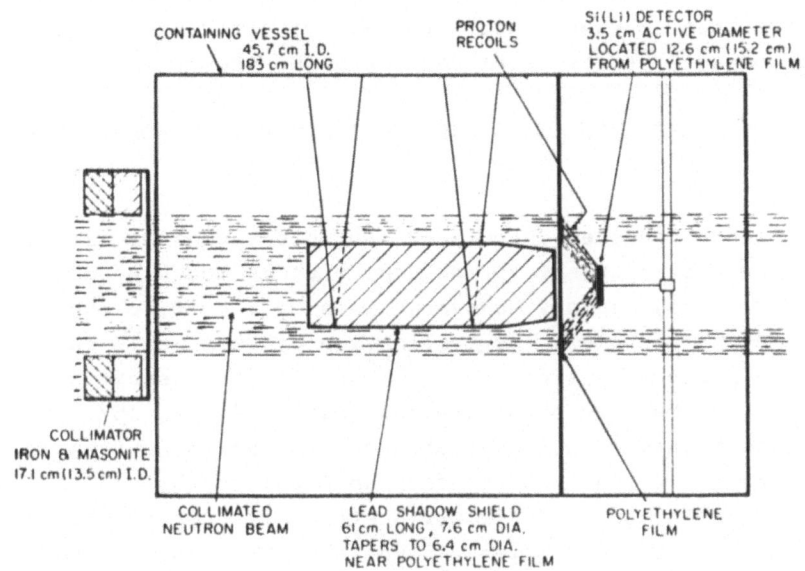

FIG. 5. Experimental geometry for annular proton telescope.

tron scattering in a thin polyethylene film were detected in a
Si(Li) detector which was shielded from the incident neutron
beam by a 61 cm long lead shadow shield. For the low-energy
measurements a polyethylene film of 0.31 mg/cm^2 was utilized.
The high-energy measurements were made with three film thicknesses (2.08, 10.7, and 44.2 mg/cm^2) in order to cover the
range from 3 to 20 MeV with satisfactory counting rate. The
measurements made with these films utilized the same geometry.
The data were then self-normalized with weights of the polyethylene films. The background was determined from a series of measurements with and without the polyethylene film in place and
with and without a tantalum cap over the Si(Li) detector. The
tantalum cap was thick enough to stop 20 MeV proton recoils.

405

The ambient background for all measurements was determined from a time window located just before the linac pulse. The net background for this detector was small, typically 2—3%. In the analysis events from $^{12}C(n,p)$ and $^{12}C(n,\alpha)$ reactions in the polyethylene foil were not counted.

This detector was used to measure the $^{235}U(n,f)$ and $^{237}Np(n,f)$ cross section in the 1—20 MeV energy interval. It is planned to repeat these measurements with this detector at a neutron source with more intensity at higher energy neutrons.

5. TIME-CORRELATED ASSOCIATED-PARTICLE METHOD FOR 14 MeV NEUTRONS

The time-correlated associated-particle technique has demonstrated the best accuracy for neutron fluence measurements in the 14 MeV energy region. However, it is difficult to apply at other neutron energies. The application of this well-known technique in nuclear physics to the $^{235}U(n,f)$ cross section measurement at 14 MeV has been very successful /4, 14, 15/. The most difficult part of all neutron cross-section measurements is the measurement of the neutron flux incident on the sample. This method replaces the difficult neutron measurement with an easier alpha-particle measurement. Since the $^{3}H(d,\alpha)n$ is a two-body nuclear reaction with a Q value of 17.6 MeV, the kinematics of the reaction define the energy and angular relationship between the emitted 14 MeV neutron and a 3.6 MeV alpha particle.

A schematic diagram of the experimental method is shown in Fig. 6. At NBS a molecular ion beam D_2^+, with 500 keV energy is produced by the 3 MV Van de Graaff accelerator and is incident on a TiT target to provide the neutron source. The number of neutrons within the indicated cone, which are incident on the ^{235}U deposit, is determined by the number of associated alpha particles, Y_α. The fission events are detected in a parallel plate ionization chamber. The number of fission events produced by the incident neutrons is given by the number of time coincident events between the alpha particle and fission events, $Y_{\alpha,f}$. The ^{235}U fission cross section is given by the simple relationship shown in the figure where n is the areal density of the

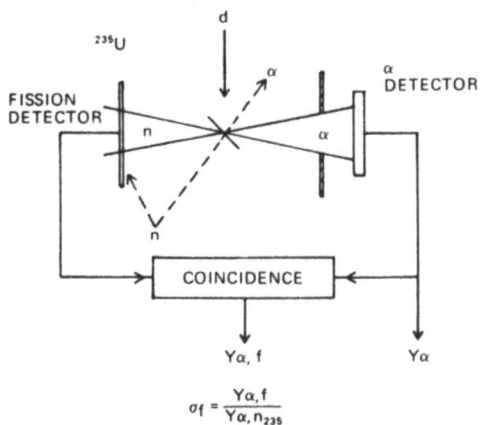

FIG. 6. Schematic diagram of the time-correlated associated-particle technique using the ³H(d,α)n reaction.

^{235}U deposit. The fission background produced by scattered neutrons, which is indicated by the dashed line in the figure, is eliminated by the coincidence requirement. This background is a major problem in most types of measurement which do not use the coincidence requirement.

The cross section does not depend on the solid angles of either the neutron beam or the collimated alpha-particle beam. The only requirement is that the ^{235}U deposit intercept the entire neutron beam. Thus, precise measurements of positions are not required. The cross section is also independent of the efficiency of the alpha-particle detector, although background events from other charged particles and neutron interactions in the detector must be evaluated. The cross section, however, does depend on the efficiency of the fission detector. The main limitation on the accuracy of fission cross section measurements at 14 MeV is the measurement of the areal density of the fission deposit.

The spectrum observed in the solid-state detector is shown in Fig. 7. This spectrum was obtained at a rate of ~ 2000 events per second for which spectral distortions due to pulse pileup were negligible. The region of the spectrum used in the measurement is indicated by the discrimination levels. The dominant feature is the alpha-particle peak from the source reaction.

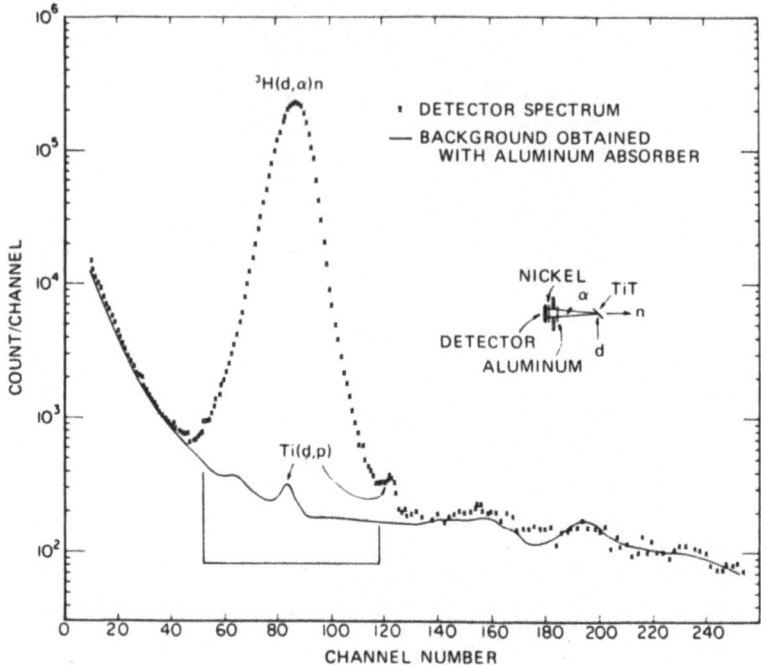

FIG. 7. Spectrum observed in the alpha-particle detector. The portion of the spectrum used for the cross-section measurements is bounded by the two vertical lines.

Since the cross-section result is independent of the fraction of the alpha-particle events utilized, the discrimination levels were selected to maximize yield and to minimize background.

The background contribution of $(0.68 \pm 0.10)\%$ indicated by the solid curve was obtained by insertion of a 0.03 mm thick aluminum foil over the surface of the detector. The absorber removed the 3.6 MeV alpha particle but not the neutrons and energetic protons. The major source of background was produced by neutron interactions in the detector while a lesser background was produced by protons from the Ti(d,p) reaction in the target materials. Detailed measurements made at several detector angles with various absorbers demonstrated that there were no other significant background sources. The use of fresh targets eliminated the possible source of 3.5 MeV alpha particles from the d(^{3}He,α)p reaction on the tritium decay products. Also, protons from the ^{2}H(d,p)^{3}H reaction were not observed.

408

The application of this method to other neutron producing
reactions is difficult because of the higher energy of the
scattered incident beam. Only the group at the Technical University of Dresden /16/ has been successful in measuring fission cross sections at 2.6 MeV using the $D(d,n)^3He$ reaction.
They have also completed measurements near 8.2 MeV. The time-
correlated associated particle technique is thus an excellent
neutron fluence measuring procedure for a few restricted neutron energies.

6. ^{3}He GAS SCINTILLATOR FOR THERMAL TO 3 MeV NEUTRONS

One of the most promising nuclear reactions for use as a neutron fluence detector in the energy region below 3 MeV is the
^{3}He(n,p)T reaction. This reaction has a positive Q value of
0.76 MeV. However, a useful detector with fast timing and adequate energy resolution to separate the (n,p) reaction from
the ^{3}He recoils from neutron elastic scattering has not been
developed. The use of this gas in the proportional counter mode
has produced excellent energy resolution although the timing
is too slow to permit use in MeV neutron time-of-flight measurements. A small high pressure gas scintillator with a resolution of 33% has been developed by Evans /17/ for delayed neutron experiments.

Our group at NBS has pushed to establish the ^{3}He(n,p)T reaction as a useful neutron standard in the keV and MeV neutron
energy regions. A prototype He-3 gas scintillator is shown in
Fig. 8. The detector consists of a 2 atmosphere He-Xe gas mixture in a 11 cm diameter by 25 cm length cylindrical volume.
The light is viewed through glass windows by two photomultiplier tubes positioned at opposite ends of the cylinder. A di-
phenylstilbene wavelength shifter was evaporated onto the inside of the windows to convert the ultraviolet light emitted
by the scintillating gas into the visible region to which the
photomultiplier tubes are sensitive. The response of the detector was measured with thermal neutron beams from the NBS reactor and higher energy neutrons from the linac facility. The
light output increased a factor of two as the Xe gas fraction

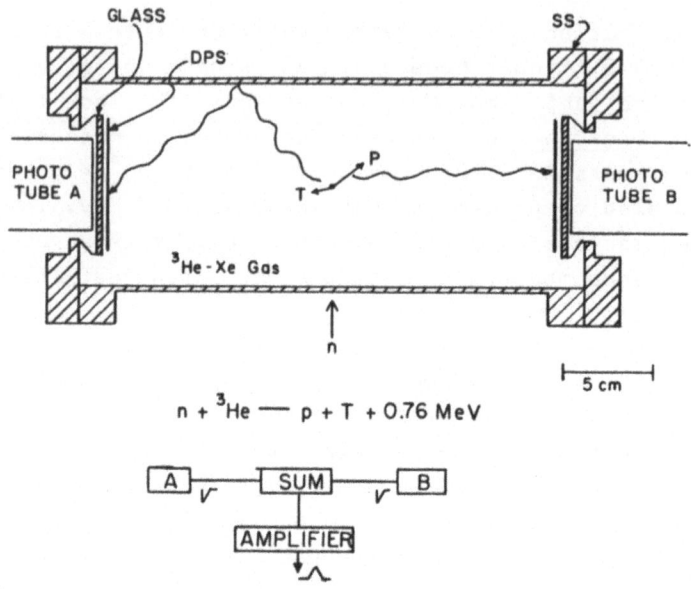

FIG. 8. ^{3}He–Xe gas scintillator.

was increased from 5% to 32% in order to reduce the range of the product particles. The spectral resolution, which was near- ly constant throughout the central 10 cm, was governed by the statistics of the number of summed photoelectrons from the two tubes (25 keV per photoelectron). The encouraging results ob- tained from these preliminary studies indicate the potential usefulness of this reaction as a neutron standard.

The author thanks his colleagues, A. D. Carlson, R. A. Schrack, R. G. Johnson, J. W. Behrens, K. C. Duvall, and M. S. Dias for their development of the detectors reported in this paper.

REFERENCES

1. A. D. Carlson and M. R. Bhat (Eds.), ENDF/B-V Cross Section Measurement Standards, BNL-NCS-51619, 1982.
2. A. D. Carlson, Prog. Nucl. Ener. <u>13</u> (1984) 79.
3. M. S. Dias, R. G. Johnson and O. A. Wasson, Nucl. Instrum. Methods <u>224</u> (1984) 532.
4. O. A. Wasson, A. D. Carlson and K. C. Duvall, Nucl. Sci. Eng. <u>80</u> (1982) 282.
5. M. M. Meier, National Bureau of Standards Internal Report (February 1978).

6. W. P. Poenitz, The Black Neutron Detector, Argonne National Laboratory Report ANL-7915, April 1972.
7. O. A. Wasson, M. M. Meier and K. C. Duvall, Nucl. Sci. Eng. 81 (1982) 196.
8. G. P. Lamaze, M. M. Meier and O. A. Wasson, Proc. Conf. on Nuclear Cross Sections and Technology, Washington, DC, 1975, Spec. Publ. 425, U.S. National Bureau of Standards, p. 73.
9. M. M. Meier, Proc. Symp. Neutron Standards and Applications, Gaithersburg, Md., 1977, Spec. Publ. 493, U.S. National Bureau of Standards, p. 221.
10. K. C. Duvall, A. D. Carlson and O. A. Wasson, Proc. 8th Conf. on Application of Accelerators in Research and Industry, Denton, Tex., 1984, in press.
11. A. D. Carlson, J. W. Behrens, R. G. Johnson and G. E. Cooper, Proc. IAEA Advisory Group Meeting on Nuclear Standard Reference Data, Geel, 1984.
12. A. D. Carlson and B. H. Patrick, Proc. Int. Conf. on Nuclear Cross Sections for Technology, Univ. of Tennessee, Knoxville, Tenn., 1979, Spec. Publ. 594, U.S. National Bureau of Standards, p. 971.
13. A. D. Carlson and B. H. Patrick, Int. Conf. on Neutron Physics and Nuclear Data for Reactors and Other Applied Purposes, Harwell, U.K., 1978, OECD Nuclear Energy Agency, 1979, p. 880.
14. M. Cancé and G. Grenier, Nucl. Sci. Eng. 68 (1978) 197.
15. R. Arlt, W. Meilung, G. Musiol, H. G. Ortlepp, R. Teichner, W. Wagner, I. D. Alkhazov, O. I. Kostockin, S. S. Kovalenko, K. A. Petrzak and V. I. Shpakov, Kernenergie 24 (1981) 48. (Earlier references listed.)
16. R. Arlt, H. U. Heidrich, M. Josch, G. Musiol, H. G. Ortlepp, R. Teichner and W. Wagner, Nucl. Instrum. Methods, in press.
17. A. E. Evans, Jr., Los Alamos National Laboratory Report LA-Q2TN-82-109, 1982.

TIME-OF-FLIGHT SPECTROMETER FOR THE MEASUREMENT OF GAMMA CORRELATED NEUTRON SPECTRA

A. V. Andriashin, B. V. Devkin, A. A. Lychagin,
J. V. Minko, A. N. Mironov, V. S. Nesterenko,
*T. Sztaricskai, *G. Pető and *L. Vasváry

Institute of Physics and Power Engineering, Obninsk, USSR
*Institute of Experimental Physics, Kossuth University,
Debrecen, Hungary

ABSTRACT. A time-of-flight spectrometer for the measurement of gamma correlated neutron spectra from $(n,xn\gamma)$ reactions is described. The operation and the main parameters are discussed. The resolution in the neutron channel is 2.2 ns/m at the 150 keV neutron energy threshold. A simultaneous measurement of the time-of-flight and amplitude distributions makes it possible to study gamma correlated neutron spectra as well as the prompt gamma spectra in coincidence with selected energy neutrons. In order to test the spectrometer, measurements of the neutron spectrum in coincidence with the 846 keV gamma line of ^{56}Fe were carried out at an incident neutron energy of 14.1 MeV.

The study of fast neutron induced neutron emitting reactions is usually performed using the neutron time-of-flight spectrometer. The utilization of gamma spectrometers in the fast neutron time-of-flight spectrometry makes it possible to study the higher excited states in (n,n) and (n,2n) reactions /1/. The original neutron spectrometer, based on the nanosecond pulsed neutron generator of the IPP /2/, was completed with a Ge(Li) spectrometer in order to be able to study in more detail the $(n,n'\gamma)$ reactions.

The neutron source in the measurement of gamma correlated neutron time-of-flight spectra was the nanosecond pulsed KG-0.3 neutron generator, producing 2.5 ns wide neutron pulses at a repetition rate of 2.5 Mcps and an average target current of 4 μA on the TiT target. To decrease the prompt gamma background of the target holder, a new holder was made of thin stainless steel tube, having wall thickness of 0.3 mm. The target was air cooled by a fan.

412

The spectrometer was constructed for simultaneous registration of the following spectra:

a) time-of-flight spectrum of direct neutrons from the target, detected by the monitor (MD-1) detector,

b) time-of-flight spectrum of scattered neutrons from the sample,

c) time-of-flight spectrum of scattered neutrons from the sample in coincidence with a selected gamma line in the gamma spectrum,

d) time-of-flight spectrum of scattered neutrons from the sample in coincidence with the background of the selected gamma line in the gamma spectrum,

e) the prompt gamma spectrum produced by the fast neutron pulses in the sample.

The detector for the registration of scattered neutrons was a $\emptyset$ 100x50 NE218 liquid scintillator mounted on an XP2041 photomultiplier, having an energy threshold of 150 keV. The efficiency of the neutron detector was determined by two methods: ^{252}Cf TOF spectrum and (n-p) scattering. The MD-1 monitor detector was a similar scintillation counter with a $\emptyset$ 70x50 mm stilbene on a FEU-30 fast multiplier. The gamma detector used in this experiment was a 50 cm^3 true coaxial Ge(Li) detector with a relatively poor energy resolution at counting rate of 5000 cps. The two

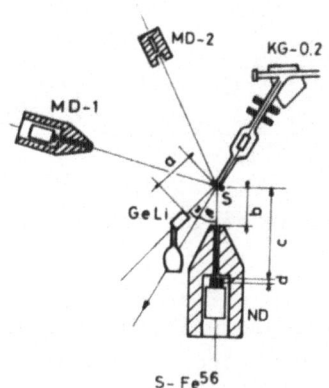

S- Fe^{56}

FIG. 1. The lay-out of the gamma correlated neutron spectrometer. α = 17°, β = 27°, a = 72.5 cm, b = 75.5 cm, c = 157.5 cm, d = 6.0 cm.

413

neutron detectors had usual shields, the gamma detector was not
shielded. The neutron yield of the generator during the experiments was measured by a long counter and it was calibrated by
activation method based on the ^{27}Al(n,α) reaction.

In an earlier version of a same-purpose TOF spectrometer 3
there were more difficulties connected with the gamma energy
peak and backgroud selection, therefore a new digital selector
was developed /4/.

Lay-out of the experimental spectrometer set-up used in the
study of (n,n´γ) reaction is shown in Fig. 1.

The detector time signals were processed by standard NIM
units using CFPH timing and n-γ discrimination. Analogue pulses
from the Ge(Li) detector were processed by CAMAC. The whole
experiment was controlled by the CAMAC DATAWAY and connected to
SM-3 minicomputer.

The block diagram of the neutron spectrometer and the measuring set-up are shown in Fig. 2.

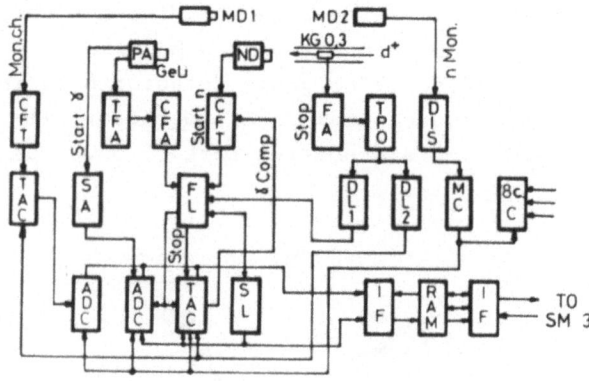

FIG. 2. The block diagram of the neutron spectrometer.

The data collection in the spectrometer was controlled by
the fast and the slow logic units of the spectrometer. These
units select the single neutron spectrum (without gamma coincidence), the prompt gamma spectrum and the neutron spectra
in coincidence with the peak and over the peak range of the
gamma spectrum.

SM-3 minicomputer which controls the optimum work of the

generator, the monitorization of the experiment and spectra
collection was also used in the evaluation of data collected.
The data and software of the experiment and evaluation were
stored on floppy discs.

The spectrometer was tested in the study of the (n,n´γ) reaction on ^{56}Fe. This reaction was investigated earlier /5, 6/.
Neutron time-of-flight spectra obtained by this gamma correlated
neutron spectrometer are shown in Fig. 3.

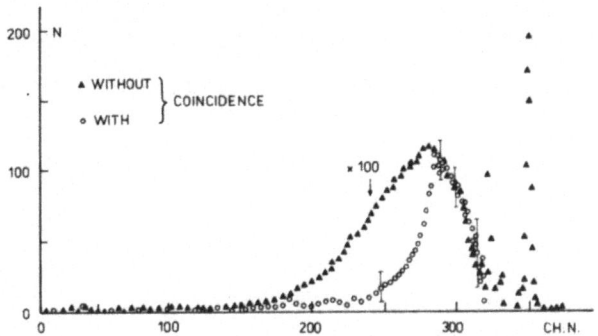

FIG. 3. The experimental time-of-flight neutron spectrum of
the ^{56}Fe(n,n´γ) reaction with and without coincidence on the
846 keV line.

The data on the measured neutron and gamma spectra, collected
in this experiment, are under evaluation in order to study the
neutron-gamma competition in the (n,n´γ) reaction.

The set-up of the spectrometer allows not only the neutron
time-of-flight measurements based on pulsed accelerators but
also the following experiments:

1. Similar measurements with associated particles (i.e.
alpha particle, fission products).

2. Simultaneous measurement of TCF and amplitude distribution
of the products in the investigated reaction.

3. Energy distribution measurement of gamma rays in coincidence with a given neutron flight time interval.

A change of the working mode of the spectrometer can be done
by the alteration of the mode that controls the fast and the
slow logics.

REFERENCES

1. S. Hlaváč and P. Obložinský, Nucl. Instr. Meth. 206 (1983)
   217.
2. V. B. Anufrienko, B. V. Devkin, Yu. A. Kulabukhov, L. A.
   Matalin, O. A. Salnikov, M. Z. Tarasko and L. A. Timokhikh,
   Report FEI-755, Obninsk, 1977.
3. B. V. Devkin, A. A. Lychagin, V. S. Nesterenko, N. N. Shadin,
   A. I. Gonchar, A. N. Mironov, S. E. Sukhin, T. Sztaricskai
   and G. Ch. Petö, Proc. 11th Int. Symp. on Interactions of
   Fast Neutrons with Nuclei, Gaussig, Report ZfK-476, 1982,
   p. 71.
4. B. V. Devkin, A. A. Lychagin, A. N. Mironov and V. S.
   Nesterenko, Report FEI-1557, Obninsk, 1980.
5. G. Stengl, M. Uhl and H. Vonach, Nucl. Phys. A290 (1977)
   109.
6. Yu. E. Kozyr and T. A. Prokopets, Yad. Fiz. 27 (1978) 616.

DESIGN OF A MULTI-PURPOSE INTENSE NEUTRON GENERATOR

P. Eckstein, F. Gleisberg, H. Helfer, R. Krause, U. Jahn,
E. Paffrath, D. Schmidt, D. Seeliger, *A. I. Glotov
and *V. A. Romanov

Technical University Dresden, GDR
 *Institute of Physics and Power Engineering, Obninsk, USSR

ABSTRACT. The concept of an intense neutron generator is outlined. Its elements are characterized in some detail.

1. INTRODUCTION

There are several nuclear reactions for producing intense neutron beams. Using an incident particle current of 1 mA neutron
strengths between 10^{11} and 10^{17} s^{-1} could be obtained. The well
known DT-reaction is not the best candidate in this sense, but
it shows two remarkable advantages: a very narrow spectral distribution near 14 MeV ("monoenergetic neutrons") and the lowest
technical requirements and costs for accelerator instrumentation.
There are some quite different possibilities for the design of
intense DT-neutron sources. The mostly used type is the pumped
small accelerator consisting of the main components source, acceleration tube, target and vacuum system. This generator type
shows the greatest variability with respect to intensity variation and adaption to specific utilization. Table 1 shows the
parameters of some typical pumped intense DT-generators which
are under operation.

The aim of the present work was the conceptual design of an
intense DT-neutron generator within the intensity range
$(1-5) \times 10^{12}$ s^{-1} basing on the experience of existing machines
of this type (as far as available from the literature), already
existing components and technological equipments in both insti-

417

TABLE 1. Some operating intense DT-generators

Generator	Location	U (kV)	I (mA)	N_n (10^{12} s^{-1})	Ref.
RTNS-II	Livermore/USA	400	150	40	/8/
RTNS-I	Livermore/USA	400	22	6	/1/
LANCELOT	Valduc/France	160	160	6	/6/
CHALK RIVER	/Canada	300	25	4	/10/
OKTAVIAN	Osaka/Japan	300	20	3	/11/
JAERI	Tokio/Japan	400	20	5	/12/
DYNAGEN	Hamburg/FRG	500	12	3	/13/
INGE-1	Dresden/GDR	300	20	Project	

tutes as well as on the possibilities of modern electronical
circuits. In the following the main components of the resulting
concept are characterized in some details. As far as the project of the Technical University Dresden is concerned the name
INGE-1 is used.

2. THE GENERAL SCHEME OF THE INTENSE NEUTRON GENERATOR INGE-1

The general arrangement proposed for the generator INGE-1 is
similar to that of the RTNS-II neutron source /1/ and other generator projects /2, 3/. The main components of the neutron generator are the following: The high voltage terminal is isolated from ground potential by ceramic isolators up to 350 kV potential difference. Within the high voltage terminal the ion
source, an ion pump and power supplies are located. The main
vacuum system is located on the ground potential. The ion beam
will be accelerated by an acceleration tube with anticorona
rings and resistors for potential dividing. Power supplies and
ion source within the terminal are controlled by opto-electro-
nic circuits in connection with glass fibres. The accelerated
beam is focussed by a lens onto the rotating target, where the
neutrons are produced by the DT-reaction. The position of the
beam axis over ground floor is about 1.6 m, the overall dimensions of the generator amount to 1.4x2.3x4.3 m^3. Figure 1 shows

the scheme of the generator components in connection with the
control elements and cooling and electrical supplies.

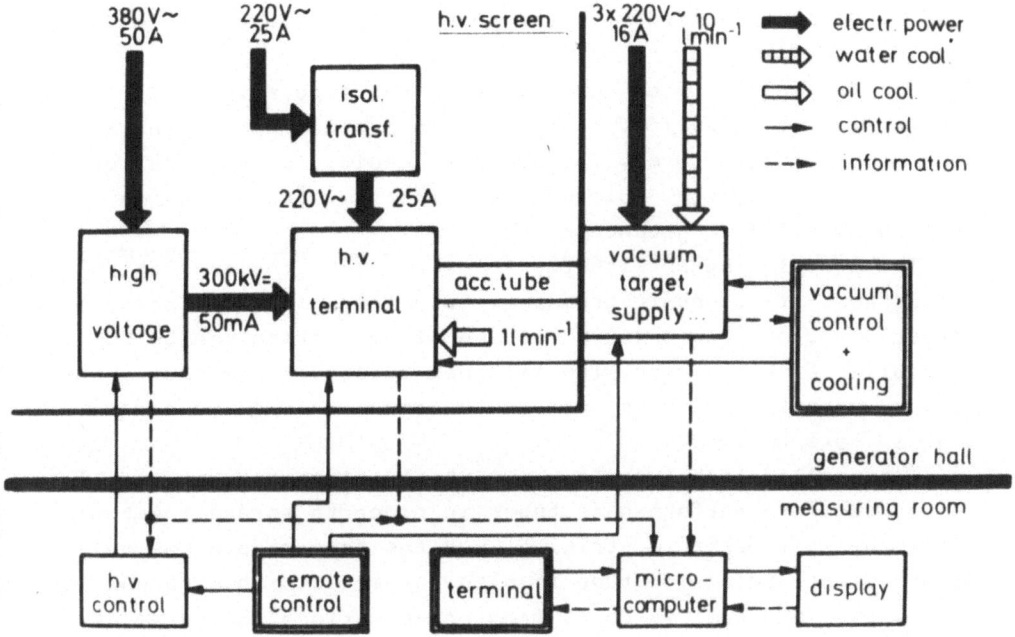

FIG. 1. Components and supply of the intense neutron generator
INGE-1.

3. DESCRIPTION OF THE MAIN COMPONENTS

The ionoptical layout of the generator INGE-1 is shown in Fig. 2

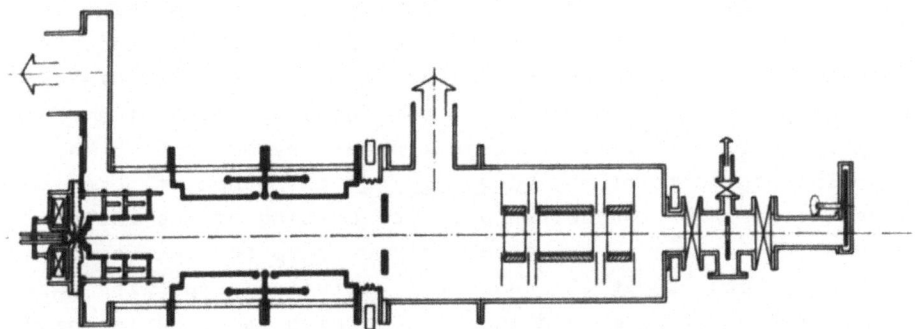

FIG. 2. Ionoptical layout of the generator (from the left: ion
source, einzel lens, acceleration tube, diaphragm, quadrupole
triplet, rotating target with valves and diaphragm).

The diaphragms at the end of the acceleration tube and also in front of the rotating target are water-cooled. The two valves near the target provide the possibility of exchange of the target.

The ion source is of the duoplasmatron type with specified extraction geometry /4/. It is able to provide ion currents in the order of some 10 mA. The cathode consists of a nickel grid with embedded barium, calcium and strontium. The source is cooled by oil from ground potential /5/.

The einzel lens is directly connected with the extraction electrode. It is constructed as a tube lens with an inner diameter of 78 mm. The maximum voltage of extraction can be 50 kV, normally 30 kV are used. The zwischen electrode has also a negative potential with respect to the anode of the ion source, but at maximum 10 kV.

The acceleration tube is a two-acceleration-gap version. The electrodes are performed as tubes in order to screen the tube walls against particle striking. For the same reason the zwischen electrode of the tube is also connected with a screening ring. The tube wall is built from ceramic rings, separated by stainless steel anticorona rings. For a definite linear potential distribution along the tube a divider is necessary. Despite of the relatively small current through the resistors (R_{total} = 7 GOhm) no high power resistors are needed because of the effective screening mentioned above.

As shown by ionoptical calculations /4/ a focusing lens after acceleration was needed. For this purpose a magnetic as well as an electrostatic quadrupole lens could be used.

The target is of the rotating type with water cooling. The active layer is 30 mm broad, therefore a water-cooled diaphragm in front of the target is arranged. By means of a rotating speed of 300 rpm the integral power may be up to 5 kW.

Main components of the vacuum system are pumps of three different types /5/: the getter pump for holding of the vacuum without operation, an oil-free orbitron pump for ion source and low-power beam operation and an oil diffusion pump (with LN_2-trap) for power beam handling. The pumping speed during beam operation must be at least 2000 l s^{-1}.

As high voltage supply the device GP 50/300 from VEB TuR
Dresden (n.e. Transformer and Roentgen Plant) is foreseen. It
can provide a maximum voltage of 300 kV with a current up to
50 mA. The power flux to the terminal is realized by an isolating transformer for 12 kW at 50 Hz.

Different types of electric power supplies are needed for
operation of the generator INGE-1. Supply units with a power
below 500 W are realized in a switching mode using frequencies
near 20 kHz /6/. For the supply units with higher power a more
conventional solution was found /7/. All supplies should be
controlled remotely and are able to provide reference signals
within a standardized level.

The general scheme of control and data acquisition is also
shown in Fig. 1. The control of the generator is foreseen by
hand at first, but the data acquisition is handled by a micro-
computer in connection with a colour display for all electrical
and other information (temperature, pressure, flow rate and
others).

4. CONCLUSIONS

The project of the intense neutron generator INGE-1 described
here seems to be suitable for a machine producing DT-neutron
intensities above 10^{12} s^{-1}. Some parts of it are realized already and show convenient properties. The concept hopefully allows
a compact design as well as reliable operation of this small
intense neutron generator.

REFERENCES

1. R. Booth et al., Nucl. Instr. Meth. 145 (1977) 25.
2. J. Pivarč et al., Report INDC(CSR)-2/L, IAEA, Vienna, 1980.
3. T. Sztaricskai, Report ZfK-459, Rossendorf, 1981, p. 176.
4. U. Jahn and J. Dietrich, Proc. 14th Int. Conference,
   Gaussig, 1984.
5. E. Paffrath et al., Proc. 14th Int. Conference, Gaussig,
   1984.
6. F. Gleisberg and H.-J. Esche, Report ZfK-503, Rossendorf,
   1983, p. 135.
7. P.Eckstein et al., Proc. 14th Int. Conference, Gaussig,
   1984.

8. D. W. Heikkinen and C. M. Logan, Report UCRL-86747, Livermore, 1982.
9. J. B. Hourst and M. Roche, Acta Phys. Slov. $\underline{30}$ (1980) No. 2.
10. J. D. Hepburn et al., IEEE Trans. Nucl. Sci. $\underline{22}$ (1975) 1809.
11. H. Ullmaier et al., Nucl. Instr. Meth. $\underline{145}$ (1977) 1.
12. T. Nakamura et al., Proc. 7th Symposium on Ion Sources, Antwerp, 1982.
13. M. R. Cleland and B. P. Offermann, Nucl. Instr. Meth. $\underline{145}$ (1977) 41.

TWO-TARGET DSAM WITH FAST REACTOR NEUTRONS IN Si

M. Georgieva, D. Elenkov, D. Lefterov and G. Toumbev

Institute of Nuclear Research and Nuclear Energy, Sofia, Bulgaria

ABSTRACT. An experimental system for DSAM (Doppler Shift Attenuation Measurement) is described using fast reactor neutrons on two targets simultaneously. The mean life times of the states 1273, 2235, 2425, 4618, 4979 and 6275 in $^{28-30}Si$ were measured.

In the recent years considerable progress has been observed in the use of fast reactor neutrons for study of nuclear states excited in $(n,n'\gamma)$ reaction. Highly intensive fluxes of fast reactor neutrons and their wide energy spectrum (0.5—6.0 MeV) allow to populate and study low-lying energy states of the stable nuclei and their parameters - population probabilities, spins, life times, etc. This reaction proved a useful tool in the extension of our knowledge obtained from accelerator experiments /1/.

An experimental apparatus for DSAM has been installed for some years on the horizontal channel of the IRT-2000 reactor in Sofia. Mean life times of excited nuclear states of over 20 nuclei in the mass region of A ~ 10—60 have been measured. We have developed the conception proposed by Warburton et al. /2/ and Antilla et al. /3/ for a simultaneous measurement using two targets and a fixed detector by applying it to the $(n,n'\gamma)$ reaction. The lay-out of the apparatus is shown in Fig. 1.

By means of filters consisting of 0.5 mm Cd, 1 cm B_4C and 4 cm Pb the flux of filtered fast neutrons hits through a system of collimators of Fe and Pb the targets T_1 and T_2. The cross-section of the flux is about 7 cm^2.

The collimators in the detector shield are of the same cross-

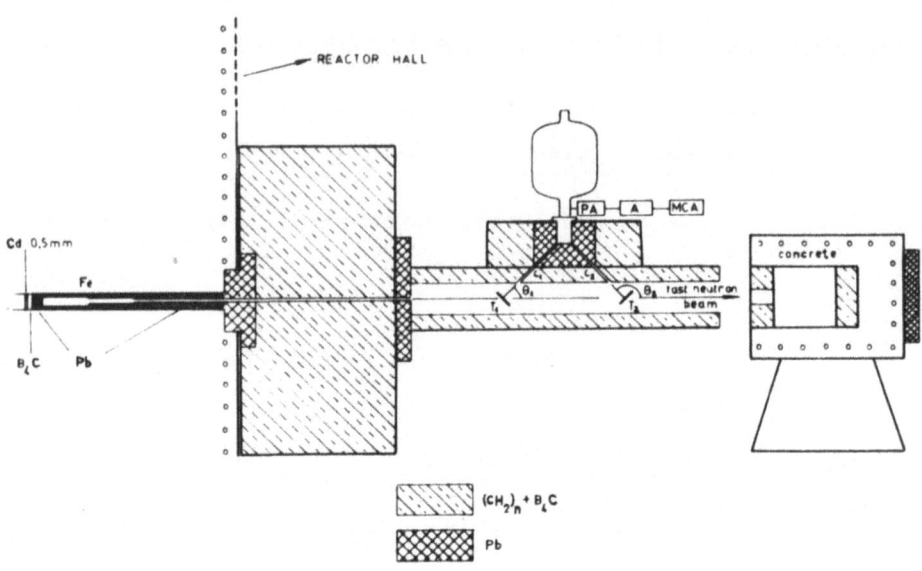

FIG. 1. Experimental apparatus for DSAM in the $(n,n'\gamma)$ reaction, using two targets simultaneously.

section and are partially filled in with Li_2CO_3. They are meant to collimate the gamma rays from the targets onto the Ge(Li)-detector. The angles between the neutron flux and the collimators are $45°$ and $135°$, respectively. A detector of effective volume of 28 cm^3 and resolution of 2.6 keV at 1.33 MeV was used. The method of data processing is described in detail in /4, 5/. Two experimental spectra of gamma transitions with energies of 2838 and 3200 keV in ^{28}Si shifted forward and backward due to the Doppler effect are shown in Fig. 2.

The numerical results on the mean life times are given in Table 1.

The above-described arrangement has some advantages compared to the one-target method: (i) the system is fixed and the data are obtained in the same background conditions; (ii) the electronic drift does not affect the distance ΔE between the two Doppler-shifted peaks; (iii) the neutron flux is practically used twice (on each target). Owing to this the experimental error is reduced by a factor of 1.4.

In the mass region $A \succ 60$ the recoil of the target nucleus

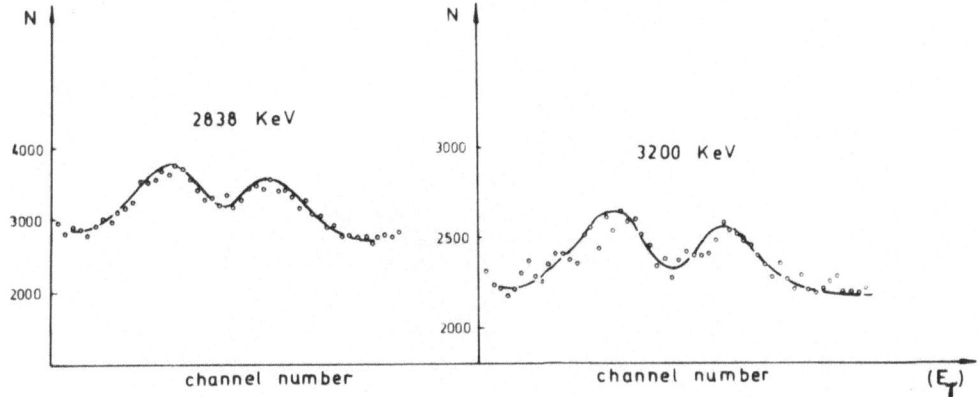

FIG. 2. Typical experimental spectra of Doppler-shifted energies of gamma transitions.

TABLE 1

E^{st} (keV)	E^{tr}_{γ} (keV)	$\tau \cdot 10^{-15}$ s	Ref. /6, 7/
1273[a]	1273	100	150 ± 50
			310 ± 100
2235[b]	2235	270 ± 50	260 ± 60
			345 ± 70
2425[a]	2425	26 ± 20	13 ± 3
			29 ± 13
4618[c]	2838	28 ± 5	58 ± 5
			39 ± 2
4979[c]	3200	65 ± 6	34 ± 12
			60 ± 20
6275[c]	4496	1900 ± 200	1300 ± 200
			1500 ± 400

a - ^{29}Si; b - ^{30}Si; c - ^{28}Si.

is smaller as a result of the insufficient kinetic momentum of the induced neutron. This disadvantage of the method could be avoided by making use of detectors of high energy resolution.

REFERENCES

1. A. M. Demidov et al.,"Atlas of Gamma-ray Spectra", Atomizdat, Moscow, 1978.
2. E. K. Warburton et al., Phys. Rev. 129 (1963) 2180.
3. A. Antilla et al., Nucl. Instr. Meth. 124 (1975) 605.
4. D. Elenkov et al., Nucl. Instr. Meth. 201 (1982) 377.
5. D. Elenkov et al., Nucl. Instr. Meth. 228 (1984) 62.
6. P. M. Endt et al., Nucl. Phys. A315 (1978) 1.
7. M. A. Meyer et al., Nucl. Phys. A250 (1975) 235.

RECENT DEVELOPMENTS IN THE MULTIPURPOSE INTENSE
14 MeV NEUTRON SOURCE AT BRATISLAVA

J. Pivarč, S. Hlaváč, P. Obložinský, V. Matoušek,
R. Lórencz, I. Turzo and L. Dostál

Institute of Physics, Electro-Physical Research Centre of
the Slovak Academy of Sciences, Bratislava, Czechoslovakia

ABSTRACT. Multipurpose intense D+T neutron source based on the
20 mA duoplasmatron ion source and the 300 kV/40 mA high voltage power supply is under construction at the Institute of
Physics in Bratislava.

1. INTRODUCTION

The work gives a short survey of the present situation in the
development of the accelerator. Several components of the
source have been developed/completed and tested successfully:
(i) The high voltage power supply together with the capacitor
battery and all supplementary units, such as water resistors,
pneumatic discharger and high voltage cable ends. (ii) All high
voltage transformers, namely the isolating transformer 300 kV/10
kVA for high voltage terminal, the isolating transformer
50 kV/2 kVA for the duoplasmatron and 50 kV dc/3.75 kVA for the
extraction power supply source.
  Under construction are the high voltage terminal, the diag-
nostic elements of ion beam as well as the rotating target.

2. NEUTRON SOURCE DESIGN

The lay-out of the source is shown in Fig. 1. The intense section should be capable of producing 300 keV/10 mA of a separated D^+ ion beam. With such a beam and a fresh target a neutron

427

yield of 10^{12} ns^{-1} for a beam spot of 1 cm^2 and a useful target life time of about 10 h can be expected.

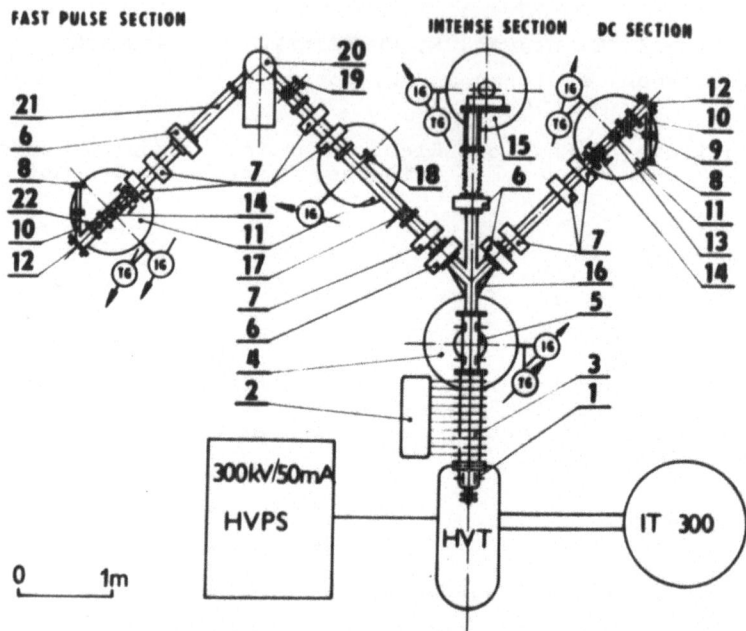

FIG. 1. The lay-out of the main components of the neutron source. 1 - ion source, 2 - column resistors, 3 - acceleration tube, 4 - main vacuum system, 5 - beam corrector, 6 - gate valve, 7 - quadrupole lens, 8 - monitor, 9 - static beam monitor, 10 - target chamber, 11 - auxiliary vacuum system, 12 - water-cooled target, 13 - water-cooled slit, 14 - beam profile monitor, 15 - rotating target chamber, 16 - analysing-switching magnet, 17 - chopper, 18 - slit, 19 - diaphragm, 20 - 90° magnet, 21 - bunching system, 22 - pick-up system, TC - thermocouple gauge, IG - ionization gauge, IT 300 - isolating transformer, HVT - high voltage terminal and HVPS - high voltage power supply.

The conventional dc section has been designed to produce a neutron yield of 10^{10} ns^{-1} and the fast pulsed section will be capable to generate a compressible D$^+$ ion beam of up to 1 ns on a target spot.

3. RECENT DEVELOPMENTS AND THE PRESENT STATE

Recent developments concentrated on the completion of drawings, assembling of a high voltage power supply, putting into operation the main vacuum system and the acceleration tube and on development of two isolating transformers as well as a transformer for the extraction source.

The optical properties of the accelerator components were calculated taking into account the effect of the beam space charge. Likewise, the electronic control system was also developed /1/.

3.1. High Voltage Power Supply

The most important components of the high voltage power supply were supplied by VEB TuR Dresden (n.e. Transformer and Roentgen Plant). The supply was constructed taking into account a long operation at 40 mA/20°C or 30 mA/35°C. The high voltage is regulated by a motor-operated regulating transformer. The ripple factor of the supply is 2.5%. The capacitor battery consisting of capacitors (1.5 μF/30 kV) has been added in order to obtain

FIG. 2. General view of the high voltage power supply.

better parameters of the system. Making use of such a capacitor bank we are able to reduce the ripple factor to 1%.

The high voltage power supply is separated from the neutron source terminal. They are interconnected through a water resistor with a high voltage cable. Suitable cable ends were constructed to prevent a breakdown between the polyethylene isolation of the cable and the lead cable covering.

The capacitor battery is connected to the ground potential through a 280 kΩ water resistor by a pneumatic discharger. This system can quickly short the capacitor battery to ground and protect the damping resistor of the high voltage power supply.

The general view of the high voltage power supply is shown in Fig. 2. It is mounted in a 6 x 6 x 3.5 m cell.

3.2. High Voltage Terminal and Ion Source

A high voltage terminal 3.4 m^2 in surface and 0.9 m high will contain the ion source and the associated equipments. Power for the high voltage terminal will be supplied by the isolating transformer IT 300 /2–4/. The general view of the transformer is shown in Fig. 3. Its development was accomplished and the transformer is at present in routine operation in an old neutron source configuration.

The ion source is a modification of the duoplasmatron source developed at Vakutronik Dresden (n.e. Vacuum Electronic Plant). The preliminary tests of the source have been performed. It was shown that the expected 20 mA beam could be achieved at an approximately 40 kV extraction voltage. The tests did not include the 20^0 double focusing magnet.

The acceleration tube was supplied by the Institute of Nuclear Research, Warsaw. In addition, the multiple series of resistor block (10 MΩ/30 kV) built into an oil-filled insulating box has been finished in the workshop of our Institute.

3.3. Ion Source Electronics and Control System

Ion extraction should be performed by the 50 kV potential. For

430

this reason two transformers were developed. One, labelled as DEZ 50 provides the 50 kV dc voltage and the 3.75 kVA power for the extraction source. The other, the isolating transformer IT 50 /3–4/ for the 50 kV potential and the 2 kVA power, should transmit the mains for the rest of the duoplasmatron power supplies. The transformers weighing about 150 kg and 80 kg, respectively, are put into insulating cylinders made of laminated paper and filled by inhibited transformer oil.

The electronic control system has been developed /1/. It should maintain the accelerator for the required operation re-

FIG. 3. General view of the IT 300 isolating transfor er.

gime. At the first stage, however, it should allow an easy setup of the sources placed on the high voltage terminal.

3.4. Vacuum System

The vacuum system is constructed of stainless steel. ConFlat flanges and cooper gasket seals are used throughout the high vacuum side. In solenoid valves Viton rings and in the target chamber aluminium wire-rings are used. The high vacuum part is bakeable.

The principal part of the main vacuum pump unit, shown in

FIG. 4. Main vacuum system.

Fig. 4, has been completed. It is based on a 2000 ls^{-1} diffusion pump with water and liquid nitrogen traps and a mechanical
pump.

The EGZ 100 and the IZ 80 ion pumps produced by High Vacuum
Dresden and Leybold Heraeus, respectively, will be used as
auxiliary vacuum units.

At present the control unit of the main vacuum system has
been designed and delivered to the workshop of the Institute.

4. CONCLUSIONS

The next steps in the construction of the accelerator should
be as follows: Putting into operation the duoplasmatron ion
source on a high vacuum level (1986); forming the intense section of the accelerator (1987) and achieving the fully operating level of the intense neutron source (1988).

The authors wish to thank Ing. Š. Luby, DrSc., Director of the
Institute of Physics, Electro-Physical Research Centre of the
Slovak Academy of Sciences, for his support and kind interest
in this work. They also wish to thank Ing. K. Málek, Ing. B.
Bajcsy, P. Rovný, E. Cerva, Ing. L. Fekete and J. Vaňová for
technical help.

REFERENCES

1. J. Pivarč, S. Hlaváč, P. Obložinský, I. Turzo, V. Matoušek,
   R. Lórencz and L. Dostál, Proc. 14th Int. Symposium on
   Nuclear Physics - Neutron Generators and Application,
   Gaussig, 1984.
2. J. Pivarč, K. Málek, B. Bajcsy and P. Rovný, Proc. 14th Int.
   Symposium on Nuclear Physics - Neutron Generators and
   Application, Gaussig, 1984.
3. K. Málek, J. Pivarč and B. Bajcsy, Cs. pat. 237148.
4. K. Málek, J. Pivarč and B. Bajcsy, Cs. pat. 237149.

SUMMARY TALK *

S. Raman

Oak Ridge National Laboratory, [+]Oak Ridge,
Tennessee, USA

Conference summaries are perhaps best given by theorists. Two
eminent theorists attending this conference managed to evade
this job, and the burden fell on me (basically an experimenta-
list), probably because I have attended all but the first of
the four conferences in this series at Smolenice. I will not
attempt to summarize the conference; instead I will give some
of my personal impressions.

The main topics were nuclear reaction theory and neutron-
induced reactions. Specialized topics such as photon strength
functions, level densities, effective interactions, clustering
phenomena, ^{252}Cf neutron emission, resonance energy shifts due
to chemical and temperature effects, and instruments were also
discussed. The organizers succeeded in their primary aim of
maximizing the interchange of ideas by fostering discussion
among the attendees and minimizing distractions.

The geographical distribution of the participants has wide-
ned with first-time participation by scientists from Belgium
and the People's Republic of China. Also, for the first time,
English was the lingua franca throughout the conference —
which must have required extra effort and sacrifices on the
part of several speakers all of whom deserve our thanks.

*This written version was prepared while the author was a
guest at the Central Bureau for Nuclear Measurements (CBNM),
Geel, Belgium. The author is grateful to CBNM for its hospitality.

[+]Operated by the Martin Marietta Energy Systems, Inc. for
the U.S. Department of Energy under Contract No. DE-AC05-
84OR21400.

A small-town newspaper in the United States of America survives by having a policy of citing the names of all townspeople once or twice in the paper each year. I will follow this policy by invoking the names of all speakers at least once during this talk, but surely my oral comments are germane only to those who actually attended the conference. This written version will therefore be very brief. I apologize in advance to any authors whose achievements do not receive adequate weight or mention in this subjective summary.

Weidenmüller explained how the mean field theories (shell model and its extensions) account nicely for average nuclear properties determined by the dynamics of the system and how the fluctuations about this average can be simulated successfully (Wigner repulsion, Dyson-Metha Δ_3 statistics, Porter-Thomas distribution) in terms of a Hamiltonian consisting of a Gaussian orthogonal ensemble. These fluctuations are generic (they are of the same type found in classical chaotic systems) and are independent of the specific dynamics of the nucleus. Then came his provocative punch line, "Fluctuations carry no information and are therefore of limited physical interest". When I first heard this statement and his "Hamiltonian with zero information", it reminded me of an oxymoron — a contradiction in terms — like, for example, a "full-length bikini". Weidenmüller, however, went on to say that although fluctuations carried no information content, it was nevertheless necessary to understand their features. His final question, "To what extent do the fluctuations prevail all the way down to the vicinity of the ground state?" led him to make a plea for more experimental data, specifically for complete sets of levels with given spin and parity. Nuclear spectroscopists should take special note of this plea.

Soloviev and Vdovin brought us back to earth by starting with a full Hamiltonian (one of the two famous Armenians invoked at the beginning of all serious theory talks, the other being Lagrangian) consisting of at least nine terms. Soloviev and his colleagues have developed an effective and a powerful method for describing the fragmentation of one-quasiparticle, two-quasiparticle, and one-phonon states within the framework of the quasiparticle-phonon model. In general, the wave function of a neutron resonance contains a large number of components — and there are many resonances in a heavy nucleus. Fortunately only a limited number of components is responsible for the absorption and emission of a neutron. Even then, the sheer amount of work involved in calculating several nuclear properties (e.g., reduced transition probabilities, spectroscopic factors, transition densities, strength functions, and cross-sections) for a wide range of nuclei must have been great. The agreement that Soloviev showed between the calculated and experimental neutron strength functions (both s- and p-wave) was quite impressive. So confident is he about his calculations that when the measured s-wave strength function in ^{124}Te appeared to be three times the calculated value, he started to question seriously the measurements!

While previous conferences in this series had dealt in great detail with the master-equation, exciton-model approach to nuclear reaction theory, other approaches and models that took direct processes explicitly into account (in addition to the precompound and compound contributions) were mentioned, but only in passing. In the current conference, the tables were turned with a presentation by Tamura of the multistep direct-reaction theory, while the exciton model appears to have taken a back seat, though, by no means, out of the picture (see, for example, Kalka's contribution). Moreover, theories dealing with preequilibrium emission is bound to resurface in future conferences as the nuclear data needs for fusion reactors increase. Towards the end of his talk, Tamura applied the direct-reaction theory to describe the fusion between heavy ions through a separation of the expression for the total reaction cross-section into a direct reaction part and a fusion part. Even though this separation was achieved by phenomenologically choosing the range of integration, the procedure appears to work well for the cases he showed (fusion of ^{40}Ar+^{122}Sn, ^{58}Ni+^{64}Ni and ^{58}Ni+^{58}Ni) and, therefore, merits further applications.

Nearly 30 years ago, Brink proposed that an E1 giant resonance can be built on every excited state of the nucleus, and this resonance is similar in energy, strength, and width to the classical E1 photoresonance built on the ground state. This idea is contained in Brink's 1955 Ph.D. thesis (Oxford University), and this thesis is probably the reference that is most often cited in nuclear physics literature without the person citing it having ever actually seen it or read it. The verification of Brink's hypothesis via the (p, γ) reaction was reported by Szeflinski, and this hypothesis was also mentioned by Lone in connection with the photon strength function. Brink's idea was incorporated long ago into neutron-resonance-capture theory, but in recent years this idea has found new applications in heavy-ion radiative capture.

Chrien's presentation at this conference together with his article in the Proceedings of the 1982 Smolenice Conference will provide an interested reader with an excellent introduction to the neutron-resonance-averaging technique. He has repeatedly stressed that a proper application of this technique requires not only a careful analysis of the data but also a proper fluctuation analysis. While useful spectroscopic information may be provided by such data for A ≺ 100 nuclei — Chrien even provided a rationale for studying a nucleus with no resonance in the averaging window — it is also clear that the extraction of reliable J^{π} information requires an s-wave average level spacing of typically 12 eV in connection with the Sc filter. Fortunately, more than 25% of the stable nuclei between Mo and Pb possess such close level spacings at their neutron separation energies.

The study by Rohr of the level density of compound resonances gives two results which may be important for cross-section predictions — the first, the pairing energy changes with excitation energy. In general, in agreement with theory, the pairing energy is lower at the neutron separation energy, but there is also an apparent breakdown of the pairing correlation for several nuclei around A ≈ 100 and A ≈ 170. The second result is that the influence of collective motion on the level density at the neutron separation energy is negligible. The latter is an empirically derived result that may be questioned because previous theoretical calculations have predicted sig-

437

nificant enhancements of the order of σ^2 (where σ is the spin cut-off factor) in the level density of deformed nuclei as compared to spherical nuclei resulting from collective effects. If Rohr is correct, his second result would simplify considerably the description of the level density at high excitation energy.

At the neutron-separation energy, it is customary to assume equal level density values for levels of a given spin but of opposite parities. In the 1982 Smolenice conference, Antalik showed, through calculations for nuclei in the f-p shell, that this assumption is valid only at energies much higher than the neutron-separation energy. During a poster session at the current meeting, he showed that this unequal parity distribution can nicely explain the gamma competition in the continuum of ^{56}Fe.

A survey of the most recent measurements of (n,p) and (n,α) reactions with thermal and resonance neutrons was given by Wagemans. An interesting general comment to be made is the shift in motivation for the various experiments. Unlike the older experiments which were motivated mainly by basic physics interests or by nuclear technology requirements, the more recent work is influenced very strongly by nucleosynthesis. In particular, the explosive nucleosynthesis calculations require data on the ^{26}Al(n,p), ^{33}S(n,α), ^{40}K(n,p), ^{40}K(n,α), ^{41}Ca(n,p), and ^{41}Ca(n,α) reactions to remove discrepancies between observed and calculated abundances of some rare isotopes.

When the exciton model is applied to composite-particle emission, the cross-section is usually underestimated. This long-standing problem has been tackled previously by the Bratislava group. Iwamoto presented a fresh approach in which he calculates the probability of cluster formation in a semiclassical way and incorporates this probability into the exciton-model expressions. In this manner, he is able to obtain much better agreement with the measured cross-sections. The generalization of the exciton model to cluster emission and also to heavy-ion reactions is a topic of considerable current interest.

Many, many examples of experimental data were shown during

the meeting. Rapaport showed some beautiful nucleon elastic-scattering data and explained how the isovector effective interactions could be deduced from them. Cvelbar's review of fast neutron capture, replete with experimental details, brought back memories of talks by Drake and Longo in the earlier Smolenice conferences. Wasson's talk on neutron detection was followed by great interest by many attendees because he was dealing with a bread-and-butter problem for experimentalists. A particular experimental result presented at this meeting however stands out clearly in my mind. Already, during the 1982 Smolenice meeting, the Dresden group had presented experimental results that indicated the presence of a hard ($\succ$20 MeV neutron energy) component of fission neutrons from ^{252}Cf. This result was greeted with some skepticism then because the measurements are difficult indeed. During the latest meeting, Seeliger presented new results which confirm the earlier observations. A clear theoretical explanation for these high-energy neutrons is still lacking. In this connection, it is interesting to note that a multiparameter measurement of the ^{252}Cf prompt fission spectrum is currently under way at Geel, Belgium, as described by Knitter. It is expected that this experiment will be able to give an independent check of the excess of fast neutrons observed in the Dresden measurements.

I have touched — and only briefly at that — on but a few of the topics discussed at the 1985 Smolenice conference. These Proceedings constitute a much more complete record of what transpired. In conclusion, I wish to express my deep appreciation, in which I am sure the conferees will all join, to the organizers, and especially to Jozef Krištiak, for making the gigantic and thoughtful effort which made this conference a very fruitful scientific gathering. I have been instructed by the organizers, in turn, to invite all of you to come back to the next Smolenice meeting in three years' time. Until then ....

LIST OF PARTICIPANTS

R. ANTALÍK, Fyzikálny ústav CEFV SAV, Dúbravská cesta 9,
    842 28 Bratislava, ČSSR

V. BARTOŠEK, ÚJV ČSKAE, 250 68 Řež u Prahy, ČSSR

F. BEČVÁŘ, MFF UK, Pelc Tyrolka, 180 00 Praha 8, ČSSR

E. BĚŤÁK, Fyzikálny ústav CEFV SAV, Dúbravská cesta 9,
    842 28 Bratislava, ČSSR

R. E. CHRIEN, Department of Physics, Brookhaven National
    Laboratory, Upton, Long Island, New York 11973, USA

F. CVELBAR, J. Stefan Institute, Jamova 39, P.O. Box 100,
    61111 Ljubljana, Yugoslavia

M. J. A. DIXMIER, 15 rue Lebrun, 75013 Paris, France

J. DOBEŠ, ÚJF ČSAV, 250 68 Řež u Prahy, ČSSR

O. DUMITRESCU, Department of Fundamental Physics, Institute
    of Physics and Nuclear Engineering, Central Institute of
    Physics, P.O. Box MG6, Magurele, Bucharest, Romania

XIANG WAN DU, Institute of Atomic Energy, Academia Sinica,
    P.O. Box 275(41), Beijing, China

T. ELFRUTH, Technische Universität Dresden, Sektion Physik,
    Mommsenstrasse 13, 8027 Dresden, DDR

M. FLOREK, Katedra jadrovej fyziky MFF UK, Mlynská dolina,
    816 31 Bratislava, ČSSR

E. FRANC, ÚJF ČSAV, 250 68 Řež u Prahy, ČSSR

V. I. FURMAN, Joint Institute for Nuclear Research, Head Post
    Office, P.O. Box 79, 101000 Moscow, USSR

Š. GMUCA, Fyzikálny ústav CEFV SAV, Dúbravská cesta 9,
    842 28 Bratislava, ČSSR

440

H. HELFER, Technische Universität Dresden, Sektion Physik,
Mommsenstrasse 13, 8027 Dresden, DDR

A. IWAMOTO, Department of Physics, Japan Atomic Energy Research
Institute, Tokai, Naka-Gun, Ibaraki 319-11, Japan

H. O. JAHN, Kernforschungszentrum Karlsruhe, Institut für
Neutronphysik und Reaktorentechnik, Postfach 3640,
7500 Karlsruhe 1, BRD

A. B. KADHIM, Nuclear Research Centre, Iraqi Atomic Energy
Commission, Baghdad, Iraq

H. KALKA, Technische Universität Dresden, Sektion Physik,
Mommsenstrasse 13, 8027 Dresden, DDR

J. KLIMAN, Joint Institute for Nuclear Research, Head Post
Office, P.O. Box 79, 101000 Moscow, USSR

H. H. KNITTER, Commission of European Communities, Gerstrstr. 8,
B-2490 Balen, Belgium

Z. KOSINA, ÚJF ČSAV, 250 68 Řež u Prahy, ČSSR

J. Krištiak, Fyzikálny ústav CEFV SAV, Dúbravská cesta 9,
842 28 Bratislava, ČSSR

J. LÁC, Fyzikálny ústav CEFV SAV, Dúbravská cesta 9,
842 28 Bratislava, ČSSR

M. A. LONE, Chalk River Nuclear Laboratories, Chalk River,
Ontario, Canada K0J 1J0, Canada

H. MALECKI, Institute of Physics, University of Lódź,
ul. Nowotki 149/153, 90236 Lódź, Poland

H. MÄRTEN, Technische Universität Dresden, Sektion Physik,
Mommsenstrasse 13, 8027 Dresden, DDR

M. K. MEHTA, Nuclear Data Section, IAEA, Wagramerstrasse 5,
P.O. Box 100, A-1400 Vienna, Austria

E. MIŠIANIKOVÁ, Katedra jadrovej fyziky MFF UK, Mlynská dolina,
816 31 Bratislava, ČSSR

P. OBLOŽINSKÝ, Fyzikálny ústav CEFV SAV, Dúbravská cesta 9,
842 28 Bratislava, ČSSR

J. ORAVEC, Katedra jadrovej fyziky MFF UK, Mlynská dolina,
816 31 Bratislava, ČSSR

Š. PISKOŘ, ÚJF ČSAV, 250 68 Řež u Prahy, ČSSR

J. PIVARČ, Fyzikálny ústav CEFV SAV, Dúbravská cesta 9,
842 28 Bratislava, ČSSR

S. RAMAN, Oak Ridge National Laboratory, P.O. Box X, Oak Ridge, Tennessee 37831, USA

J. RAPAPORT, Department of Physics and Astronomy, Ohio University, Athens, Ohio 45701, USA

R. REIF, Technische Universität Dresden, Sektion Physik, Mommsenstrasse 13, 8027 Dresden, DDR

G. H. ROHR, CEC-JRC, Central Bureau for Nuclear Measurements, Steenweg naar Retie, B-2440 Geel, Belgium

J. RONDIO, Institute for Nuclear Studies, Hoża 69, 00681 Warsaw, Poland

E. RURARZ, Institute for Nuclear Studies, Department PII, 05-400 Świerk near Otwock, Poland

D. RYCKBOSCH, Nuclear Physics Laboratory, University of Gent, Proeftuinstraat 86, B-9000 Gent, Belgium

O. A. SALNIKOV, Institute of Physics and Power Engineering, 249020 Obninsk, USSR

P. SCHILLEBEECKX, Nuclear Physics Laboratory, University of Gent, Proeftuinstraat 86, B-9000 Gent, Belgium

D. SEELIGER, Technische Universität Dresden, Sektion Physik, Mommsenstrasse 13, 8027 Dresden, DDR

K. SEIDEL, Technische Universität Dresden, Sektion Physik, Mommsenstrasse 13, 8027 Dresden, DDR

Yu. A. SELITSKY, V. G. Khlopin Radium Institute, ul. Roentgena 1, 197022 Leningrad, USSR

A. I. SERGACHOV, Institute of Physics and Power Engineering, 249020 Obninsk, USSR

V. G. SOLOVIEV, Laboratory of Theoretical Physics, Joint Institute for Nuclear Research, Head Post Office, P.O. Box 79, 101000 Moscow, USSR

M. STEMPINSKI, Institute of Physics. University of Lódź, ul. Nowotki 149/153, 90236 Lódź, Poland

S. SUDÁR, Institute of Experimental Physics, Kossuth University, Bem Tér 18/A, P.O. Box 105, 4001 Debrecen, Hungary

I. SZARKA, Katedra jadrovej fyziky MFF UK, Mlynská dolina, 816 31 Bratislava, ČSSR

Z. SZEFLIŃSKI, Institute of Experimental Physics, Warsaw University, Hoża 69, 00681 Warsaw, Poland

T. SZTARICSKAI, Institute of Experimental Physics, Kossuth
University, Bem Tér 18/A, B.O. Box 105, 4001 Debrecen,
Hungary

J. ŠÁCHA, Fyzikálny ústav CEFV SAV, Dúbravská cesta 9,
842 28 Bratislava, ČSSR

O. ŠAUŠA, Fyzikálny ústav CEFV SAV, Dúbravská cesta 9,
842 28 Bratislava, ČSSR

F. ŠTĚRBA, MFF UK, Pelc Tyrolka, 180 00 Praha 8, ČSSR

T. TAMURA, Department of Physics, University of Texas, Austin,
Texas 78712, USA

G. TOUMBEV, Institute of Nuclear Research and Nuclear Energy,
Boul. Lenin 72, 1184 Sofia, Bulgaria

E. TRUHLÍK, ÚJF ČSAV, 250 68 Řež u Prahy, ČSSR

A. VDOVIN, Joint Institute for Nuclear Research, Head Post
Office, P.O. Box 79, 101000 Moscow, USSR

C. WAGEMANS, Laboratory of the SCK-CEN, Boeretang 200,
B-2400 Mol, Belgium

O. A. WASSON, National Bureau of Standards, Gaithersburg,
Maryland 20832, USA

H. A. WEIDENMÜLLER, Max-Planck-Institut für Kernphysik,
Postfach 103980, 6900 Heidelberg 1, BRD

I. WILHELM, Nukleární centrum MFF UK, Pelc Tyrolka,
180 00 Praha 8, ČSSR

TIAN-YUAN ZHANG, Institute of Applied Physics and Computational Mathematics, P.O. Box 8009, Beijing, China